Challenges of Latino Aging in the Americas

William A. Vega • Kyriakos S. Markides
Jacqueline L. Angel • Fernando M. Torres-Gil
Editors

Challenges of Latino Aging in the Americas

 Springer

Editors
William A. Vega
University of Southern California
Los Angeles
California
USA

Jacqueline L. Angel
The University of Texas at Austin
Austin
Texas
USA

Kyriakos S. Markides
University of Texas Medical Branch
at Galveston
Galveston
Texas
USA

Fernando M. Torres-Gil
School of Public Affairs
University of California. Los Angeles
Los Angeles
California
USA

ISBN 978-3-319-12597-8   ISBN 978-3-319-12598-5 (eBook)
DOI 10.1007/978-3-319-12598-5
Springer Cham Heidelberg New York Dordrecht London

Library of Congress Control Number: 2014955342

Printed on acid-free paper

Springer is part of Springer Science+Business Media (www.springer.com)

Preface

The Conference Series on Aging in the Americas (CAA) was launched in 2001 to promote interdisciplinary collaboration on social science issues research on health of the growing population of older Hispanics in the United States and Mexico. The current volume contains scholarly research presented during the sixth installment of the series focused on "Aging in the Americas" and was sponsored by the National Institute on Aging of the National Institutes of Health (R-13 5R13AG029767). The conference series was initiated because the conference organizers believed that demographic transitions in aging occurring across the Americas in the first half of the twenty-first century would be transformative and create unprecedented social, economic, and political challenges. The populations of Mexico and the United States were a focal interest for the conference series due to the historic interlocking of the two societies, the long contested nature of migration and the border, and the dynamic population growth of the Mexican-origin population in the United States. Conference organizers were aware of a significant number of active and concerned scholars who were using unprecedented amounts of information from field surveys and administrative data collected in Mexico and the United States. This research had importance for advancing knowledge in social science, health and public policy, and an outlet was needed for organized presentation and publication to further develop the field.

This volume is an invitation to readers to participate in unraveling a very complicated story of Hispanic population diaspora and health. A comprehensive understanding of issues affecting health and aging of Hispanic populations in the Americas is a global research challenge. This volume has a two-nation perspective, Mexico- U.S., which nonetheless incorporates many general themes of relevance to global aging research. The topics covered in this volume span demography, migration, economics of sustainable aging, social networks and support, determinants of health risk, and medical and social services. Considerable attention is given to key policy issues in recognition that pose current and emerging problems that require effective societal responses.

The approach taken in the volume is rooted in social science methods and supplemented by other disciplines as required to provide comprehensive perspectives. There are four sections designed by our Co-Editors; Sect. 1: Demographic and Economic Implications for Health and Well-being in Mexico and the USA (J. Angel); Sect. 2: New Data and Methodological Approaches on Aging in Mexico and the United States (K. Markides); Sect. 3: Binational, Transnational Migration Perspectives: Mexico, Latin America, and the USA (W. Vega); and, Sect. 4; Cost and Coverage: Fiscal Impacts (F. Torres-Gil).

The ethnic descriptors "Hispanic" and "Latino" are used interchangeably in this volume to inadequately capture the rich histories and distinctive nationality of Spanish-speaking people with origins in the Western Hemispheres. Hispanics (or Latinos) have a great linguistic diversity including dialects of Spanish and indigenous (native) tongues and equally diverse physical—phenotypic features derived from mixed African, Native Indian, and European heritage. This rich diversity is reflected in the Mexican-origin population which is the point of interest in this volume. The ethnic diversity of the Americas is relevant to patterns of migration, resettlement, and social behavior in the United States. These aspects of social life and social adjustments have been underappreciated and understudied in the past, especially as they affect aging people, and are essential for accurately characterizing critical health-related processes.

The tools and methods for doing transnational research are improving steadily, and we now have more than two generations of experienced investigators in the field. More importantly, there is a greater receptivity and capacity in the academic research world to provide accurate and actionable information for improving public and private services benefitting the well-being of older people. Today social science researchers can have greater confidence that the implications of their work, when appropriately focused for community, organizational, and governmental audiences, will be understood and appreciated. It is now evident in both Mexico and the United States that public policy and health policy must be designed to address social determinants of health to keep aging people functioning at the highest level for as long as possible, and to provide Hispanic elders with information to live well and become effective advocates for their own health—not simply consumers of late life medical and social care.

William A. Vega, Lead Editor
williaav@usc.edu

Kyriakos S. Markides, Co-Editor
kmarkide@UTMB.EDU

Jacqueline L. Angel, Co-Editor
jangel@austin.utexas.edu

Fernando M. Torres-Gil, Co-Editor
torres@luskin.ucla.edu

Acknowledgements

Conference Series on Aging in the Americas
2013 ICAA
The University of Texas at Austin

We would like to thank each individual who contributed to the success of the 2013 ICAA including the invited speakers, emerging scholar poster presenters, presiders, discussants, rapporteurs, as well as those who provided staff assistance. The conference was funded in part by the (R13) Scientific Meeting Grant from the National Institute on Aging (NIA) No. AG029767-01A2. Deserving of profound appreciation are our partners and supporters. They include: the LBJ School of Public Affairs and the Population Research Center at The University of Texas at Austin, University of Southern California, Edward R. Roybal Institute on Aging, University of Texas Medical School at Galveston, University of California, Los Angeles, Irma Rangel Public Policy Institute, National Institute on Aging, Office of Special Populations Planning Grant, AARP Texas, The University of Texas at Austin, Graduate School, George W. Jalonick, III and Dorothy Cockrell, Jalonick Centennial Lectureship, U.T. Coop, and the Austin Community Foundation.

Additional thanks must be given to our peer-reviewers:
Emma Aguila, University of Southern California
Flavia Andrade, University of Illinois at Urbana-Champaign
Ronald Angel, The University of Texas at Austin
Rosina Becerra, University of California, Los Angeles
Guilherme Borges, Instituto Nacional de Psiquiatría, Mexico
Richard Cervantes, Behavioral Assessment, Inc.
Carlos Díaz-Venegas, University of Texas Medical Branch
Chenoa Flippen, University of Pennsylvania
Hector Gonzalez, Michigan State University
Roberto Ham-Chande, El Colegio de la Frontera Norte, Mexico
Donald Lloyd, University of Southern California
Silvia Mejía-Arango, El Colegio de la Frontera Norte, Mexico
Carolyn Mendez-Luck, Oregon State University
Stipica Mudrazija, University of Southern California

Manuel Pastor, University of Southern California
Jennifer Salinas, The University of Texas Health Science Center at Houston
Valentine Villa, University of California, Los Angeles
Kathleen Wilber, University of Southern California
Special thanks to Stipica Mudrazija for expert assistance in the production of all phases of this volume.

Contents

Contributors

Emma Aguila Sol Price School of Public Policy, University of Southern California (USC) and RAND, Los Angeles, CA, USA

Jacqueline L. Angel The University of Texas, Austin, TX, USA

Ronald J. Angel The University of Texas at Austin, Austin, TX, USA

Ester Carolina Apesoa-Varano University of California, Davis, CA, USA

María P. Aranda School of Social Work, University of Southern California, Los Angeles, CA, USA

USC Edward R. Roybal Institute on Aging, University of Southern California, Los Angeles, CA, USA

Robert P. Berrens Department of Economics, MSC 05 3060, University of New Mexico, Albuquerque, NM, USA

Jorge Bravo Population Division, New York, NY, USA

Henry S. Brown School of Public Health Austin Regional Campus, The University of Texas Health Science Center, Houston, TX, USA

Raúl Hernán Medina Campos Instituto Nacional de Geriatría, Distrito Federal, Mexico

Mariana Campos Horta Office of Population Research & Department of Sociology, Princeton University, Princeton, NJ, USA

Michael Cline Hobby Center for the Study of Texas, Rice University, Houston, TX, USA

Marie-Laure Coubès El Colegio de la Frontera Norte, Tijuana, Baja California, Mexico

Courtney M. Demko University of California, Los Angeles, CA, USA

Gretchen Donehower Center for the Economics and Demography of Aging of the Department of Demography, University of California, Berkeley, CA, USA

Flávia Cristina Drumond Andrade University of Illinois at Urbana-Champaign, Champaign, IL, USA

Chenoa Flippen University of Pennsylvania, Philadelphia, PA, USA

Raquel Fonseca Département des sciences économiques, Université du Québec a Montréal and RAND, Montréal, QC, Canada

San Juanita García The Ohio State University, Department of Sociology, Columbus, CA, USA

Zachary D. Gassoumis Davis School of Gerontology, University of Southern California, Los Angeles, CA, USA

Yarin Gomez University of California, Davis, CA, USA

Erick G. Guerrero School of Social Work, University of Southern California, Los Angeles, CA, USA

Roberto Ham-Chande El Colegio de la Frontera Norte, Tijuana, Baja California, Mexico

Ladson Hinton University of California, Davis, CA, USA

Tenie Khachikian School of Social Work, University of Southern California, Los Angeles, CA, USA

Yinfei Kong School of Social Work, University of Southern California, Los Angeles, CA, USA

Nicole Mun Sim Lai Population Division, New York, NY, USA

Mariana López-Ortega National Geriatrics Institute, National Institutes of Health, Mexico City, Mexico

USC Edward R. Roybal Institute on Aging, University of Southern California, Los Angeles, CA, USA

Silvia Mejía-Arango El Colegio de la Frontera Norte, Tijuana, Baja California, Mexico

Ivan Mejia-Guevara Harvard Center for Population and Development Studies, Harvard University, Cambridge, MA, USA

Verónica Montes-de-Oca Universidad Nacional Autónoma de México, Mexico City, Mexico

Marie T. Mora University of Texas–Pan American, Edinburg, TX, USA

Stipica Mudrazija University of Southern California, Los Angeles, CA, USA

Steve H. Murdock Hobby Center for the Study of Texas, Allyn and Gladys Cline Professor Sociology, Rice University, Houston, TX, USA

Olufolake O. Odufuwa Department of Economics, MSC 05 3060, University of New Mexico, Albuquerque, NM, USA

Javier Pereira Universidad Católica del Uruguay, Montevideo, Uruguay

Nekehia Quashie College of Population Studies, Chulalongkorn University, Bangkok, Thailand

Telésforo Ramírez Researcher of CONACYT at the CRIM-UNAM, Tijuana, Mexico

Luis Miguel Gutiérrez Robledo National Geriatrics Institute, National Institutes of Health, Mexico City, Mexico

Rogelio Sáenz University of Texas at San Antonio, San Antonio, TX, USA

University of Texas, San Antonio, USA

Nadia Santillanes Centro de Investigaciones y Estudios Superiores en Antopologia Social (CIESAS-DF), Tlalpan, Mexico

Marta Tienda Office of Population Research, Woodrow Wilson School of International and Public Affairs and Department of Sociology, Princeton University, Princeton, USA

Fernando M. Torres-Gil Luskin School of Public Affairs, University of California, Los Angeles, CA, USA

R. Burciaga Valdez Department of Economics, MSC 05 3060, University of New Mexico, Albuquerque, NM, USA

Alma Vega RAND, Santa Monica, CA, USA

William A. Vega University of Southern California, Los Angeles, CA, USA

School of Social Work, University of Southern California, Los Angeles, CA, USA

Peter M. Ward The LBJ School of Public Affairs and The Department of Sociology, The University of Texas, Austin, TX, USA

Gregory B. Weeks University of North Carolina, Charlotte, NC, USA

John R. Weeks San Diego State University, San Diego, CA, USA

Kathleen H. Wilber Davis School of Gerontology, University of Southern California, Los Angeles, CA, USA

Kimberly J. Wilson School of Public Health Austin Regional Campus, The University of Texas Health Science Center, Houston, TX, USA

Mary Zey Department of Demography and Organizational Studies, University of Texas at San Antonio, San Antonio, TX, USA

Part I
Demographic and Economic Implications for Health and Well-being (Perspective) in Mexico and the USA: An Overview

Chapter 1
The Growing Importance of Educational Attainment and Retirement Security of Mexican-Origin Adults in the U.S. and in Mexico

Marie T. Mora

Introduction and Background

Hispanic population growth has fueled much of the U.S. population growth in recent years. To illustrate, between 2000 and 2012, the total U.S. population increased by 27.1 million people, 17.1 million (62%) of whom were Hispanic.[1] Indeed, the 48% growth in the Hispanic population was 11.5 times larger than the 4.2% growth in the non-Hispanic population during this time.[2] Moreover, much of this growth was driven by Mexican Americans, as this population increased by 63.6% (from 20.6 to 33.7 million), compared to a 27.2% growth in the non-Mexican-American Hispanic population. As such, Mexican Americans—who were already the largest Hispanic group—represented three-quarters of the total Hispanic population growth during this time period. As Steve Murdock, Michael Cline, and Mary Zey indicate in their upcoming chapter, relatively high fertility rates and immigration explain this large shift.

These changes have resulted in an increase in the presence of Mexican Americans, from 58 to 64% among Hispanics, and from 7.3 to 10.9% among the total U.S. population over the 12-year period. It follows that the socioeconomic and demographic characteristics of Mexican Americans have become particularly important in driving the socioeconomic and demographic outcomes in the U.S. overall. As Gregory Weeks and John Weeks argue in Chap. 3, the demographic projections

---

[1] I obtained these estimates using data from the U.S. Census Bureau (2001, 2013); these data exclude the Commonwealth of Puerto Rico and other outlying U.S. territories.

[2] Such dramatic demographic changes have not been ignored by scholars and policymakers; examples of recent studies analyzing socioeconomic and demographic outcomes of Hispanic population include those published in edited volumes such as Rodríguez et al. (2008); Leal and Trejo (2011); Verdugo (2012); Gastic and Verdugo (2013); and Mora and Dávila (2013).

M. T. Mora (✉)
University of Texas—Pan American, Edinburg, TX, USA
e-mail: mtmora@utpa.edu

© Springer International Publishing Switzerland 2015
W. A. Vega et al. (eds.), *Challenges of Latino Aging in the Americas,*
DOI 10.1007/978-3-319-12598-5_1

of the growth of Mexican Americans will reshape the landscape for many states' political futures as well. When combined with the 120.8 million people living in Mexico (World Bank Group 2014), the 154.5 million Mexican-origin people thus represent a significant determinant of the socioeconomic and demographic future of the Americas.

Age Distributions in the U.S. and Mexico

One demographic characteristic of interest is the relative youth of the Mexican-origin population. As seen in Table 1.1,[3] the average age of Mexican Americans in 2012 was 28.1 years, compared to 41.1 years among non-Hispanic whites. Moreover, 95 % of all Mexican Americans were under the age of 65 (12 % points higher than the corresponding share among non-Hispanic whites), and more than one-third were under 18 years of age (about twice the share for non-Hispanic whites). Among Mexicans living in Mexico, the average resident was 29.2 years old, nearly 94 % were younger than 65, and over one-third were younger than 18. The Mexican American population thus more closely resembles Mexico's population than non-Hispanic whites with respect to age. Issues related to aging and retirement among Mexican-origin adults, regardless if they reside in the U.S. or Mexico, should therefore affect both countries along the same timeline.

Education, Assets, and Retirement Income

In terms of aging and retirement, Jorge Bravo and his colleagues in their chapter point out that Mexican Americans have a higher life expectancy than non-Hispanic whites. At the same time, these authors note that Mexican Americans appear to be less economically secure in their old-age. Given their findings about asset income becoming the major source of retirement income in the U.S. as well as Mexico, the asset accumulation of Mexican-origin adults is of policy concern for both countries.

As with other socioeconomic resources discussed by Murdock, Cline, and Zey, one basic predictor of asset accumulation is education. Consider that in 2012, 26.4 % of adults of traditional working ages (25–64) who had a bachelor's degree or higher reported positive interest, dividend, or investment income, compared to

---

[3] The estimates in Table 1.1 come from two main sources: (1) the 2012 American Community Survey (ACS), made available by Ruggles et al. (2014) in the Integrated Public Use Microdata Series (IPUMS), and (2) the 2010 Population and Housing Census of Mexico, made available by the Minnesota Population Center (2013) in the IPUMS-International. The 2012 ACS contains approximately 1 % of the total U.S. population, and the Mexican census data includes 10 % of Mexico's population. The IPUMS-provided sampling weights were employed in all of the analyses to maintain the national representation of the data.

Table 1.1 Selected demographic and socioeconomic characteristics of Mexican Americans and non-Hispanic whites in the U.S. in 2012, and Mexicans in Mexico in 2010. (Source: Author's estimates using data from the 2012 ACS in the IPUMS, and from the 2010 Mexican census available in the International-IPUMS (see Note 3))

Characteristic	Mexican Americans in U.S. (2012)	Non-Hispanic whites in U.S. (2012)	Mexicans in Mexico (2010)
Age data (for entire population)			
Age (in years)	28.1	41.1	29.2
Less than 65 years	95.1%	82.7%	93.5%
Less than 18 years	35.8%	19.7%	35.1%
Adults ages 25–64			
Did not complete secondary school	39.5%	6.5%	68.0%
Secondary school, but not college grad.	50.0%	58.6%	18.8%
College graduate	10.4%	34.9%	12.8%
Has positive interest, div., or investment income	2.3%	12.9%	–
Average interest, dividend, or inv. income	$ 9383	$ 11,163	–
Owns home	52.1%	74.4%	–
Impoverished	21.1%	9.5%	–
Adults ages 65+			
Did not complete secondary school	60.5%	14.9%	89.8%
Secondary school, but not college grad.	33.5%	60.0%	5.6%
College graduate	6.0%	25.2%	3.9%
Has positive interest, div., or investment income	7.9%	31.9%	–
Average interest, dividend, or inv. income	$ 11,668	$ 17,084	–
Has retirement income	23.2%	39.6%	–
Average retirement income	$ 15,562	$ 19,759	–
Owns home	73.4%	84.8%	–
Impoverished	18.5%	7.2%	–
Has social security income	73.8%	84.8%	–
Average soc. sec. inc.	$ 10,491	$ 13,522	–

All differences between Mexican Americans and non-Hispanic whites in the U.S. are statistically significant at the 1% level. The various average income levels are only estimated for those who had such income.

8.9% of those who had less than a bachelor's degree.[4] Given that Mexican Americans have less education than the national average, the relationship between human capital and financial capital suggests they will lag behind the rest of the nation with respect to asset accumulation as they approach retirement.

Indeed, Table 1.1 indicates that only one out of ten of Mexican Americans ages 25–64 had graduated from college, while nearly 4 out of 10 had not completed high school. In contrast, over one-third of non-Hispanic whites had at least a bachelor's degree, and only 6.5% did not have a high school diploma in 2012. It is therefore not surprising that a considerably lower share of Mexican Americans than non-Hispanic whites in this age range had assets, including those that generated interest, dividend, or investment income (2.3 versus 12.9%, respectively). Among those with this type of income, Mexican Americans accrued less on average than non-Hispanic whites ($ 9,400 versus $ 11,200). Moreover, half of Mexican Americans of traditional working ages owned their homes, compared to three-quarters of non-Hispanic whites.

Similar gaps also exist between Mexican Americans and non-Hispanic whites with respect to education, asset accumulation, and retirement income when focusing on older adults (those above the age of 64). Over 60% of older Mexican Americans had not completed high school, which was ten times the share of those who had finished college. In contrast, a smaller share of older non-Hispanic whites did not have a high school diploma versus a college diploma (14.9 versus 25.2%).

These differences in education are consistent with asset and retirement income differentials between these groups. Similar to their younger counterparts, Table 1.1 indicates that older Mexican Americans were less likely than non-Hispanic whites to have interest, dividend, or investment income, and a lower proportion owned their homes. Moreover, a smaller percentage of older Mexican Americans than non-Hispanic whites had retirement income (23.2 versus 39.6%); among those who did, the average retirement income was significantly lower on average for the former group than the latter ($ 11,700 versus $ 17,100). The relative economic vulnerability of older Mexican Americans is further emphasized by their 18.5-% poverty rate, which exceeded the poverty rate of older non-Hispanic whites by more than 11% points. Coupled with the dramatic increase in the size of the Mexican American population, the increased dependence on asset accumulation for retirement security reported by Bravo and his colleagues sends a warning signal for the near future unless something is done to improve their relatively low education levels.

This does not mean that Mexican Americans have failed to make progress in attaining higher levels of schooling over time. As discussed by Mora and Rodríguez (2014), the average education of U.S.-born Mexican Americans, while below that of non-Hispanic whites, has risen almost every year since 2000. Moreover, Table 1.1 shows that Mexican Americans tend to be more educated than their counterparts in Mexico, as two-thirds of Mexican residents ages 25–64 in 2010 had not completed secondary school. Still, Table 1.1 indicates that progress in educational attainment

---

[4] Unless otherwise noted, the statistics cited are based on my estimates using data from the 2012 ACS (see Note 3).

along the age dimension has also occurred in Mexico, as nine out of ten Mexicans ages 65 and older had not completed secondary school in 2010. College graduates represented only 3.9 % of Mexicans in this age range, which was less than one-third of the representation of college graduates among younger Mexicans.

Because higher educational attainment translates into more asset accumulation (and thus more retirement security), if these patterns continue, Mexican Americans in the U.S. should be better off than Mexicans in Mexico as they age, other things the same. At the same time, they will likely continue to trail behind non-Hispanic whites in the U.S. in this regard. It follows that policies designed to increase schooling access and attainment in both countries should lead to improved socioeconomic outcomes, including more retirement security. Given the growth in the Mexican American population in the U.S., and given the role of international migration in shaping the socioeconomic and demographic futures of both Mexico and the U.S., some of these policies might be more effective if they can be coordinated between the two countries.

Pensions and International Migration

One issue related to international migration and retirement security among the Mexican-origin population is how such migration, even when temporary, affects access to pensions. The last two rows in Table 1.1 indicate that a smaller share of older Mexican Americans than non-Hispanic whites in the U.S. received social security income (76 versus 89 %) in 2012. Among those who did, Mexican Americans received less than non-Hispanic whites (approximately $ 10,500 versus $ 13,500).

Immigration relates to some of these differences, as Mariana Horta and Marta Tienda point out in their chapter. The Mexican American/non-Hispanic white social-security gaps narrow when focusing exclusively on U.S.-born Mexican Americans. To illustrate, approximately 85 % of U.S.-born Mexican Americans received social security (compared to 68 % of older Mexican immigrants), averaging $ 11,400 (compared to $ 9300 among Mexican immigrants) in 2012. Part of these differences likely relate to cases where some older Mexican immigrants did not contribute enough credits into social security to be eligible, particularly for those who migrated later in their life cycles as Horta and Tienda suggest. However, it is also likely that some immigrants had been unauthorized workers who used false social security numbers, and thus cannot *collect* the benefits (despite paying into the system) unless they legally resided in the U.S. (U.S. General Accounting Office 2003). According to the Social Security Administration, Office of the Chief Actuary (2013), moreover, recent legislation requires that immigrants receiving social security numbers after 2003 cannot receive benefits unless they had legal work authorization at some point *before* retiring. The issue is increasingly important given the rise of late-age immigration and changes in the employment patterns of older migrants during the past few decades, as discussed by Horta and Tienda.

In fact, the elder migration flow from Mexico to the U.S. is investigated in the chapter by Silvia Mejía-Arango, Roberto Ham-Chande, and Marie-Laure Coubes, who report a significant increase of aging people in Mexico migrating northward in recent years. Many of their study's respondents expressed the intention to spend an extended period of time residing in the U.S. This finding is counterintuitive to the broader reduction of border crossings widely reported over the same time frame. While the reasons are not confirmable at this point, the possibilities include decreased security in Mexico, the inability of adult children to return to Mexico to visit elder parents, or perhaps the inadequacy of personal support and networks in Mexico. It is especially notable that many elderly Mexicans are legally entering the United States, suggesting greater flexibility for family engagement and a higher quality of life rather than economic necessity as the central driver for migration.

Mexican emigration to the U.S. has further pension-access implications for those who eventually return to Mexico. As noted by Ronald Angel and Javier Pereira later in this volume, Mexico now has mandatory Individual Retirement Accounts for all private sector employees, and the government guarantees minimum pensions for workers who contribute for a given number of weeks. It remains unclear, however, how the temporary emigration of Mexicans to the U.S. affects these accounts. Presumably, many of these workers will not be eligible for Mexican pensions because they fail to meet the minimum amount of contribution time.

It follows that the mandatory minimum contributions to public pension programs might preclude return migrants from accessing these programs in Mexico. The socioeconomic implications of this reduced pension access are compounded when further considering Mexicans who migrated to the U.S. at older ages and thus lack the credits for social security as well as those who migrated illegally and cannot access the social security benefits they earned. It therefore appears that a non-trivial number of Mexican international migrants are vulnerable with respect to their retirement security, which has socioeconomic implications for both countries as the population ages. This raises the issue of coordinating retirement and pension-related policies between the U.S. and Mexico for international migrants, including moving forward with a "totalization agreement" between the U.S. and Mexico and making the pension-access process more flexible and transparent.[5]

International Coordination of Education and Retirement Policies

The international coordination of policies related to education, asset accumulation, and retirement programs is not as unlikely as the issue might first appear. Indeed, many communities along the U.S.-Mexico border already have coordinated their

---

[5] The U.S. has Totalization Agreements with two dozen countries for the purpose of avoiding double taxation of income with respect to social security taxes; Mexico is not currently one of them (Internal Revenue Service 2013).

efforts to promote human capital acquisition, employment, and economic growth on both sides of the border. As such, these communities might serve as "experiments" for broader international policies and programs.

One example is that certain institutions of higher education in Texas-Mexico border communities, including the University of Texas—Pan American (UTPA) and the University of Texas at El Paso (UTEP) have cross-border access policies in which Mexican residents near the border pay in-state tuition. Moreover, UTPA and Universidad Autónoma de Tamaulipas (UAT) have articulation agreements with respect to tuition and transferring coursework between the institutions (Espinoza 2012). Given the interdependence of border communities, such policies are designed to increase the access to institutions of higher education in the region, regardless in which country the students reside.

Consider also that residents of border communities access labor markets, product markets, and health care services on both sides of the border. In terms of labor markets, local corporations and government agencies have coordinated their efforts to stimulate employment on both sides of the border.[6] A fact that few people realize outside of the border region is that a non-trivial number of workers (including U.S.-born workers) live on one side of the border but work on the other side (e.g., Mora and Dávila 2011; Alegría 2002). There are already cross-border income tax implications for such workers. Perhaps these types of tax implications can be considered while designing more flexible and transparent cross-border pension and retirement programs. In all, policymakers and scholars of the implications of aging populations may wish to devote more attention to studying human capital and financial capital accumulation in dynamic border regions, as the Mexican-origin population continues to become increasingly important for the future socioeconomic direction of the Americas.

Concluding Remarks

Recent demographic changes have led to a significant increase in the presence of Mexican Americans in the U.S. In addition to relatively high fertility rates, migration from Mexico has fueled this increase. As such, despite their relatively youth, these demographic changes in both countries indicate that issues related to education and retirement security among Mexican-origin individuals have become increasingly important to the socioeconomic direction of both countries. It follows that a better understanding of factors and policies that affect the human capital and financial capital accumulation among members of this population should have a real impact on the future economic prosperity of the Americas.

---

[6] For example, the McAllen Economic Development Corporation (MEDC, http://www.mcallenedc.org) is a not-for-profit corporation under contract with the City of McAllen, Texas, that wants "to create jobs by attracting new industry and helping existing companies expand within the city and in Reynosa, Mexico". It works with Mexican agencies, including the Reynosa Maquila Association, as part of its strategy.

References

Alegría, T. (2002). Demand and supply of Mexican cross border workers. *Journal of Borderlands Studies, 17*(1), 37–55.

Espinoza, J. (2012). UTPA and UAT renew articulation agreement, The University of Texas Pan American. http://www.utpa.edu/news/index.cfm?newsid=4694. Accessed 5 Sept 2012.

Gastic, B., & Verdugo, R. R. (Ed.). (2013). *The education of the Hispanic population: Selected essays*. Charlotte: Information Age Publishing.

Internal Revenue Service. (2013). Totalization agreements. Washington, DC. http://www.irs.gov/Individuals/International-Taxpayers/Totalization-Agreements. Accessed 2 Aug 2013.

Leal, D. L., & Trejo, S. J. (Eds.). (2011). *Latinos and the economy: Integration and impact in schools, labor markets, and beyond* (pp. 153–168). New York: Springer.

Minnesota Population Center. (2013). *Integrated public use microdata series, international: Version 6.2 [Machine-readable database]*. Minneapolis: University of Minnesota. http://www.ipums.org. Accessed 03 May 2014.

Mora, M. T., & Dávila, A. (2011). Cross-border earnings of U.S. Natives along the U.S.-Mexico border. *Social Science Quarterly, 92*(3), 850–874.

Mora, M. T., & Dávila, A. (Eds.). (2013). *The economic status of the Hispanic population: Selected essays*. Charlotte: Information Age Publishing.

Mora, M. T., & Rodríguez, H. (2014). *Hispanic student success in higher education: Implications for earnings, health, and other socioeconomic outcomes*. Working paper, The University of Texas—Pan American.

Rodríguez, H., Sáenz, R., & Menjívar, C. (Eds.). (2008). *Latinos in the United States: Changing the face of América*. New York: Springer.

Ruggles, S., Alexander, J. T., Genadek, K., Goeken, R., Schroeder, M. B., & Sobek, M. (2014). *Integrated public use microdata series: Version 5.0 [Machine-readable database]*. Minneapolis: University of Minnesota. http://www.ipums.org. Accessed 03 May 2014.

Social Security Administration, Office of the Chief Actuary. (2013). Effects of unauthorized immigration on the actuarial status of the social security trust funds. Actuarial Note Number 161, Baltimore, MD, April 2013.

U.S. Census Bureau. (2001). The Hispanic population: Census 2000 brief. Washington, DC. http://www.census.gov/prod/2001pubs/c2kbr01-3.pdf. Accessed 04 May 2014.

U.S. Census Bureau. (2013). The Hispanic population in the United States: 2012. Washington, DC. http://www.census.gov/population/hispanic/data/2012.html. Accessed 04 May 2014.

U.S. General Accounting Office. (2003). Social security: Proposed totalization agreement with Mexico presents unique challenges. Report to congressional requesters, GAO 03–993, Washington, DC, Sept 2003.

Verdugo, R. R. (Ed.). (2012). *The demography of the Hispanic population: Selected essays*. Charlotte: Information Age Publishing.

World Bank Group (2014). Data: Mexico. Washington, DC. http://data.worldbank.org/country/mexico. Accessed 03 May 2014.

Chapter 2
The Future of Hispanics May Determine the Socioeconomic Future of the United States

Steve H. Murdock, Michael Cline and Mary Zey

The growth of Latino populations is widely recognized (Bean and Tienda 1987; Bergad and Klein 2010; Murdock 1995; Murdock et al. 2003, 2013, 2014; Ramos 2005) and this growth and the health, education and socioeconomic characteristics of their households and families are receiving increased and much needed attention (Angel and Angel 2009; Murdock et al. 2012). Long seen as a major component of the populations and rates of population growth in states in the southwestern United States, such as California, Texas, Arizona, and New Mexico, the growth of the Latino population has become pervasive across the United States. In fact, some have argued (see Murdock et al. 1997, 2003, 2014; Murdock 1995; Ramos 2005) that in large part the future of the United States overall is increasing linked to the future of its Latino population.

However, with rates of poverty that are generally two to three times as high as those for non-Hispanic whites and incomes that are only 60–75 % (depending on the area) of those for non-Hispanic white, and with educational levels that are the lowest of any major racial/ethnic group in the United States (e.g., in 2010, 38 % of Hispanics in the United States had less than a high school level of education compared to 18 % of African Americans and 9 % of non-Hispanic whites), it is not clear what the socioeconomic future will be for Latinos in the United States. It is equally unclear whether their impacts (in the absence of increased opportunities for educational attainment) will lead to a younger more vibrant United States in the future or to a nation that is poorer and less competitive. This chapter examines issues related

S. H. Murdock (✉)
Hobby Center for the Study of Texas, Allyn and Gladys Cline Professor Sociology,
Rice University, Houston, TX, USA
e-mail: shm3@rice.edu

M. Cline
Hobby Center for the Study of Texas, Rice University, Houston, TX, USA

M. Zey
Department of Demography and Organizational Studies, University of Texas at San Antonio,
San Antonio, TX, USA

© Springer International Publishing Switzerland 2015
W. A. Vega et al. (eds.), *Challenges of Latino Aging in the Americas,*
DOI 10.1007/978-3-319-12598-5_2

to the growth of Latino populations and their potential socioeconomic impacts on the United States.

The analysis presented is informed by literature that asserts that demographic structure can largely determine the socioeconomic characteristics of states and the Nation as a whole and that the major link between demographic structure and socio-economic change is mediated through education (Lutz 2013; Cuaresma et al. 2014). We argue with Lutz (2013) and Murdock et al. (2014) that demographic structure may have aggregate effects that have no micro-level analogue. We further argue in agreement with Cuaresma et al. (2014) that differences in educational attainment are the major factors affecting the relative socioeconomic conditions of major population groups such as Hispanics and non-Hispanic Blacks when compared to non-Hispanic whites.

These effects are documented in an extensive literature. As delineated below there are substantially lower levels of socioeconomic resources in Hispanic and non-Hispanic black households compared to non-Hispanic white households but, as noted in the quote provided below, these differences may be related to higher levels of education for non-Hispanic whites. In fact, this literature suggests that not only for individuals, but for populations as a whole (see Cuaresma et al. 2014), educational attainment may lead to population-wide benefits and be largely responsible for the societal-level economic benefits once attributed to the age structure of western populations, i.e., the often noted benefits of the "demographic dividend".

For example, Murdock et al. (2014, pp. 7–8) have noted the strong relationship that exists between education and income (Romer 1990) in multiple countries (Mincer 1974; Card 1999; Ashenfelter and Rouse 1999; Abdullah et al. 2011) across time periods to the present (Becker 1967; Hanoch 1967; Schultz 1968; U.S. Bureau of Labor Statistics 2011). Such studies show that increased education results in increased income. Murdock et al. (2014) note that work by Cheeseman-Day and Newburger (2002) documents that college graduates, on average, make more than $ 1 million more lifetime income than high school dropouts. They further point out that analyses by Julian and Kominski (2011) show that non-Hispanic whites males with graduate degrees make more than $ 1.9 million more in their lifetime than non-Hispanic whites with less than a high school degree and indicate that lifetime differences of $ 1.6, $ 1.4, and $ 2.3 million are evident for Hispanic males, non-Hispanic black males and non-Hispanic Asian males with the same educational differences. They conclude that although racial/ethnic differences clearly exist, education pays for all racial/ethnic groups.

Murdock and his colleagues (2014) also cite literature that indicates that individuals with higher levels of education experience less unemployment and work in higher paying occupations than their less well-educated counterparts. They cite works by Cohn and Addison (1998) and Abdullah et al. (2011) that indicate that children's future income is largely determined by the education they obtain. Because poor children are less likely to do well in school they tend to attain low levels of educational attainment resulting in employment in occupations that are lower paying and, as a result, have lower incomes.

When examined at the macro level (i.e., for states and nations) empirical findings (see U.S. Bureau of Labor Statistics 2011) also indicate strong relationships between a nation's general level of education and aggregate income. Studies by numerous scholars (Denison 1962; Bowman and Anderson 1963; Schultz 1963) have estimated the contribution of education expenditures to national income and how state educational expenditures affect state income. These analyses have consistently found that increased education leads to higher state and national income (Tolley and Olson 1971; Davern and Fisher 2001). Murdock et al. (2014) cite a study by McKinsey and Company (2009) that concludes that if the educational achievement gap between Hispanics and non-Hispanic blacks and non-Hispanic white students in the United States continue it will have the effect of a permanent national recession. McKinsey and Company (2009) maintain that had the achievement and related economic gap not existed so that all groups earned at the level of non-Hispanic whites, GDP in 2008 would have been $ 310–$ 525 billion higher and, if the gaps between all minorities and non-Hispanic whites had been eliminated, the GDP would have been between $ 400 and $ 675 billion higher. These represent increases in GDP of 2–4 % and 3–5 % respectively. Yet another analysis of 64 empirical studies (Abdullah et al. 2011) examining 868 estimates of the effects of education on income found that education was an important factor for reducing income differences among racial/ethnic groups.

What such analyses demonstrate is that education may be the key factor for closing socioeconomic differences among racial/ethnic groups. As such, education may not only largely explain socioeconomic disparities among racial/ethnic groups but may also provide the major means for eliminating such disparities (Cohen and Soto 2007; Bloom et al. 2009).

Based on the premises above, we provide descriptions of three sets of conditions related to the demographic and socioeconomic future of Latino populations and their socioeconomic impacts on the United States. These are:

1. The extent to which Hispanic population growth is, and will continue to, dominate population growth not only in "Hispanic" states but in the United States as a whole;
2. The level of socioeconomic and educational disadvantage in the Latino population;
3. The role that improved education may have in changing the socioeconomic future of Latinos and through them the socioeconomic future of Hispanic states such as Texas and the United States as a whole.

In addressing these factors we provide an overview of Hispanic population growth in the United States and their current socioeconomic characteristics and, using the example of Texas, delineate the potential socioeconomic impacts of such growth depending on the extent to which Latino populations are provided with the opportunities to obtain the skills and education they need to be competitive. The chapter is descriptive of both what is, and what could be the future of Latinos in both states, such as Texas, and the United States as a whole, and of what the effects of this population may be on the overall socioeconomic characteristics of both areas in the coming decades.

Table 2.1 Total population and percent population change in Texas and the United States, 1850–2010. (Source: Derived from the U.S. Census Bureau decennial census April 1 of reported year)

Year	Total population		Percent change	
	Texas	U.S.	Texas	U.S.
1850	212,592	23,191,876	–	–
1860	604,215	31,443,321	184.2	35.6
1870	818,579	39,818,449	35.5	26.6
1880	1,591,749	50,155,783	94.5	26.0
1890	2,235,527	62,947,714	40.4	25.5
1900	3,048,710	75,994,575	36.4	20.7
1910	3,896,542	91,972,266	27.8	21.0
1920	4,663,228	105,710,620	19.7	14.9
1930	5,824,715	122,775,046	24.9	16.1
1940	6,414,824	131,669,275	10.1	7.2
1950	7,711,194	150,697,361	20.2	14.5
1960	9,579,677	179,323,175	24.2	19.0
1970	11,196,730	203,302,031	16.9	13.4
1980	14,229,191	226,545,805	27.1	11.4
1990	16,986,510	248,709,873	19.4	9.8
2000	20,851,820	281,421,906	22.8	13.2
2010	25,145,561	308,745,538	20.6	9.7

Current and Future Patterns of Population Growth in Texas and the United States

Current Patterns of Population Growth and Diversification

Population growth in both the United States as a whole and in key Hispanic states, such as Texas, has been extensive. In the 50 years from 1960 to 2010 (see Table 2.1), the United States population increased by 129.4 million people, by 72.2% and Texas population increased by 15.6 million or 162.5% In recent decades this growth has been increasingly dependent on minority population growth, particularly Hispanic population growth.

Tables 2.2 and 2.3 provide data on the growth of racial/ethnic populations in Texas and the United States from 2000 to 2010. The data in Table 2.2 show that Texas' population increase of about 4.3 million or 20.6% during the decade from 2000 to 2010 was largely a result of minority, particularly Hispanic, population growth. The non-Hispanic white population increased by only 4.2% and accounted for only 10.8% of the total net growth in Texas population from 2000 to 2010 whereas Hispanic population growth was 41.8% and accounted for 65% of net growth, non-Hispanic black populations increased by 22.1% and accounted for 12.2% of the net growth, and non-Hispanic Asian populations increased by 71.1% and accounted

Table 2.2 Population, population change, and proportion of the total population by race/ethnicity for Texas, 2000 and 2010. (Source: Murdock 2014)

Race/ethnicity[a]	Population		Population change			Percent of total population	
	2000	2010	Numeric	Percent	Percent of total change	2000	2010
NH white	10,933,313	11,397,345	464,032	4.2	10.8	52.4	45.3
Hispanic (all races)	6,669,666	9,460,921	2,791,255	41.8	65.0	32.0	37.6
NH black	2,364,255	2,886,825	522,570	22.1	12.2	11.3	11.5
NH Asian	554,445	948,426	393,981	71.1	9.2	2.7	3.8
NH other	330,141	452,044	121,903	36.9	2.8	1.6	1.8
Total	20,851,820	25,145,561	4,293,741	20.6	100.0	100.0	100.0

[a] Hispanic includes persons of all races. All other race/ethnicity categories shown here are non-Hispanic. non-Hispanic other includes persons identifying themselves as non-Hispanic American Indian or Alaska Native, non-Hispanic Native Hawaiian or Pacific Islander, non-Hispanic some other race, or non-Hispanic and a combination of two or more races

Table 2.3 Population, population change, and proportion of the total population by race/ethnicity for the United States, 2000 and 2010. (Source: Murdock 2014)

Race/ethnicity[a]	Population		Population change			Percent of total population	
	2000	2010	Numeric	Percent	Percent of total change	2000	2010
NH white	194,552,774	196,817,552	2,264,778	1.2	8.3	69.1	63.7
Hispanic (all races)	35,305,818	50,477,594	15,171,776	43.0	55.5	12.5	16.3
NH black	33,947,837	37,685,848	3,738,011	11.0	13.7	12.1	12.2
NH Asian	10,123,169	14,465,124	4,341,955	42.9	15.9	3.6	4.7
NH other	7,492,308	9,299,420	1,807,112	24.1	6.6	2.7	3.0
Total	281,421,906	308,745,538	27,323,632	9.7	100.0	100.0	100.0

[a] Hispanic includes persons of all races. All other race/ethnicity categories shown here are non-Hispanic. non-Hispanic other includes persons identifying themselves as non-Hispanic American Indian or Alaska Native, non-Hispanic Native Hawaiian or Pacific Islander, non-Hispanic some other race, or non-Hispanic and a combination of two or more races

for 9.2% of the net growth in the total population. All other non-Hispanic populations increased by 36.9% and accounted for 2.8% of net growth. As a result of such changes, Texas became a majority minority state by 2010 (along with California, Hawaii and New Mexico) with non-Hispanic whites decreasing from 52.4% of the total population in 2000 to 45.3% in 2010 while Hispanic populations accounted for 37.6% non-Hispanic black populations for 11.5% non-Hispanic Asian populations for 3.8% and all other populations for 1.8% of the total population of Texas in 2010.

Although extensive and diverse, such a pattern of population growth would not be seen as unusual by many because Texas is a "Hispanic" state, with patterns similar to those in Arizona, California, Florida, and New Mexico. Many residents of the United States believe that such patterns of minority growth are largely limited to such diverse states and these state are believed by many to have patterns that differ substantially from those for the Nation as a whole.

The data in Table 2.3 suggest that rather than being unique such patterns are increasingly pervasive across the Nation. As shown in this table, in the United States as a whole, the non-Hispanic white population increased by only 1.2% from 2000 to 2010 and accounted for only 8.3% of the total net change in the population in the decade. Although the total population remained 63.7% non-Hispanic white in 2010, from 2000 to 2010, 55.5% of the net growth in the United States as a whole was due to Hispanics; 13.7% was due to non-Hispanic black; 15.9% due to non-Hispanic Asians; and 6.6% was a result of growth in other non-Hispanic minority populations. The overall effect of minority population growth, particularly Hispanic population growth, on total population growth in the United States was larger than the effect of such growth on total population change in Texas.

The extent and impact of minority population growth is thus evident in both states, such as Texas, and in the Nation as a whole. Analyses not shown here (due to space limitations) indicate that minority growth is increasingly the largest source of population growth not only in large central city counties but also in suburban and nonmetropolitan counties in the Nation and is increasingly the dominant source of growth in the Northeast, Midwest, Southern, and Western regions (as defined by the United States Census Bureau) of the United States (see Murdock et al. 2014).

The dominance of minority population growth is also evident in an examination of county population growth from 2000 to 2010. For example, if one examines counties in Texas with at least 100 persons in a racial/ethnic group (thereby eliminating counties in which such a group plays a very limited role in overall patterns of change in the county) the dominance of minority population growth is evident. Such data on population change in racial/ethnic groups from 2000 to 2010 show that of the 252 (of 254 total) counties in Texas with sufficient numbers (i.e., 100 or more) of non-Hispanic white population members there were increases in non-Hispanic white populations in 91 counties and declines in non-Hispanic white populations in 161 counties. The number of increasing relative to declining counties (for counties with 100 or more population group members) for Hispanics was 228 increasing and 22 decreasing; for non-Hispanic Black populations 83 counties showed increases and 102 showed declines; and for non-Hispanic Asian and other populations there were 97 increasing and 7 declining counties from 2000 to 2010. In percentage terms this indicates that there were increases in non-Hispanic whites in only 36.1% of Texas relevant (252) counties, increases in Hispanic populations in 91.2% of relevant (250) counties, increases in non-Hispanic black populations in 55.1% of relevant (185) counties, and increases in non-Hispanic Asian and other populations in 93.3% of relevant (107) counties from 2000 to 2010.

If one examines these patterns for the United States, as a whole, there were similar patterns for the 2000 to 2010 decade. There were increases in non-Hispanic white populations in 46.7% of the relevant (3133) counties, increases in the number of Hispanics in 96.5% of the relevant (2797) counties, increases in 67.3% of the relevant (2212) counties for non-Hispanic black populations, and increases in 95.1% of the relevant (1660) counties for non-Hispanic Asian and other populations. It is evident that recent population growth in the United States, as in Texas, has been largely due to minority, particularly Hispanic, population growth.

This pattern of more rapid growth in minority, particularly Hispanic, than non-Hispanic white populations is particularly evident in the child population. Analyses of the same form of data as shown in Table 2.2 and 2.3 for adults, for children indicated that there was an absolute numerical decline in the number of non-Hispanic white children in both the State of Texas and the United States as a whole from 2000 to 2010. In fact, an examination of 2000 to 2010 population data for the United States, indicated that if the decline of more than 4.3 million non-Hispanic white and the decline of 248,000 non-Hispanic black children from 2000 to 2010 had not been offset by an increase of nearly 4.8 million Hispanic children, the United States would have experienced one of the largest declines in its child population in its history. Total population growth, particularly for younger populations, is increasingly dependent on minority, particularly Hispanic, population growth. Since patterns of change in child populations are indicative of the future of population change overall, they strongly suggest that future populations may be even more determined by Hispanic population growth not only in states such as Texas but in the United States overall.

Projections of the Population of Texas and the United States

Projections for Texas

Projections of the population of Texas and the United States indicate increasing minority populations in both Texas and the United States as a whole, and increasing contributions of minority, particularly Hispanic, populations to total population growth in both areas. As indicated in Table 2.4, Texas is projected to show substantial population growth in the coming decades. Under the scenario that assumes that levels of growth of the 2000 to 2010 time period continue, Texas would have a population of 55.2 million people in 2050 up from 25.1 million in 2010. This increase of more than 30 million people would include an increase of only 628,000 non-Hispanic whites accounting for just 2.1% of net population growth from 2010 to 2050. By comparison, the non-Hispanic black population is projected to increase by 2.3 million accounting for 7.7% of the net increase, non-Hispanic Asians and others are projected to increase by about 5.9 million persons and account for 19.5% of the net increase and Hispanics are projected to increase by 21.2 million and account for 70.7% of the net increase from 2010 to 2050.

Table 2.4 Population in Texas by race/ethnicity in 2010 and projections of the population in Texas by race/ethnicity from 2020 to 2050 under alternative assumptions of age and race/ethnicity-specific rates of net migration

Year	NH[a] white	NH black	Hispanic	NH Asian and other[b]	Total
Assuming zero net migration					
2010	11,397,345	2,886,825	9,460,921	1,400,470	25,145,561
2020	11,576,595	3,122,637	11,137,672	1,536,693	27,373,597
2030	11,501,020	3,280,941	12,869,753	1,638,249	29,289,963
2040	11,182,576	3,355,500	14,570,851	1,714,232	30,823,159
2050	10,766,622	3,366,528	16,191,150	1,728,206	32,052,506
Assuming net migration equal to one-half of 2000–2010					
2010	11,397,345	2,886,825	9,460,921	1,400,470	25,145,561
2020	11,752,530	3,295,198	12,031,059	1,825,130	28,903,917
2030	11,850,180	3,658,997	15,082,058	2,309,763	32,900,998
2040	11,676,157	3,951,909	18,489,803	2,881,525	36,999,394
2050	11,376,576	4,182,155	22,268,390	3,483,178	41,310,299
Assuming net migration equal to 2000–2010					
2010	11,397,345	2,886,825	9,460,921	1,400,470	25,145,561
2020	11,931,815	3,477,928	13,003,159	2,170,409	30,583,311
2030	12,211,664	4,080,453	17,702,132	3,288,536	37,282,785
2040	12,194,151	4,653,725	23,514,974	4,953,861	45,316,711
2050	12,024,913	5,195,861	30,701,208	7,283,548	55,205,530

[a] NH refers to non-Hispanic; values shown are only for the non-Hispanic persons in each race category. Hispanic includes Hispanics of all races
[b] NH Asian and other category includes non-Hispanic persons who identify themselves as belonging to two or more race groups

Under the scenario that assumes growth levels only one-half of those of the 2000 to 2010 decade, the increase of 16.1 million people from the 25.1 million in 2010 to a population of 41.3 million in 2050, the growth of minority populations would play an even larger role in the determination of overall levels of growth. This projection, would involve an absolute **decline** of roughly 21,000 non-Hispanic whites and increases of 1.3 million non-Hispanic black, 2.1 million non-Hispanic Asian and others and an increase of 12.8 million Hispanics. The three minority populations would account for 8.0, 12.9 and 79.2% respectively, of the net increase in the total population (with the non-Hispanic white population accounting for −0.1% of the net change).

Thus, under the two scenarios noted above non-Hispanic whites would account for between 22 and 28% of the total population in 2050, non-Hispanic blacks for 9–10% of the population, non-Hispanic Asians and others for 8–13% of the total population and Hispanics for 54–56% of the total population by 2050. When examined by age group (see Table 2.5), by 2050 Hispanics would be the largest single population group in every age group and would be less than 50% only in the age

Table 2.5 Percent of the population by age group and race/ethnicity in 2010 and projected percent of the population by age group and race/ethnicity from 2030 to 2050 assuming age and race/ethnicity-specific net migration equal to 2000–2010 for the State of Texas (Scenario 1.0)

Age	NH[a] white	NH black	Hispanic	NH Asian and other[b]	Total
2010					
< 18	20.4	28.1	35.1	29.6	27.3
18–24	8.7	11.2	11.7	10.2	10.2
25–44	25.7	29.1	30.1	32.3	28.2
45–64	29.8	24.0	17.5	21.4	24.0
65+	15.4	7.6	5.6	6.5	10.3
2030					
< 18	19.3	23.4	29.7	21.7	24.9
18–24	7.5	9.6	10.8	11.5	9.6
25–44	23.0	27.2	29.9	31.6	27.5
45–64	24.3	24.2	19.8	24.1	22.2
65+	25.9	15.6	9.8	11.1	15.8
2050					
< 18	18.2	20.7	27.3	21.9	24.0
18–24	7.4	8.9	10.6	8.6	9.4
25–44	23.0	25.5	27.8	30.5	26.9
45–64	23.7	25.6	21.4	24.9	22.8
65+	27.7	19.3	12.9	14.1	16.9

[a] NH refers to non-Hispanic; values shown are only for the non-Hispanic persons in each race category. Hispanic includes Hispanics of all races
[b] NH Asian and other category includes non-Hispanic persons who identify themselves as belonging to two or more race groups

group of 65 years of age or older (for which they would represent 42.5 % of the total population of that age, a larger percentage of persons in that age group than that represented by any other racial/ethnic group).

For a population in a very diverse state such as Texas there is little doubt that minority populations today will be the demographically dominant populations of the future. The future welfare of such states will be determined by the education and socioeconomic opportunities and accomplishments of its minority, and particularly, its Hispanic population.

Projections for the United States

Recent projections of the United States population verify similar patterns (United States Census Bureau 2012) to those for states such as Texas for the United States as a whole. As shown in Table 2.6, under this projection, the non-Hispanic white population is projected to decline from 196.8 million in 2010 to 178.9 million in

Table 2.6 U.S. population by race and Hispanic origin: census 2010 and projected for 2020–2060. (Source: U.S. Census Bureau 2010 Census and 2012 National Projections)

Year	(Resident population as of July 1, number in thousands)					
	Non-Hispanic				Hispanic (any race)	Total
	White alone	Black alone	Asian alone	All other races		
2010	196,818	37,686	14,465	9,299	50,478	308,746
2020	199,312	41,776	18,246	10,778	63,784	333,896
2030	198,818	45,452	22,044	13,502	78,655	358,471
2040	193,887	48,769	25,881	16,603	94,876	380,016
2050	186,334	51,988	29,583	20,166	111,732	399,803
2060	178,952	55,302	33,106	24,128	128,780	420,268

2060 while the non-Hispanic black population increases from 37.7 to 55.3 million, the non-Hispanic Asian population increases from 14.5 to 33.1 million, all other non-Hispanic populations would increase from 9.3 to 24.1 million. However, the Hispanic population is projected to increase from 50.5 to 128.8 million persons from 2010 to 2060. All of the net increase in the total population of the United States from 2010 to 2060 will be due to populations other than non-Hispanic whites (because non-Hispanic whites are projected to decline) and, of the projected **net increase** in the total population, 60.5 % is projected to be due to the Hispanic population, 13.6 % to the non-Hispanic black population and 25.9 % to non-Hispanic Asian and other non-Hispanic populations.

The projected population will also show a general aging but much less so for minority than for non-Hispanic white populations. An examination of projections data from the United States Census Bureau indicates that the overall population will age such that by 2060 nearly 22 % of the entire population of the United States will be 65 years of age or older while the percentages in this age group made up of persons from different racial/ethnic groups would vary substantially. In 2010, the majority in all age groups were non-Hispanic white but for the elderly population (persons 65 years of age or older) 80 % of this population in the United States was non-Hispanic white while 8.4 % were non-Hispanic black, 4.7 % were members of Asian and other non-Hispanic groups and 6.9 % were Hispanic. By 2050, the only age group in which a majority of the population would be non-Hispanic white would be the age group 65 years of age or older. Even in this age group there would be significant change with the percent of all elderly who would be non-Hispanic white declining to 55.8 % while non-Hispanic blacks account for 12.5 % Asian and other non-Hispanic population groups for 10.4 % and Hispanics for 21.3 % of the elderly population in 2060. By 2042, no single racial/ethnic group will account for 50 % or more of the total population and, by 2060, 42.6 % of the total population would be non-Hispanic white, 13.2 % non-Hispanic black, 13.6 % non-Hispanic Asian and other, and 30.6 % would be Hispanic. In both major Hispanic States, such as Texas and California, and in the Nation as a whole, minority, particularly Hispanic, populations will be of increasing importance in determining the future population of all ages.

Overall, then, the data here suggest that the United States as a whole, and particularly the southwestern and other states with large minority populations, will increasingly depend on change in their Hispanic populations to determine their demographic futures. But do such changes have implications beyond the demographic? What are the implications of America's changing population base for its socioeconomic resource base? In the next section we address this issue by examining disparities in the socioeconomic and educational resources of racial/ethnic groups in Texas and the Nation.

The Socioeconomic and Educational Characteristics of Hispanic and Other Populations in the Southwest and the United States

Although population growth has substantially increased the numerical presence of Hispanic and other minority populations, other data continue to show patterns of substantial socioeconomic and educational disadvantage for non-Hispanic black and Hispanic populations compared to non-Hispanic white and non-Hispanic Asian and other populations. Table 2.7 presents data on the recent socioeconomic characteristics of the Texas' and United States' population by race/ethnicity in 2010. The data in this table indicate that Hispanic, and non-Hispanic black levels of income are between 60 and 75% of the income of non-Hispanic whites both in the United States and in Texas. At the same time, the percent of non-Hispanic black and Hispanic households in poverty are two to two and one-half times as high as those for non-Hispanic white populations. In 2010 non-Hispanic whites in Texas had a median household income of $ 59,517, while Hispanics had median household incomes of $ 37,087, and non-Hispanic black households had a median household income of $ 35,674. The percent of persons in poverty was 9.5% for non-Hispanic white but 26.8% for Hispanic and 24.7% for non-Hispanic black populations. These values for the United States

Table 2.7 Total and percent of the population in poverty and median household income by race/ethnicity for the United States and Texas, 2010. (Source: U.S. Census Bureau 2006–2010 American Community Survey)

Race/ ethnicity	United States			Texas		
	Population in poverty	%	Median house-hold income[a]	Population in poverty	%	Median house-hold income[a]
NH white	21,831,928	10.8	$ 53,988	1,111,251	9.5	$ 59,517
Hispanic (all races)	12,583,542	24.8	$ 40,165	2,555,080	26.8	$ 37,087
NH black	10,270,229	27.1	$ 33,568	717,458	24.7	$ 35,674
NH Asian	1,806,217	12.4	$ 67,142	120,601	12.6	$ 64,191
Total	47,330,502	15.3	$ 50,046	4,521,023	17.9	$ 48,615

Not all race/ethnic groups are shown here but are included in the total
[a] In 2010 US dollars

Table 2.8 Educational attainment for the population age 25 and over by race/ethnicity for the United States and Texas, 2010. (Source: U.S. Census Bureau 2010 American Community Survey)

Race/ethnicity	United States		Texas	
	Less than high school	Bachelor degree or more	Less than high school	Bachelor degree or more
NH white	9.3	31.4	8.0	34.1
Hispanic[a]	37.8	13.0	40.4	11.6
NH black	17.9	17.9	13.7	19.7
NH Asian	14.5	50.0	14.3	52.0
NH other	15.2	21.3	9.3	20.8
Total	14.4	28.2	19.3	25.9

[a] Hispanic includes persons of all races. All other race/ethnicity categories shown here are non-Hispanic. non-Hispanic other includes persons identifying themselves as non-Hispanic American Indian or Alaska Native, non-Hispanic Native Hawaiian or Pacific Islander, non-Hispanic some other race, or non-Hispanic and a combination of two or more races

were $ 54,671 for non-Hispanic white, $ 39,923 for Hispanic, and $ 33,923 for non-Hispanic black populations. Poverty rates for households in these three groups were 10.0, 23.5, and 25.8% respectively, in 2010. Unfortunately, these race/ethnicity differences are pervasive. Analyses not shown here, indicate that such differences between non-Hispanic whites and Hispanics and non-Hispanic black populations are evident across each of the four major regions of the United States–the Northeast, Midwest, West, and South–and across metropolitan central city, metropolitan suburban, and nonmetropolitan counties in virtually every state with significant numbers of minority population members. The reality is that Hispanic and non-Hispanic black populations are substantially disadvantaged economically.

Closely related to socioeconomic resources are differentials in levels of education (see Murdock et al. 2014). Higher levels of education are related to increased socioeconomic advantage. Table 2.8 provides data indicating that both in Texas and the Nation, minorities, particularly Hispanics, have substantially lower levels of educational attainment. Nationally, in 2010, 37.8% of Hispanics had less than a high school level of education and only 13.0% had a bachelor's degree or higher level of education. Similarly, in Texas, 40.4% of Hispanics had less than a high school level of education and 11.6% had bachelor's degree or higher levels of education. These compare to less than high school attainment levels for non-Hispanic whites of 9.3% in the United States and 8.0% in Texas and bachelors or higher levels of attainment of 31.4% in United States and 34.1% in Texas.

Education is an important factor differentiating levels of socioeconomic advantage and impacting the levels of socioeconomic advantage and disadvantage at both the individual and aggregate level. As shown in Fig. 2.1, both levels of

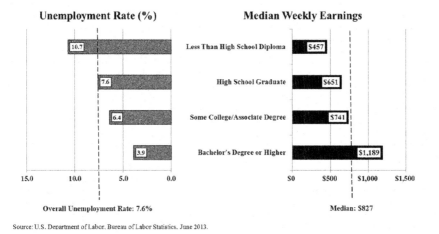

Source: U.S. Department of Labor. Bureau of Labor Statistics. June 2013.

Fig. 2.1 Relationship between educational attainment and unemployment and earnings

unemployment (and hence employment) and income are related to levels of educational attainment. As shown in this figure those with some college or higher levels of education have lower rates of unemployment than those with lower levels of education with 10.7 % of those with less than a high school level of education being unemployed (in 2013) compared to only 3.9 % with a bachelor's degree or higher level of education who were unemployed. Similarly, those with a bachelor's degree or higher level of education had median weekly earning of $ 1,189 compared to median weekly earnings of $ 457 for those with less than a high school level of education.

The interrelationship between education and socioeconomic resources is also apparent when detailed data on their interrelationships are examined. Table 2.9 shows such interrelationships between income and education across occupations and racial/ethnic groups in Texas. What is evident in these data is that no matter what the occupation or the racial/ethnic group, as education increases so does average income. Although differences among racial/ethnic groups remain, education pays for all racial/ethnic groups. For example, Hispanic households with a householder who had less than a high school level of education had a median income of $ 38,221 compared to median household incomes of $ 108,473 for those with a graduate or professional degree. Similarly, for Hispanics employed in operative and laborer positions, increased levels of education are related to increased income with those with less than a high school level of education having an average income of $ 41,273 compared to 57,806 for those with a graduate degree. Although not the only factor, education is a major factor in improving socioeconomic resources for all groups.

Table 2.9 Mean household income by race/ethnicity, educational attainment, and occupation in Texas in 2010. (Source: Ruggles et al. 2010; U.S. Census Bureau, American Community Survey 2010 Public Use Microdata Sample, 2011)

Occupations	Less than high school	High school/ GED	Bachelor's degree	Graduate/ prof degree
NH white				
Managerial and professional	$ 72,446	$ 84,504	$ 120,792	$ 149,819
Technical, sales, and admin	52,930	64,179	112,843	130,471
Precision prod., craft, and repairers	55,079	68,247	94,875	119,057
Operatives and laborers	48,819	61,211	77,581	89,720
Total	50,533	64,975	114,698	144,714
NH black				
Managerial and professional	$ 47,483	$ 52,186	$ 84,758	$ 99,073
Technical, sales, and admin	33,980	43,933	75,416	80,852
Precision prod., craft, and repairers	42,767	53,192	82,885	69,805
Operatives and laborers	37,811	46,296	66,181	63,707
Total	31,072	42,465	78,933	94,490
Hispanic				
Managerial and professional	$ 52,178	$ 61,992	$ 88,036	$ 116,400
Technical, sales, and admin	40,131	50,127	81,005	91,155
Precision prod., craft, and repairers	42,302	51,527	63,695	63,345
Operatives and laborers	41,273	47,455	54,282	57,806
Total	38,221	47,760	81,548	108,473
NH Asian and other				
Managerial and professional	$ 84,172	$ 73,016	$ 103,737	$ 130,603
Technical, sales, and admin	52,764	60,719	86,039	105,469
Precision prod., craft, and repairers	54,594	63,451	70,778	92,041
Operatives and laborers	47,939	49,015	60,895	65,074
Total	50,815	55,496	93,474	123,035

The Potential Impact of Increased Socioeconomic Attainment for Hispanic and Other Minority Populations on the Socioeconomic Future of Texas and the United States

The data presented above indicate that there are substantially lower levels of socioeconomic resources in Hispanic and non-Hispanic black households compared to non-Hispanic white households and these differences (as noted in the literature summarized above) are related to higher levels of education for non-Hispanic whites. In fact, as noted above, a substantial body of literature has developed to suggest that, not only for individuals; but, for populations as a whole (see Cuaresma et al. 2014; Murdock et al. 2014; Leal and Trejo 2010; Angel and Angel 2009) educational attainment may lead to *The Hispanic population of the United* socioeconomic benefits.

Thus across geographic areas within the United States incomes are substantially lower and poverty higher for both Hispanic and non-Hispanic black populations but improve substantially for both groups with higher levels of education. This strongly suggests that increases in education in both groups will lead to increased socioeconomic resources. However, because Hispanics, will provide the major source of population growth in both states such as Texas and the Nation as a whole, it is evident that improvements in education and other socioeconomic resources for Hispanic populations could have particularly substantial impacts on the overall economic conditions of states with large Latino populations, and in the long term, for the Nation.

As one mean of examining the potential impacts of closing the socioeconomic gaps between Hispanic and other minority populations, and non-Hispanic whites, we present the data in Figs. 2.2, 2.3, 2.4, 2.5. The data displayed in these figures indicate the total population-related values for selected socioeconomic factors in Texas attained by closing the differentials between non-Hispanic white and minority populations for the socioeconomic factors indicated. The first bar in the charts shows the 2010 value for the socioeconomic factor indicated and subsequent bars provide projections of the values for these same factors in 2050 under three different assumptions.

The second bar in each figure provides a 2050 projection (for the factor being examined) if the 2010 race/ethnicity specific values of each factor examined do not change (and thus race/ethnicity specific socioeconomic gaps do not change) but the population comes to have the size and race/ethnicity characteristics projected for

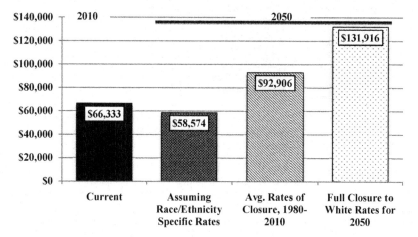

Fig. 2.2 Mean household income in 2010 and projected for 2050 under alternative assumptions of socioeconomic closure between minority and non-Hispanic white households

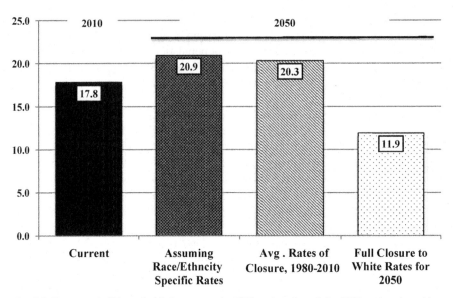

Fig. 2.3 Percent of all households in poverty in 2010 and projected for 2050 under alternative assumptions of socioeconomic closure between minority and non-Hispanic white households

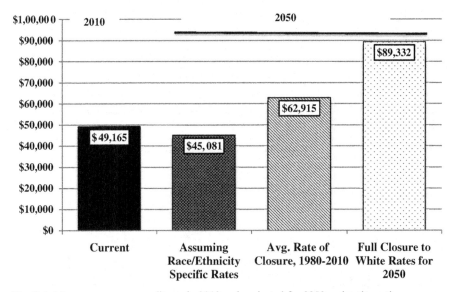

Fig. 2.4 Mean consumer expenditures in 2010 and projected for 2050 under alternative assumptions of socioeconomic closure between minority and non-Hispanic white households

2050 under the assumption of a continuation of 2000–2010 race/ethnicity specific rates of population growth. This bar is labeled the "Assuming Race/Ethnicity Specific Rates".

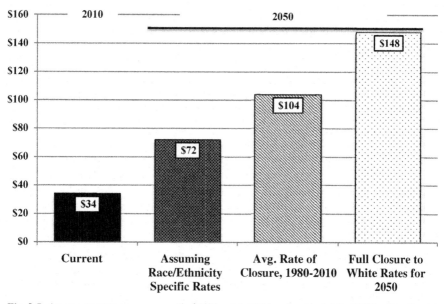

Fig. 2.5 Aggregate state tax revenues (in $Billions) in 2010 and projected for 2050 under alternative assumptions of socioeconomic closure between minority and non-Hispanic white households

The third bar shows 2050 values for the socioeconomic characteristic assuming that the mean (average) rates of closure (i.e., mean change across the 1980–1990, 1990–2000 and 2000–2010 decades) in the difference in the value of the factor for each minority population and that of the non-Hispanic white population continue for each of the decades from 2010 to 2050. This is done for all racial/ethnic groups with 2010 values lower than those for the non-Hispanic white population. The use of the average of three decades of change as the basis for projecting 2010–2050 levels of closure combines data for decades such as 1980–1990 and 1990–2000, which provided indications of some progress in closing gaps between non-Hispanic whites and minorities, with the period (the 2000–2010 decade) in which there was little if any closure. It thus represents a moderate level of change. For racial/ethnic minority groups with higher values than those for non-Hispanic whites for the factor of interest their rates for 2010 were assumed to continue through all decennial periods to 2050.

The fourth and final set of bars in each figure for each socioeconomic factor indicates the results of a projection for 2050 that assumes that by 2050 there is complete closure in socioeconomic differences between racial and ethnic groups (for which there were lower levels of each socioeconomic factor than those for non-Hispanic whites in 2010) to the levels for socioeconomic factors for whites in 2010 by 2050. For any other racial and ethnic group with values that are higher than those for non-Hispanic whites in 2010 a continuation of their 2010 values for the factor were again assumed to continue to 2050.

The values in Fig. 2.2 show that the effect of closure (through education and other factors) in the socioeconomic characteristics of Hispanic and other minority populations with reduced resources to the levels of non-Hispanic whites through increased education would substantially improve the overall socioeconomic characteristics of a population within an area (in this case the State of Texas). If complete closure were to occur mean household income in Texas would increase from the $ 66,333 that it was in 2010 to $ 131,916 in 2050. On the other hand, failure to achieve any closure could lead to a less prosperous population with the average household income of Texans (in the absence of change in socioeconomic differences among Texas rapidly growing Hispanic and other populations) being $ 7,759 lower in 2050 than the average income in 2010 (in 2010 constant dollars).

Similarly, as shown in Fig. 2.3, poverty rates for households, in the absence of change in socioeconomic differentials, could increase from 17.8 % in 2010 to 20.9 % by 2050 or, with complete closure to non-Hispanic income levels, could be reduced to 11.9 % in 2050. The values in Fig. 2.4 indicate that mean consumer expenditures would decrease by nearly $ 4,000 per household in the absence of minority-non-Hispanic white closure or, under complete closure, increase by more than $ 40,000 by 2050 (in 2010 constant dollars). Finally, the values in Fig. 2.5 show that such closure could substantially improve the State of Texas' fiscal resources. Although population growth alone will lead to an increase in total state tax revenues from $ 34 billion in 2010 to $ 72 billion in 2050, if complete socioeconomic closure were to occur the state's annual tax revenues could be $ 148 billion in 2050, an increase of an additional $ 76 billion per year in 2050 compared to the 2050 value assuming no closure (in 2010 constant dollars). Obviously, improvements in the socioeconomic resources of the rapidly growing Hispanic and other minority populations would not only benefit these populations but the overall economy of the area experiencing socioeconomic change. This would be true not only for Texas but for the Nation as a whole.

Conclusions

The data in this chapter suggest that the future both in states, such as Texas, and the United States overall, will be increasingly tied to their minority populations, particularly their Hispanic and Latino populations. Not only will Hispanic population growth, coupled with growth in other minority populations, be responsible for nearly all population change (for 2010–2050) in key states such as Texas, Arizona, California, and New Mexico but increasingly for growth across all regions and metropolitan and nonmetropolitan areas in the Nation. In fact, the projected future population of the United States (for 2060) shows a smaller non-Hispanic white population in 2060 than in 2010. For the Nation, population growth will be entirely due to minority populations with more than 60 % of the net growth from 2010 to 2060 being due to Hispanic population growth alone.

The critical question for such populations in key Hispanic states and the Nation as a whole is what their increased demographic dominance will mean relative to their socioeconomic futures and the socioeconomic future of their states and the United States. With current Hispanic income levels only 60–75 % of those for non-Hispanic whites and with poverty rates that are two to three times as high as those for non-Hispanic whites their increased numerical dominance could, in fact, make them and the Nation poorer and less competitive. On the other hand, if the educational progress of Hispanic and other minority populations can be enhanced, the growth of the Hispanic population could be the key to increased socioeconomic resources not only for this population but also for the Nation as a whole.

What the data in this chapter and the sources from which it is drawn (see Murdock et al. 2014) make evident is that states, such as Texas, and the United States as a whole have a clear choice. These areas can take the steps necessary to close the educational gaps between Hispanic and other minority populations and their non-Hispanic white populations and reap a demographic dividend in socioeconomic resources not evident in other states and nations with older and/or less educated populations or they can become more socioeconomically disadvantaged as a result of population growth. The data presented here show that there are marked advantages to be obtained not only for Hispanic and other minority populations, but for all populations in all geographic regions and states and the Nation overall.

What kind of future it will be is dependent on what all of us do, or fail to do. The challenge is to ensure that all people in the United States have the skills and education they need to be competitive in the world's increasingly international economy. It is a challenge that if met will improve the socioeconomic future not only for Hispanic populations but also for all populations in states such as Texas and for the United States.

References

Abdullah, A. J., Doucouliagos, H., & Manning, E. (2011). *Education and income inequality: A meta-regression analysis.* Paper presented at MAER-NET Colloquium, Wolfson College, University of Cambridge, UK, 16–18 Sept 2011.

Angel, R., & Angel, J. (2009). *Hispanic families at risk: The new economy, work, and the welfare state.* New York: Springer.

Ashenfelter, O., & Rouse, C. (1999). *Schooling, intelligence and income in America: Cracks in the Bell Curve* (Working Paper 6902, Jan). Cambridge: National Bureau of Economic Research.

Bean, F., & Tienda, M. (1987). *The Hispanic population of the United States.* New York: Russell Sage.

Becker, G. S. (1967). *Human capital and the personal distribution of income.* Ann Arbor: University of Michigan Press.

Bergad, L. W., & Klein, H. (2010). *Hispanics in the United States: A demographic, social and economic history, 1980–2005.* London: Cambridge University Press.

Bloom, D. E., Canning, D., Fink, G., & Finlay, J. (2009). Fertility, female labor force participation and the demographic divide. *Journal of Economic Growth, 12,* 79–101.

Bowman, M., & Anderson, C. (1963). Concerning the role of education in development. In C. Geetz (Ed.), *Old societies and new states*. Glencoe: Free Press.

Card, D. (1999). The casual effect of education on earnings. In O. Ashenfelter & D. Card (Eds.), *Handbook of labor economics* (Vol. 3, pp. 1801–1834). Amsterdam: Elsevier Science.

Cheeseman-Day, J., & Newburger, C. (2002). *The big payoff: Educational attainment and synthetic estimates of work-life earnings*. Current Population Reports (July). U.S. Department of Commerce Economics and Statistics Administration. Washington, DC: U.S. Census Bureau.

Cohen, D., & Soto, M. (2007). Growth and human capital: Good data, good results. *Journal of Economic Growth, 12*, 51–76.

Cohn, E., & Addison, J. (1998). The economic returns to lifelong learning. *Education Economics, 6*(3), 253–308.

Cuaresma, J., Lutz, W., & Sanderson, W. (2014). Is the demographic divide an education divide. *Demography, 51*, 299–315.

Davern, M. E., & Fisher, P. J. (2001). *Household net worth and asset ownership: 1995*. Current Population Reports (Feb). U.S. Department of Commerce, Economics and Statistics Administration. Washington, DC: U.S. Census Bureau.

Denison, E. (1962). *The sources of economic growth in the United States and alternatives before us* (*Supplementary Paper No. 13*). New York: Committee on Economic Development.

Hanoch, G. (1967). An economic analysis of earning and schooling. *Journal of Human Resources, 2*(Summer), 310–329.

Julian, T., & Kominski, R. (2011). *Education and synthetic work-life earnings estimates* (American Community Survey Reports ACS-14). Washington, DC: U.S. Census Bureau. Accessed April 2012.

Leal, D., & Trejo, S. (2010). *Latinos and the economy*. New York: Springer.

Lutz, W. (2013). Demographic metabolism: A predictive theory of socioeconomic change. *Population and Development Review, 38*(Suppl.), 283–301.

McKinsey and Company. (2009). *The economic impact of the achievement gap in America's schools, summary of findings*. New York: Social Sector Office.

Mincer, J. (1974). *Schooling, experience and earnings*. New York: Columbia University Press.

Murdock, S. (1995). *An America challenged: Population change and the future of the United States*. Boulder: Westview Press

Murdock, S. (2014). *Population Change in Texas: Implications for Education and the Socioeconomic Future of Texas*. Available at http://www.thecb.state.tx.us/download.cfm?downloadfile=95499447-F8E1-C85A-D3E9B300CFA20F3E&typename=dmFile&fieldname=filename.

Murdock, S., Hoque, N., Michael, M., White, S., & Pecotte, B. (1997). *The Texas challenge: Population change and the future of Texas*. College Station: Texas A & M University Press.

Murdock, S., Hoque, N., White, S., Pecotte, B., You, X., & Balkan, J. (2003). *The new Texas challenge: Population change and the future of Texas*. College Station: Texas A & M University Press.

Murdock, S., Cline, M., & Zey, M. (2012). The children of the Southwest: Demographic and socioeconomic characteristics impacting the future of the Southwest and the United States. *Big ideas: Children in the Southwest* (pp. 11–29). Washington, DC: First Focus.

Murdock, S., Cline, M., & Zey, M. (2013) The history of Texas population. In S. Haynes (Ed.), *Major problems in Texas history* (2nd ed.). Arlington: University of Texas-Arlington Press.

Murdock, S., Cline, M., Zey, M., Jeanty, P., & Deborah, P. (2014). *Changing Texas: Implications of addressing or ignoring the Texas challenge*. College Station: Texas A & M University Press.

Ramos, J. (2005). *The Latino wave: How Hispanics are transforming politics in America*. New York: Harper Collins Publishers.

Romer, P. (1990). Human capital and growth: Theory and evidence. *Carnegie-Rochester Series on Public Policy, 32*, 251–286.

Schultz, T. (1963). *The economic values of education*. New York: Columbia University Press.

Schultz, T. (1968). Resources for higher education: An economist's view. *Journal of Political Economy, 76,* 327–347.

Tolley, G., & Olson, E. (1971). The interdependence between income and education. *Journal of Political Economy, 79,* 460–480.

U.S. Bureau of Labor Statistics (2011). *Current population survey.* Washington, DC: U.S. Bureau of Labor Statistics.

U.S. Census Bureau. (2012). *2012 population national projections [MRDF].* Washington, DC: U.S. Census Bureau.

Chapter 3
The Train Has Left the Station: Latino Aging in the New South

Gregory B. Weeks and John R. Weeks

Introduction

Nowhere in the United States has seen a more rapid growth of the Latino population in recent years than the South. In the 1990s the region, and especially Georgia and North Carolina, became a popular destination for many different people because of its booming economy, low cost of living, and temperate climate. References to the "new" South reflect this dynamism, moving away from traditional images of economic backwardness and racial segregation (Weeks et al. 2007; Smith and Furuseth 2008). Since it is a new migrant destination, the region's Latino population is quite young and to this point has little direct political influence. Indirectly, influx of Latinos has created an initial backlash, largely due to the fact that a high proportion of adults are undocumented immigrants. The politics will almost certainly shift over time, however, as the Latino population ages and continues to increase as a share of the total population. In this analysis we demonstrate what we mean by "the train has left the station" by employing a unique set of population projections stretching to the year 2040.

Using U.S. Census data and other sources to develop projections, the core of our argument is that as the cohort of older (65 years and over) Latinos grows in North Carolina, there will be concomitant political shifts. Children who are citizens will eventually become eligible to vote, legislative districts will be transformed, and Hispanic adults will be taking care of a growing elderly population. Although much has been written about Latinos and politics, almost no work has addressed the issue of aging more specifically.

The percentage of the Latino population eligible to vote will be much larger by 2040 than it is currently, which will slowly—and imperfectly—increase that popu-

G. B. Weeks (✉)
University of North Carolina, Charlotte, NC, USA
e-mail: gbweeks@uncc.edu

J. R. Weeks
San Diego State University, San Diego, CA, USA

© Springer International Publishing Switzerland 2015
W. A. Vega et al. (eds.), *Challenges of Latino Aging in the Americas*,
DOI 10.1007/978-3-319-12598-5_3

lation's political leverage. At the same time, we examine how electoral politics, especially gerrymandering, suggest that the South will have a different experience than other parts of the United States because of race. The chapter will first discuss the methods used to generate the population projections, analyze the results, and proceed to examining the potential political ramifications of the likely demographic changes that the New South will experience over the next several decades.

Background

As we have argued elsewhere, understanding migration patterns from Latin America to the United States, and to the southeastern part of the country specifically, requires examining demographic "irresistible forces" (Weeks and Weeks 2010). This refers to the demographic transformations taking place in both Latin America and the US that are drivers of migration, mitigated, of course, by economic and political factors. Although at times it can seem like irresistible forces of demography are meeting politically immovable objects, we contend that they are powerful enough that they cannot be legislated away. In other words, policy makers can respond to demographic forces, but cannot eliminate them or wish them away.

What does that mean? Although it is slowly coming to an end, the boom years of the 1990s and early 2000s were marked by a "demographic fit," whereby a youth bulge in Latin America coincided with a youth "dent" in the United States as the population has been growing increasingly older. Just as jobs—many of them physically demanding—were plentiful in the United States, there were too few young people to take them. Meanwhile, in Latin America (and, of course, particularly Mexico) there were too many young workers for too few jobs. The result was a migration surge, and the New South was a major destination.

Policy reactions in the United States can be best understood from a political demography standpoint. The influx of migrants fostered hostility in some quarters, which led to increased federal enforcement budgets along with state and local restrictions on undocumented immigrants. Paradoxically, the increased border security since 9/11 has served to "fence in" undocumented immigrants who previously would have circulated between Mexico and the US. The increased risks and costs of crossing the border have led to people staying once they safely cross into the US, thus giving rise to the unprecedented increase in the number of undocumented immigrants (Massey 2008). Given the demographic reality, however, there was no way to stop the forces underway. Those irresistible forces are literally changing the face of the South.

In this chapter, we extend our previous study from the historical and contemporary perspective to consider the future. Over the past two decades, economic growth in the South combined with the demographic fit drove large scale in-migration of relatively young workers and their families, often with children born after the parents' arrival in the US and who are thus United States citizens. That has brought with it a host of economic and political challenges that include a strain on public

Table 3.1 Latino population in the New South, 2010. (Source: U.S. Census Bureau 2010 Census of Housing and Population)

State	Population in 2010		Percent Latino
	Total	Total Latino	
Georgia	9,687,653	853,689	8.8
North Carolina	9,535,483	800,120	8.4
Virginia	8,001,024	631,825	7.9
Arkansas	2,915,918	186,050	6.4
South Carolina	4,625,364	235,682	5.1
Tennessee	6,346,105	290,059	4.6
Louisiana	4,533,372	192,560	4.2
Alabama	4,779,736	185,602	3.9
Mississippi	2,967,297	81,481	2.7
Total	53,391,952	3,457,068	6.5

schools, use of federal and state social services, and the appropriate application of law enforcement, not to mention myths and outright untruths about the economic impact of undocumented immigrants. As we look ahead to 2040, however, the issues shift as the older population grows and those who are currently children simultaneously become political participants and caregivers to their aging parents.

For the purposes of this study, we define the "South" as states that were part of the Confederacy, but that also previously did not have any experience with Hispanic migration. Therefore Florida and Texas are excluded, leaving Alabama, Arkansas, Georgia, Louisiana, Mississippi, North Carolina, South Carolina, Tennessee, and Virginia. These are all states that seceded from the Union but were also historically non-Hispanic. That is no longer true, as can be seen in Table 3.1, which shows the total and Latino populations for each state as of the 2010 Census. There were nearly 3.5 million Latinos in the nine southern states in 2010, accounting for 6.5 % of the population. The three most populous states of Georgia, North Carolina, and Virginia are disproportionately home to Latinos, accounting for 50 % of the total population in the nine states, but 66 % of the Latinos. The state we will examine in most detail is North Carolina which, as we show below, is projected to have the highest increase in its Latino population over the next three decades.

Population Projections Taking Hispanic Population Growth into Account

The U.S. Census Bureau no longer produces state-level population projections, so we have generated our own projections for each of the nine states of the "New South." We have projected the population out to the year 2040 on the assumption that policy initiatives should reflect the possible demographic changes taking place in the next few decades. Too short a time frame (e.g., a single decade) is unlikely to capture the trajectory of change that should be accounted for, whereas projections over too long a time frame (e.g., a half-century) will have such high levels of uncertainty that they may not be useful for policy purposes.

Our projections utilize a sophisticated cohort-component method of projection (Smith et al. 2001; Swanson and Tayman 2012), with which we are able to model the projections separately for Latinos and non-Latinos, and separately for males and females within each group, on the basis of the beginning age structure in each state according to the 2010 Census, taking into account the different patterns of fertility and mortality for the two groups, and then making differing assumptions about the future pattern of migration into each state. For both Latinos and non-Latinos we held the mortality and fertility schedules constant over time at the circa 2010 levels, applying national level data to each state, in the absence of state level data for all states. The projections are based on 5-year age groups, and so the population is projected forward 5 years at a time.

Mortality data were drawn from the 2008 life tables by race/ethnicity produced by the US Centers for Disease Control (Arias 2012). These were the most recent life tables available at the time we began our analysis. Life expectancy at birth for Latino females in 2008 was 83.3 years, compared to 80.7 for non-Latino white women and 80.6 for women of all race/ethnic groups combined. We used the latter as a proxy for the non-Latino population, recognizing that it is a conservative approach since life expectancy would be somewhat lower for this group had we been able to take out Latinos from its calculation. For Latino males, life expectancy in 2008 was 78.4, compared to 75.9 for non-Latino white males and 75.6 for all males. These data show that we can expect a higher fraction of Latinos to survive into old age than is true for the rest of the population, so that alone has implications for future patterns of aging.

Fertility data refer to 2010 and are also from the US Centers for Disease Control (Martin et al. 2012). The total fertility rate (expected number of lifetime births based on current age-specific fertility rates) for Latino women in 2010 was 2.35 children. For non-Latino whites, the rate was 1.79, for all groups combined the rate was 1.93, and for non-Latinos it was 1.83. In this case, we do have non-Latino birth rates, so we used them instead of the figure for all groups combined. As of 2010, Latino women were having 0.5 more children each than were non-Latino women, and this implies that the Latino population will be an ever increasing share of the younger ages in the population out to 2040, where our projection ends. It is likely that Latino fertility will drop over time at the national level, especially considering that the total fertility rate in Mexico is now down to 2.2 children per woman. However, to the extent that Latino migration into the New South is driven especially by undocumented immigrants (Weeks and Weeks 2010), the Latino fertility in this region is likely to remain above the average for Latinos nationally, and thus above the non-Latino level.

The combination of mortality and fertility levels implies that the number and percentage of Latinos at both the younger and older ages will increase over time in each state, even if there were no continued in-migration of Latinos. In fact, the decade between 2000 and 2010 witnessed high levels of migration into all of the states of the New South. What if this was to continue unabated out to 2040? That possibility represents our "high" projection scenario, in which the 5-year net migration rates (in-migrants minus out-migrants) experienced by each age, sex and ethnic group (Latino or non-Latino) in the 5-year period before the 2010 Census are held constant for each subsequent 5-year period up to 2040. The "low" scenario maintains the rates for non-Latinos, but then applies the non-Latino migration rates to Latinos. Thus, in the low scenario we assume that Latinos had a one-time increase in migration between 2000 and 2010, and then reverted to the level of non-Latinos beyond 2010. The net migration data by age, sex, and ethnic group are drawn from the work of researchers at the Center for Demography and Ecology, University of Wisconsin-Madison (Winkler et al. 2013), which are available for download on the internet at http://www.netmigration.wisc.edu.

We believe that the high projection has a low probability of occurring, since it represents the likely upper limit of the impact of Latinos in the New South. The region saw tremendous Latino growth in the 2000–2010 period, at a rate that is difficult to imagine continuing. Its value lies in showing us what is at least possible, even if not likely. The low projection has a higher probability of occurring than does the high projection, but we believe that the low projection is an underestimate of what policy-planners should expect. Serendipitously, as we were undertaking our population projections, a group of researchers at the University of Virginia completed and made available their set of state-by-state population projections (Demographics and Workforce Group 2013). These projections are also out to 2040, and are based on the Hamilton and Perry (1962) method, which is "a reduced form of the cohort-component method that requires less detailed data and captures the major components of population change (births, deaths, and migration) in a general way" (Demographics and Workforce Group 2013, p. 1). They also produced projections for Hispanic and non-Hispanic populations in each state, allowing us to compare their projections (which we refer to as the "medium" projections) with our high and low scenarios. Their "medium" projections fall within our high and low projections (with one exception discussed below) but they are not consistently closer to either our high or low projections, probably because they chose to mitigate the state-level impact of the high 2000–2010 migration rates of immigrant groups by applying ratios for data at the national level to the state-level data, whereas we used state-specific migration rates for Latinos and non-Latinos.

What Do the Population Projections Tell Us?

The major conclusion from the three sets of population projections is that, as we imply in the title of the paper, the train has left the station when it comes to the Latinoization of the New South. This is especially noticeable at the younger ages, since migrants tend to be young adults of child-bearing age, and they are indeed having children. But the Latino population will almost certainly become an increasingly larger fraction of the older population in the South, with a burgeoning group of younger people, most of them citizens due to their birth in the US, who may be politically active on behalf of themselves and the elderly within the Latino community. Table 3.2 summarizes the three projections—high, medium, and low—for the total population in 2040.

It is striking that by 2040 even the low projection suggests that nearly one in ten residents of the New South will be Latino, with the percentages even higher than that in North Carolina (the highest at 12.1%, Georgia, and Virginia. Only Alabama and Mississippi are substantially below the regional average). The high projections show that in North Carolina nearly one in four residents could be Latino by 2040, and more than one in five in Georgia and Virginia could be Latino. The medium projections also show the power of the growth in the Latino population, with 17% of the population of both Georgia and North Carolina projected to be Latino. Five of the nine states are above 10% even in the medium projections. Again, only Alabama and Mississippi are bucking the regional trend, but the Latino population will still be a higher fraction of the total population in 2040 than it was in 2010.

Turning now to specific age groups of interest, Table 3.3 summarizes the results for each of the nine state of the New South, with states ordered by the proportion of the younger population that was Latino as of the 2010 Census. North Carolina and Georgia have been the most attractive destinations for Latinos, as we discussed earlier, with Virginia close behind. North Carolina and Georgia boomed in the 1990s, the former with growth related to banks and the latter sparked by the Olympics. Those created jobs and then a host of ancillary low-wage employment. In these three states, even the medium projections suggest that at least one in five people under the age of 20 in 2040 will be Latino. Furthermore, by 2040, even the low projections in each of those three states suggest that one in 15 older people in 2040 will be Latino.

It can be seen in Tables 3.1 through 3.3 that the least attractive state in the South among Latinos is Mississippi, and this state provides the one exception to the generalization that the University of Virginia projections lie between our high and low projections. At ages 65 and older, our high projection suggests that 2.5% will be Latino, while our low projection is at 2.3% and the "medium" projection is 2.7%. The difference is caused by the fact that between 2000 and 2010 older people of all ages, but especially Latinos, were experiencing net out-migration from Mississippi, a phenomenon not picked up by the University of Virginia projection method.

The decision to leave a state is based on a number of factors, and one that is particularly difficult to predict is political. Along with Alabama, Mississippi enacted some of the harsher state-level laws aimed at undocumented immigrants, such as requiring immigration checks of anyone who is arrested. It is difficult to measure the emigration impact of such laws, but reasonable to suggest that a reputation for hostility (combined with a long history of discriminatory practices) could well act as a prompt to migrant residents to leave as well as a deterrent to would-be migrants. Our projections seek to take demography into full account, but changes in political climate can happen quickly (or, indeed, be reversed).

Table 3.2 Projections of the total Latino population (all ages) to 2040. (Source: See text for derivation)

State	Projection of the total population			Projection of Latino population			Percent Latino		
	High	Med	Low	High	Med	Low	High	Med	Low
Georgia	13,929,731	13,599,293	12,376,706	3,003,522	2,309,259	1,450,497	21.6	17.0	11.7
North Carolina	14,213,002	12,892,556	12,235,221	3,455,767	2,190,853	1,477,987	24.3	17.0	12.1
Virginia	10,585,660	10,415,576	9,470,130	2,130,326	1,554,133	1,014,796	20.1	14.9	10.7
Arkansas	3,725,042	3,465,661	3,311,585	718,728	452,488	305,270	19.3	13.1	9.2
South Carolina	6,379,894	5,991,058	5,600,679	1,171,901	612,245	392,687	18.4	10.2	7.0
Tennessee	8,509,284	7,918,884	7,374,310	1,615,802	756,735	480,828	19.0	9.6	6.5
Louisiana	5,107,369	4,751,516	4,672,177	1,216,589	446,524	290,808	23.8	9.4	6.2
Alabama	6,241,523	5,538,151	5,315,742	677,748	355,264	242,556	10.9	6.4	4.6
Mississippi	3,402,457	3,337,170	3,140,740	372,923	178,247	111,206	11.0	5.3	3.5
TOTAL	72,093,961	67,909,865	63,497,290	14,363,305	8,855,748	5,766,635	19.9	13.0	9.1

Table 3.3 Percent Latino by age group in 2010 and in 2040: three projections

State	Percent Latino among less than 20 years of age				Percent Latino among 20–64				Percent Latino among 65+			
	2010	2040 high	2040 medium	2040 low	2010	2040 high	2040 medium	2040 low	2010	2040 high	2040 medium	2040 low
North Carolina	13.1	27.6	23.1	15.1	7.8	28.1	17.1	12.8	1.4	8.5	7.5	6.7
Georgia	12.3	23.5	22.3	14.3	8.4	24.8	16.7	12.3	1.9	8.6	8.1	6.8
Virginia	10.8	22.4	20.0	12.9	7.8	22.9	14.9	11.2	2.0	8.6	7.8	6.8
Arkansas	10.3	23.7	18.3	11.9	5.9	21.7	13.0	9.7	1.0	6.5	6.0	5.0
South Carolina	7.4	17.9	14.2	8.6	5.0	23.4	10.4	7.6	1.0	8.9	4.7	3.8
Tennessee	7.0	20.6	13.4	8.3	4.3	23.2	9.6	7.0	0.8	4.4	4.0	3.3
Alabama	5.8	17.9	11.3	6.8	3.8	25.3	8.2	6.0	0.8	3.6	3.3	2.8
Louisiana	4.8	14.7	9.9	6.0	4.4	15.6	7.3	5.4	2.1	4.8	4.5	3.8
Mississippi	3.5	11.2	7.2	4.1	2.8	13.9	5.3	3.8	0.7	2.5	2.7	2.3

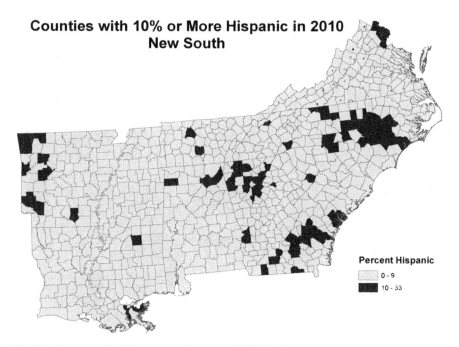

Fig. 3.1 Counties with 10 % or more Hispanic in 2010 New South

Figure 3.1 uses Census data to show where Latinos have settled in the New South, based on 2010 Census data. They are dispersed, though there is a clear preference for North Carolina and Georgia, which have the greatest number of census tracts with populations that are 10 % or more Hispanic. In general, there are more in the northeast part of the region and fewer in the southwest part.

Figure 3.2 shows the spatial clustering of Latinos by county in the New South in 2010, based on the same data as Fig. 3.1. We used the Getis-Ord Gi* statistic to measure clustering (Getis and Ord 1992; Mitchell 2005). Hot spots are clusters of counties in which the proportion Latino among neighboring counties is higher than would be expected by chance alone, whereas cold spots are counties in which the proportion Latino among neighboring counties is lower than would be expected by chance. Thus, positive values of Gi* that exceed a z-score of 1.96 (the 0.05 level of statistical significance) indicate spatial association of high values (hot spots) and negative values of Gi* that are less than -1.96 indicate spatial association of low values (cold spots). Hot spots are especially noticeable through the entire central section of North Carolina, as well as in the area surrounding Atlanta (both of which we would expect), a swath of agricultural counties in southeast Georgia and the southern tip of South Carolina, and agricultural counties in western Arkansas. Cold spots are in the Appalachian range in western Virginia, as well as in parts of Alabama, Mississippi, and Louisiana (with the exception of New Orleans, which is a Latino hot spot). These hot and cold spots are generally consistent with the idea that

Spatial Clusters of Hispanics by County 2010
New South

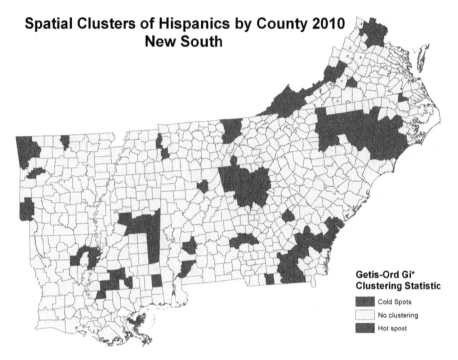

Getis-Ord Gi*
Clustering Statistic

■ Cold Spots
□ No clustering
■ Hot spost

Fig. 3.2 Spatial clusters of Hispanics by county 2010 New South

Latinos are migrating to places with above average economic opportunities. This matches anecdotal evidence about the popularity of these locations for Hispanic migrants.

The Implications of an Aging Latino Population

Currently, the Latino population is quite young. As we can see in Fig. 3.3, the small retirement-age population is concentrated largely in pockets in Virginia, Georgia, and Louisiana (specifically New Orleans). Nonetheless, like the general population there is also dispersion, with smaller concentrations of older Latinos found all across the region. Given our projections, over time there will be many more areas with relatively high numbers of Latinos aged 65 and older, which can then become a political issue. However, this has yet to become a topic of scholarly analysis.

There is a growing literature on the Latino population in the South, which has been concentrated primarily in Geography. It has tended to center largely on challenges related to newly established migrant communities. There is, for example, considerable discussion of whether Latinos are viewed primarily as "other" (Winders 2011) or if there also evidence that assimilation is taking place, particularly in rural areas (Marrow 2011). Either way, this population is now permanent and will

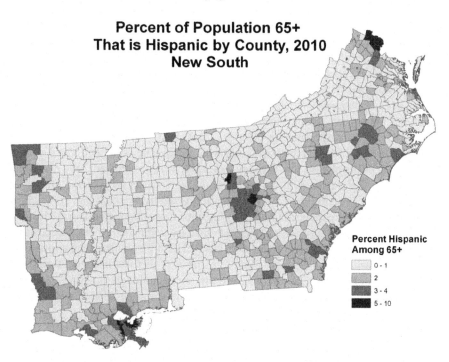

Fig. 3.3 Percent of population 65 + that is Hispanic by county, 2010 New South

continue growing, and as yet very little attention has been paid to the demographic implications.

Further, little is written about Latino politics in the South. Numerous general studies on the topic, however, provide clues about how politics will likely play out. For example, given the relatively high number of non-citizens, political influence is low, and even naturalization does not necessarily increase political participation because of the relatively low socio-economic status of Latinos (Levin 2013). Thus, even an increase of voters does not immediately translate into political influence, which we will examine below.

At the same time, certain issues do help spur political mobilization. This can include media coverage of the high social and national security costs of immigration. Merolla et al. (2013) note the future strain on social welfare programs in the United States as the Latino population in the U.S. ages, but in this chapter we examine the more specific question of the political, rather than policy, implications. That is, we know more much about the social aspects of Latino aging than we do about the political.

What these studies ignore, however, is aging. By 2040, the Hispanic population will no longer be "new." What will be the political implications? A recent study outlines the literature on the policy implications of Latino aging (Torres-Gil et al. 2013), noting that the rapid aging of the Hispanic population in the United States as well as that of Mexico has started to generate fruitful research on policy options. Policy, though, is not synonymous with politics. The fact that certain policies might be optimal from a public health standpoint does not necessarily translate into the political support required to obtain the votes to make them pass.

One important issue for the future political influence of the Latino population relates to congressional districts. Historically, of course, the primary division in the South has been binary: black and white. When they draw districts each decade with the new Census, state legislatures have done their best within the constraints of the Voting Rights Act to use those two categories to separate white and black voters. Districts are political tools that in many cases practically guarantee that a candidate of one party or the other will win because the voters lean so overwhelmingly in one direction. Currently, the Latino population is relatively small and so much of the South has not experienced the development of more majority-minority districts at the federal level. These refer to districts that are comprised of populations that nationally are racial or ethnic minorities, but which in given districts comprise the majority. In turn, other districts—often a majority of districts in a given southern state—are majority-majority and are similarly impenetrable to the opposing party. However, some states, such as North Carolina, are seeing the development of such districts at the state and local levels (Peralta and Larkin 2011).

But in which districts will these changes occur? Our research on Charlotte, North Carolina shows that Latinos are moving to the suburbs and are not concentrated in enclaves (Weeks et al. 2007). A similar dynamic holds for Durham, NC (Flippan and Parrado 2012). This suburbanization is evident elsewhere in the South as well. Since the population is dispersed, it will hold a greater share even in districts that are currently predominantly white. What this means is that districts that currently are majority white will possibly become majority-minority. When combined with districts that are majority black, the implications are potentially quite large.

This is also true for state-level legislative districts, and even the local level. The suburbanization, combined with dispersion, means that politically drawn boundaries that were once predictably leaning toward one party or another because of racial composition will inevitably change. Irresistible forces have important political consequences.

According to the Pew Hispanic Center, in the 2012 presidential election Latinos voted for Barack Obama over Mitt Romney by a 71–27% margin (Lopez and Taylor 2012). There is a dip for Latinos aged 65 and over, but even then Obama received 65%. Meanwhile, about 93% of African American voters chose Obama in 2012. Overall, these mirror support for the Democratic Party over the Republican Party. Combined, then, the two groups represent a solid Democratic force. There is no guarantee that the two would necessarily share commitment to the same interests—particularly as immigration reform is not high on the list of priorities for African American voters—but they would still be affecting the electoral foundation of the Republican Party.

In the long-term, then, the share of the white Republican vote will decrease, while the Democratic vote will increase because the total number of majority white districts in the South—which is the most Republican region of the country—will be

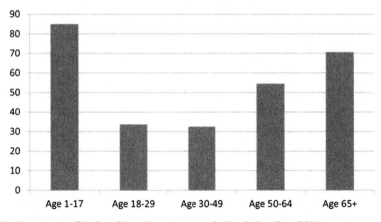

Fig. 3.4 Percentage of Latino citizens by age groups in North Carolina, 2010

on the decline. Some states that for years have been solidly red, like North Carolina, are already "purple," having gone for a Democratic candidate in the 2008 presidential election for the first time since 1964, and then for Republicans in the 2012 election. As the number of Latinos in southern states grow, they will also lose their distinctively deep red label.

Across all groups, older citizens vote in greater numbers than younger. For example, according to the Census Bureau, in the in 2012 election the percentage of eligible adults who voted ranged from 41.2 % for 18- to 24-year-olds, to a high of 71.9 % for those 65 and older (File 2013).Therefore, we can expect that the voting share of the Latino population will increase as it ages, and that it will be voting predominantly for the Democratic Party.

At the same time, Latinos nationwide tend to vote less than whites or blacks. In the 2012 presidential election, a record 11.2 million Latinos voted but turnout was 48 % of registered voters, which was actually a decrease from 49.9 % in 2008 (Lopez and Gonzalez-Barrera 2013). In the case of North Carolina, as of 2013 Latinos constitute only 1.8 % of all registered voters (http://www.app.sboe.state.nc.us/webapps/voter_stats/results.aspx?date=10–19-2013). The political changes that the region will experience are therefore much more relevant for 2040 than they are for the present day.

It is important to note that as a relatively new migrant destination, there is a sizeable undocumented population in the South. However, there is a large cohort of young Latinos that is not yet of voting age, but by 2040 will be adult. The cohort that in 2010 was aged 1–17 (that is, those are not yet of voting age) will in 2040 be between the ages of 31 and 47, and as citizens they will be voting. Figure 3.4 shows the distribution of citizenship among age groups of Latinos in North Carolina. The proportion of undocumented immigrants will almost certainly decrease since we anticipate that the demographic fit between the US and Mexico, in particular, will have ended well before that date.

As the region experiences demographic shifts, political interests will shift as well. In particular, the needs of the older population will become more prominent.

At the moment, working age Latinos in the New South are not supporting a large number of aging parents. Since the New South has been a destination for only a bit more than a decade, their parents are still likely to be living back home, whether that be another state (generally western) or in another country. As yet, studies of Hispanic public opinion—already an underdeveloped area of study—have not examined aging at all (e.g. see Abrajano and Alvarez 2011). As more Latinos bear the costs associated with aging parents, the issue will become more salient.

Conclusions

In the South, as in much of the United States, the demographic train has left the station. For over a decade the region has been attractive to migrants leaving either a Latin American country or areas of the United States with weaker economies and/or higher costs of living. Our projections going out to 2040 show continued growth under virtually all assumptions, signaling a permanent shift in what had traditionally not been a destination for Hispanics. This demographic shift has important political repercussions, in large part because currently Latino voters overwhelmingly support the Democratic Party.

Further, even under assumptions of low future growth the size of the older Latino population will be significant. It is likely that the 2040 population will be something that falls within our high and low projections. The key point, however, is that either way it will grow substantially. As noted, there are many different public health policy implications, but politics will also change. The number of Latino voters, which at this point is still relatively low, will go steadily upward. Older voters go to the polls more frequently than younger, so older voters as well as their caregivers will gradually constitute a greater share of the electorate.

There are several strands of future research that merit greater attention. First, what is the relationship between voting behavior and age of parents? In particular, there is the question of whether caring for an older parent affects both propensity to vote and the likelihood of voting for a specific party. Second, what are the political dynamics between Latinos and African Americans in the New South? To what degree can they coordinate as minority-majority districts grow? The train has left the station, but it is not yet clear exactly where it is headed.

References

Abrajano, M., & Alvarez, R. (2011). Hispanic public opinion and partisanship in America. *Political Science Quarterly, 126*(2), 255–285.

Arias, E. (2012). United States life tables, 2008. *National Center for Health Statistics, Vital Health Reports, 61*(3).

Demographics and Workforce Group. (2013). State and national projections methodology: Weldon Cooper center for public service, University of Virginia.

File, T. (2013). The diversifying electorate—voting rates by race and Hispanic origin in 2012 (and other recent elections). *US Census Bureau Population Characteristic* P20–568.

Flippan, C., & Parrado, E. (2012). Forging Hispanic communities in new destinations: A case study of Durham, North Carolina. *City & Community, 11*(1), 1–30.

Getis, A., & Ord, J. K. (1992). The analysis of spatial association by use of distance statistics. *Geographical Analysis, 24*(3), 189–206.

Hamilton, C. H., & Perry, J. (1962). A short method for projecting population by age from one decennial census to another. *Social Forces, 41,*163–170.

Levin, I. (2013). Political inclusion of Latino immigrants: Becoming a citizen and political participation. *American Politics Research, 41*(4) 535–568.

Lopez, M., & Gonzalez-Barrera, A. (2013). Inside the 2012 Latino electorate. *Pew Research Hispanic Trends Project*, June.

Lopez, M., & Taylor, P. (2012). Latino voters in the 2012 election. *Pew Research Hispanic Trends Project*, November.

Marrow, H. (2011). *New destination dreaming: immigration race, and legal status in the rural American South*. Stanford: Stanford University Press.

Martin, J. A., Hamilton, B. E., Ventura, S. J., Osterman, M. J. K., Wilson, E. C., & Mathews, T. J. (2012). Births: Final data from 2010. *National Center for Health Statistics, Vital Health Reports,* 61(1).

Massey, D. (Ed.). (2008). *New faces in new places: The changing geography of American immigration*. New York: Russell Sage Foundation.

Merolla, J., Pantoja, A., Cargile, I., & Mora, J. (2013). From coverage to action: The immigration debate and its effects on participation. *Political Research Quarterly, 66*(2), 322–335.

Mitchell, A. (2005). *The ESRI guide to GIS analysis, Volume 2: Spatial measurements & statistics*. Redlands: ESRI Press.

Peralta, J., & Larkin, G. (2011). Counting those who count: The impact of Latino population growth on redistricting in southern states. *PS: Political Science & Politics, 44*(3), 552–561.

Smith, H. A., & Furuseth, O. J. (2008). The 'Nuevo South': Latino place making and community building in the middle-ring suburbs of Charlotte. In A. Singer, S. W. Hardwick, & C. B. Brettell (Eds.), *Twenty-first century gateways: Immigrant incorporation in Suburban America*. Washington, DC: Brookings Institution Press.

Smith, S., Tayman, J., & Swanson, D. (2001). *Population projections for states and local areas: methodology and analysis*. New York: Plenum Press.

Swanson, D. A., & Tayman, J. (2012). *Subnational population estimates*. Dordrecht: Springer.

Torres-Gil, F., Suh, E., & Angel, J. (2013). Working across borders: The social and policy implications of aging in the Americas. *Journal of Cross-Cultural Gerontology, 28,* 215–222.

U.S. Census Bureau (2010). Census of Housing and Population. www.census.gov. Accessed 2013.

Weeks, G. B., & Weeks, J. 2010 *Irresistible forces: Latin American migration to the United States and its effects on the south*. Albuquerque: University of New Mexico Press.

Weeks, G. B., Weeks, J. R., & Weeks, A. J. (2007). Latino immigration to the U.S. south: 'Carolatinos' and public policy in Charlotte, North Carolina. *Latino(a) Research Review, 6*(1–2), 50–72.

Winders, J. (2011). Representing the immigrant: Social movements, political discourse, and immigration in the New South. *Southeast Geographer, 51*(4), 596–614.

Winkler, R. L., Johnson, K. M., Cheng, C., Voss, P. R., & Curtis, K. J. (2013). County-specific net migration by five-year age groups, Hispanic origin, race, and sex 2000–2010. In CDE Working Paper No. 2013-04. : Center for Demography and Ecology, University of Wisconsin, Madison, WI.

Chapter 4
Of Work and the Welfare State: Labor Market Activity of Mexican Origin Seniors

Mariana Campos Horta and Marta Tienda

Several developments motivate our study of Mexican origin seniors' labor market activity. First, the Mexican origin population has grown in size even as it has become more diverse by birthplace, legal status and economic standing. Between 1970 and 2010 the U.S. Mexican origin population almost quadrupled, rising from approximately 9 million to roughly 34 million (Bean and Tienda 1987; Gonzalez-Barrera and Lopez 2013). Immigration figured prominently in the growth of the U.S. Mexican origin population, especially during the 1980s and 1990s (Tienda and Mitchell 2006). Although U.S.-bound migration from Mexico slowed after the Great Recession, Mexico remains the largest single source of U.S. immigrants. Mexicans account for 13 % of all legal permanent residents admitted (LPRs) since 2010 (Monger and Yankay 2012); nearly one in five new LPRs admitted between 2000 and 2005 (U.S. Department of Homeland Security 2006); and over half (52 %) of the unauthorized immigrant population in 2012 (Passel et al. 2013). Currently over one-third of the U.S. Mexican origin population is foreign-born[1].

Second, the foreign-born population is growing older due both to aging of long-term immigrants and to the rise of late-age immigration (Carr and Tienda 2013; Batalova 2012; Terrazas 2009; Leach 2008; Wilmoth 2012). Although the Hispanic population, and Mexicans in particular, are notable for their youthfulness relative to

---

[1] Most researchers use the terms foreign-born and immigrant interchangeably but technically only persons granted legal permanent residents are immigrants. Following convention, we use the terms immigrant and foreign-born interchangeably, with the caveat that estimates of the foreign- born population based on census and survey data such as those we use include nontrivial numbers of temporary residents and unauthorized migrants.

M. C. Horta (✉)
Office of Population Research & Department of Sociology,
Princeton University, Princeton, USA
e-mail: mcampos@princeton.edu

M. Tienda
Office of Population Research, Woodrow Wilson School
of International and Public Affairs & Department of Sociology,
Princeton University, Princeton, USA

© Springer International Publishing Switzerland 2015
W. A. Vega et al. (eds.), *Challenges of Latino Aging in the Americas,*
DOI 10.1007/978-3-319-12598-5_4

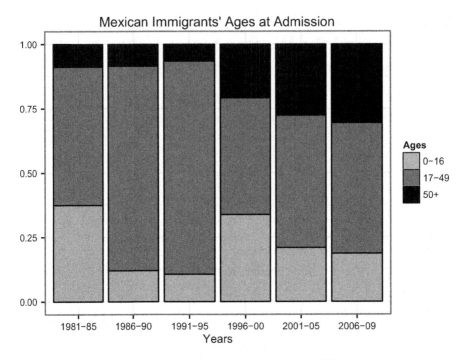

Fig. 4.1 Ages at admission for Mexican legal permanent residents, 1981–2009. (Source: Authors'
tabulations from Immigrants Admitted to the United States 1981–2000 data files and Special Tabulations provided by the U.S. Department of Homeland Security for 2000–2009)

the white majority, Gonzalez-Barrera and Lopez (2013) report that by 2011, nearly
one-quarter of foreign-born Mexicans were ages 50 and older, compared with 13 %
in 1990. Most Mexican seniors have aged in place, but there is some evidence that
late-age immigration is increasing. Based on census data, Leach (2008) points out
that the proportion of seniors among recent immigrants rose from 3 % in 1970 to
11 % in 2005. These stock measures can only approximate the influx of late-life
immigrants because census data include temporary visitors as well as unauthorized
residents; however, administrative data about immigrant flows also confirms a rise
in late-age immigration—from 11 % of new LPRs admitted between 1981 and 1985
to 17 % of admissions between 2006 and 2009 (Carr and Tienda 2013)[2].

Figure 4.1, which uses administrative data from the Department of Homeland
Security, shows how the age composition of legal permanent residents (LPRs) from
Mexico changed for 5-year cohorts admitted between 1981 and 2009. Three points
are noteworthy. First, cohort sizes fluctuated considerably over time owing to the
large numbers that benefitted from status adjustment under IRCA. Both the 1986–
1990 and 1991–1995 cohorts exceeded 1 million LPRs, with a significant majority
representing the legalized population from Mexico (Carr and Tienda 2013). Second,

---

[2] The vast majority of late-life Mexican immigrants are parents of U.S. citizens who activate their
family unification entitlements authorized by the 1965 Amendments to the Immigration and Nationality Act (Carr and Tienda 2013; Wilmoth 2012; Leach 2008).

although the majority of new LPRs from Mexico were in prime working ages, the share aged 50 and over rose sharply after 1995—precisely after eligibility for SSI was tightened. Third, even though only 7–8% of Mexican LPRs admitted between 1986 and 1995 were aged 50 and older, the outsized cohorts imply large numbers of Mexican LPRs with relatively short exposure to U.S. labor markets. To illustrate, nearly 90,000 Mexicans aged 50 and over were granted LPR status between 1991 and 1995 compared with 118,000 between 2001 and 2005.

A third development that motivates our study derives from recent trends in early retirement, which reversed as growing numbers of seniors—and immigrants in particular—joined the labor force (Kaushal 2010). After bottoming out during the late 1980s, labor force participation rates of men ages 55 and over rebounded, although not uniformly among demographic groups (Fullerton 1999; Toossi 2012). In 1950, 87% of men ages 55–64 were economically active compared with 68% in 1998; the comparable period participation rates for men ages 65 and over were, respectively, 46 and 16% (Fullerton 1999; Bureau of Labor Statistics 2008). According to Toossi (2012), only persons ages 55 and over are currently experiencing a rise in labor market activity: persons ages 55 and older represented 12% of the labor force in 1990 and 19% in 2010; moreover, she projects that by 2020 one-in-four workers will be ages 55 and over.

Toossi's projections do not distinguish between native and foreign-born workers, but Mosisa (2013) reports that the foreign-born share of the U.S. workforce also has been rising since the mid-1990s, which coincides with the enactment of legislation restricting new immigrants' access to means-tested benefit programs for 5 years (Medicaid) or until they become citizens (Supplemental Security Income). In addition to provisions to promote self-sufficiency among immigrants, the Personal Responsibility and Work Opportunity Reconciliation Act of 1996 (PRWORA) restricted Supplemental Security Income benefits to U.S. citizens and rendered immigrants with less than 5 years of U.S. residence ineligible for Medicaid[3]. In response to these sweeping reforms, many studies focused on changes in immigrants' receipt of means-tested benefits and social insurance (Borjas 2003; Van Hook 2000, 2003; Angel 2003; Nam and Jung 2008; Nam 2011; Nam and Kim 2012); however, except for Borjas (2011) and Kaushal (2010), past studies did not analyze complementary changes in employment behavior and earned benefits.

These three macro trends—the rapid growth of the Mexican origin population; the rise in late-life immigration and aging of the foreign-born population; and the rise in late-age labor force activity—raise several questions that motivate our analyses of Mexican origin seniors' labor force activity. First, how has the labor market activity of Mexican origin seniors changed in the wake of mass immigration? Second, are there discernible variations in labor market activity of Mexican origin seniors according to age at arrival? Finally, what do changes in late-age immigration and labor force activity portend for Social Security participation among Mexican origin seniors? Specifically, we assess Borjas' (2011) claim that immigrant men

---

[3] The Balanced Budget Act of 1997 restored SSI access to immigrants who were receiving benefits prior to August 1996 and made it easier for immigrants whose benefits were restored to qualify for Medicaid (Binstock and Jean-Baptiste 1999, p. 32).

work longer in order to qualify for Social Security benefits by considering variations by age at arrival and years since immigration for Mexican origin men.

Evaluating the claim that late-life Mexican immigrants work longer in order to qualify for the Social Security entitlement is not straightforward because of countervailing trends toward early retirement and working longer since 1980, and because of restrictions on immigrants' access to Supplemental Security Income (SSI) imposed by PRWORA (Kaushal 2010). Therefore, we consider how age at migration, length of U.S. residence, and agricultural employment altered Mexican origin men's labor force behavior before and after PRWORA went into effect. Our results indicate that Mexican immigration after age 50 is associated with higher rates of late-age labor force participation relative to statistically comparable Mexican origin natives only after the 1996 welfare legislation went into effect.

Unauthorized migrants are ineligible for social insurance benefits (Social Security) or means-tested benefits (Medicaid and SSI); therefore, the restrictions imposed by PRWORA technically did not affect this segment of the Mexico-born population. Although approximately 52 % of all Mexican origin workers are estimated to be unauthorized, fewer than 7 % of unauthorized migrants are men ages 50 and over (Passel et al. 2013; Passel and Cohn 2009)[4]. One reason for the relatively low share unauthorized among immigrant seniors is that many adjusted their status under the Immigration Reform and Control Act of 1986 (IRCA). Our tabulations from the 1990 census reveal that nearly 90 % of Mexican origin men ages 50 and over arrived in the United States on or before 1981, and thus would have qualified for status adjustment. For the 2000 and 2010 periods, 74 and 65 %, respectively, of Mexican origin men ages 50 and over likely qualified for status adjustment under IRCA. Furthermore, most late age arrivals since 1980 are family-reunification migrants (Carr and Tienda 2013), and thus eligible for welfare benefits without restrictions if they arrived prior to August 1996, or in accordance with the restrictions imposed by PRWORA if they arrived after this date.

We focus on men for two reasons. First, overall trends in women's labor force activity differ from those of men (Fullerton 1999), and so also do those of noncitizen elderly women (Kaushal 2010). Toossi (2012) notes that among the four major ethno-racial groups, Hispanic women have the lowest labor force participation rate whereas Hispanic men have the highest. Therefore, female labor market activity warrants a separate analysis. Second, we seek to compare our results with those of Borjas (2011) and Kaushal (2010), who analyze the labor market activity of retirement and near-retirement age foreign-born workers; both studies consider men, but neither distinguishes between Mexican and other foreign-born workers. We use these studies as a yardstick against which to evaluate whether and how the labor market behavior of older Mexican origin men, and immigrants in particular, differs from those of all foreign-born men.

To begin we identify changes in the policy environment that altered incentives for early retirement and late-age labor force participation. Subsequently we describe recent trends in the employment activity of Mexican origin men, with atten-

---

[4] Retirement-age men are under-represented among unauthorized migrants, who are concentrated in the prime working ages. Whereas approximately 4.5 % of unauthorized migrants were estimated to be men ages 50–59 in 2008, only 1.2 % were men ages 65 and over (Passel and Cohn 2009).

tion to variations by nativity, age at arrival, and industry sector distribution in order to focus on jobs with limited Social Security coverage. Finally, we evaluate how the lifecycle timing of migration, length of U.S. residence, and disproportionate representation in the agricultural sector affects late-age labor force behavior and receipt of Social Security income. The concluding section discusses the policy implications of late-age immigration in light of the diluted social safety net since 1996 and the spotty Social Security coverage for agricultural workers.

Policy Background and the Age Composition of Mexican Immigration

Immigration at older ages has direct implications for employment behavior and economic well being because late-life immigrants are less likely than foreign-born seniors who aged in place to master English; to qualify for work pensions; and to achieve economic independence (Population Reference Bureau 2013; Wilmoth 2012; O'Neil and Tienda forthcoming). In particular, as the share of late-age-immigrants among the Mexico-born population increases, employment differentials between native and foreign-born men should rise. Several academic papers and published reports track differentials in labor force participation by age and Hispanic origin (Fullerton 1999; Toossi 2012) or by age and nativity (Mosisa 2013), but few consider how labor force participation rates differ by age at immigration. Notable exceptions are recent studies by (2011, 2003) and Kaushal (2010), who explore nativity differentials in employment status; however, neither disaggregates the foreign-born workforce by national origin.

Using census data from 1960 to 2000, Borjas (2011) analyzes the labor supply behavior of older native and foreign-born men. He argues that access to Social Security and Medicare entitlements is a major reason for the higher labor force participation rates of foreign-born men compared with their native-born age counterparts. Once the 10-year work rule is satisfied, according to Borjas, participation rates of immigrants decline (Borjas 2011)[5]. This simple logic is complicated by changes in rules governing pension and disability payments for all workers, as well as by legislation that restricted immigrants' access to means-tested transfer income (O'Neil and Tienda forthcoming). Supplemental Security Income (SSI), which is a means-tested program initially created for citizens and legal permanent residents who satisfy age or disability requirements, was especially impacted by changes in eligibility criteria after 1996 (Kaushal 2010; Daly and Burkhauser 2003)[6].

---

[5] Less than 10 years of covered employment is required for receiving disability and survivor benefits (Nuschler and Siskin 2010). Although most immigrants are unaware of how benefits are calculated, Gustman and Steinmeier (2000) point out that immigrants receive higher benefits than their U.S.-born earnings counterparts because the benefit formula transfers benefits toward retirees with low lifetime earnings and because years spent outside of the United States are averaged as zeros in the average calculation.

[6] SSI was created in 1972 to replace federal grants to states for aid to aged, blind and disabled (Daly and Burkhauser 2003); however, growing utilization by immigrants resulted in gradual tightening of eligibility criteria, such as raising the residency requirement from 3 to 5 years in 1994.

Gendell (2006) and others argue that the availability of pensions and disability payments contributed to the secular trend toward early retirement during the 1960s and 1970s, but subsequent legislative provisions eliminated work disincentives for retirement-age seniors. Specifically, the 1977 and 1983 amendments to the Social Security Act that increased the full-benefit age from 65 to 67 years beginning in 2000, accounted for some of the increases in labor market activity of retirement-age seniors, as did the removal of earnings ceilings among Social Security beneficiaries in 2000.

Changes in immigrants' access to SSI also contributed to the rise in labor force activity of foreign-born seniors since the late 1990s (Van Hook 2003; Broder et al. 2005; Kaushal 2010). PRWORA made eligibility for SSI more similar to those for Social Security by imposing a minimum 10-year work requirement (Daly and Burkhauser 2003), thus eliminating its characterization as a workless pension especially among late-life immigrants. Kaushal (2010) finds that the policy was associated with a 9 % increase in labor market activity of foreign-born elderly men. Whether a similar work effect obtains among Mexican origin men is uncertain because their usage rate was lower than that of other nationalities even before changes in eligibility were imposed and because unauthorized immigrants were never eligible for benefits (Dunn 1995). We address this question for Mexican origin seniors following a description of data and operational definitions.

Data, Definitions and Descriptive Results

We use data from the 2009–2011 American Community Survey (ACS) and the 1980–2000 census microdata files, both obtained through IPUMS-USA (Ruggles et al. 2010). We merged the 3 years of ACS data, which corresponds to a 3 % sample of the U.S. population with 5 % samples for census 1980, 1990, and 2000 to account for nativity differentials in employment and Social Security receipt among Mexican origin seniors. We restrict the analysis to U.S.-born men of Mexican ancestry and Mexico-born men ages 50–80, which yields a sample of 163,115 observations—20,887 for 1980; 31,318 for 1990; 53,177 for 2000; and 57,733 for 2010. Persons born abroad to American citizens are classified as natives, and we make no distinction by naturalization status for the foreign-born because citizenship is not required either for labor market entry or to become eligible for Social Security.

By convention, employment status is measured either with an indicator variable designating whether respondents worked during the week prior to a census or survey interview, or the number of weeks worked in the previous year. Because seniors often reduce their market commitment in the years before retirement, the fraction of the year employed (i.e., weeks worked $\div$ 52) takes account of systematic variation in labor supply among older workers. Our descriptive and multivariate analyses use the week-adjusted measure to represent annual employment rates. Industry sector data is provided for individuals who have worked in the last 5 years, which includes both current workers and recent retirees.

Table 4.1 Employment as fraction of the year worked by Mexican origin men ages 50–80. (Source: Authors' tabulations of IPUMS USA 5% census samples from 1980, 1990, and 2000 and ACS 2009–2011 3% sample, person weights applied)

Ages	1980		1990		2000		2010	
	Foreign	Native	Foreign	Native	Foreign	Native	Foreign	Native
50–54	0.80	0.81	0.77	0.80	0.77	0.80	0.82	0.73
55–59	0.74	0.73	0.71	0.71	0.70	0.73	0.75	0.66
60–64	0.63	0.58	0.58	0.52	0.55	0.55	0.60	0.51
65–69	0.31	0.29	0.29	0.25	0.31	0.27	0.35	0.32
70–74	0.16	0.17	0.17	0.16	0.17	0.17	0.20	0.18
75+	0.09	0.12	0.12	0.10	0.13	0.10	0.09	0.09
All	0.53	0.63	0.56	0.55	0.59	0.54	0.63	0.53

Census surveys record information about year of arrival as well as birth year, which are used to derive approximate age at arrival. Specifically, we classify respondents into three broad groups representing youth, prime-age workers, and late-life workers. Receipt of Social Security benefits is defined as receiving any Social Security income, which includes entitlements such as pensions, survivors benefits, permanent disability insurance and U.S. government railroad retirement insurance, but excludes means-tested assistance, such as SSI benefits. We use 5-year age groups to compute age-specific employment and Social Security participation rates.

Employment of Mexican Origin Seniors

Table 4.1 reports age-specific employment measured as the fraction of year worked by Mexican origin men according to birthplace. The Bureau of Labor (2012, p. 3) reports a 10 percentage-point difference in 2012 labor force participation rates for native and foreign-born men—68 versus 78%, respectively, but not for workers over age 50. In 2010, nativity differentials in labor force activity among Mexican origin seniors were consistent with the national average in labor force participation, but this was not the case between 1980 and 2000. Rather, the nativity gap in employment reversed over the observation period. In 1980 the employment rate for native men was 10 percentage points higher than that for Mexico-born men, but the average rates converged in 1990 and began to diverge thereafter, favoring the foreign-born by 5 percentage points in 2000 and 10 percentage points in 2010.

Between 1980 and 2000, nativity differentials in employment were relatively modest among 50–59 year old men, but in 2010 employment rates of 50–64 year old foreign-born men exceeded those of natives by 9 percentage points, and by only 2–3 percentage points before age 75. For both native- and foreign-born Mexican origin men, employment rates decline monotonically with age, and the nativity differential narrows at later ages, as is evident in Fig. 4.2.

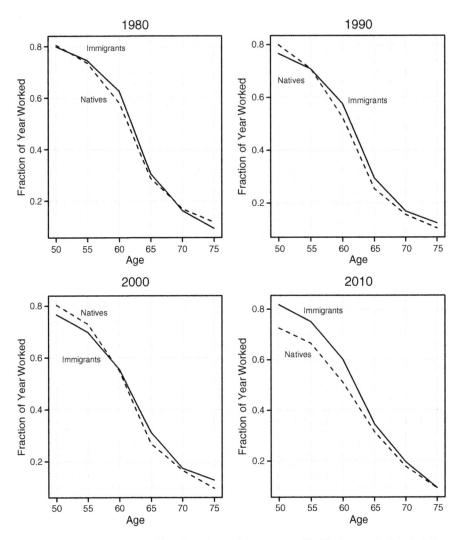

Fig. 4.2 Age-Employment profiles of Mexican origin men ages 50–80. (Source: Authors' tabulations of IPUMS USA 5% census samples from 1980, 1990, and 2000 and ACS 2009–2011 3% sample, person weights applied)

Another notable trend in the graphical representation is the steep drop in market engagement after age 64, which corresponds with the normative retirement age and the age at which workers were eligible for full benefits. This changed in 2000, when the Social Security full-benefit age was raised to 67 years. Although there is no prima facie evidence that Mexican origin men delayed retirement to qualify for full benefits in 2000, the uptick in employment among foreign-born Mexican men in 2010 may be a response to the change in eligibility for full benefits, the restriction on immigrants eligibility for SSI imposed in 1996, a rise in late-life immigration, or a combination of all three factors.

Based on data for the year 2000, Borjas (2011) shows a crossover in the employ-ment rates of immigrants around the age of Social Security eligibility, which he claims reflects the desire to accumulate the required 40 quarters of covered em-ployment to qualify for benefits. Our analyses of Mexican origin men replicate his age-employment curves only for the year 2000 (Fig. 4.2), which show immigrants' employment rates converging with and then overtaking those of natives around re-tirement age. This age-specific employment pattern does not obtain for other pe-riods, however. Rather, in 1980 the employment rates of older Mexico-born men consistently exceed those of their U.S.-born compatriots at every age until 70, when labor market engagement falls to around 15 %. Mexico-born men's age-specific employment curve in 1990 is slightly higher than that of native-born men between ages 55 and 70, and the crossover occurs well before eligibility for Social Security benefits. In 2010, nativity differentials in employment are notably larger at younger ages, but the rates converge gradually over successive ages[7].

If program eligibility is a primary driver of foreign-born Mexican seniors' mar-ket activity, it is conceivable that changes in SSI eligibility since 1996 (as Kaushal maintains) rather than a desire to qualify for Social Security benefits (as Borjas claims), is responsible for temporal variation in the timing of the nativity employ-ment crossover. Kaushal (2010) argues that SSI eligibility is the core driver of changes in labor supply because eligibility for Social Security remained relatively stable over the period, with the proviso that the increase in the age for full benefits from 65 to 67 years went into effect in 2000. This change, however, affected both natives and immigrants.

The 1990 and 2000 age-employment curves for immigrants do not reveal a clear trend towards higher rates of labor force activity. Nevertheless, it is plausible that the higher immigrant employment rates observed in 2010 compared with 1990 and 2000 for Mexican men ages 50–74 are market responses to the transfer income policy changes enacted in 1996. Kaushal (2010) argues that the effect of the 1996 policy on SSI receipt should increase as immigrants reach retirement ages because many low-wage workers who could have claimed SSI benefits under the old rules are denied the means-tested benefit under the current regime. This implies that the impact on labor supply of the SSI eligibility restrictions will cumulate over time, resulting in larger effects of the 1996 policy in 2010 compared with 2000.

Two plausible reasons why the age-employment profiles of Mexican origin se-niors reported in Fig. 4.2 differ from those reported by Borjas (2011) and Kaushal (2010) for all immigrants are the growth in late-life immigration and their dispro-portionate concentration in low-skill jobs (Bean and Tienda 1987; Duncan et al. 2006). Although sponsored Mexican seniors are likely to co-reside with their citizen relatives (Angel et al. 2000), the low earnings thresholds required to sponsor rela-

---

[7] As a sensitivity test, we plotted the age-specific employment rates of native and foreign-born Mexican origin men ages 30 and over to ensure that the crossover observed in 2000 did not occur at earlier ages in 2010. These curves are available from the authors.

Table 4.2 Employment as fraction of the year worked by Mexican immigrant men ages 50–80 by age at arrival. (Source: Authors' tabulations of IPUMS USA 5 % census samples from 1980, 1990, and 2000 and ACS 2009–2011 3 % sample, person weights applied)

Age at arrival	1980	1990	2000	2010
< 20	0.69	0.71	0.62	0.66
20–49	0.54	0.58	0.61	0.65
50 +	0.48	0.43	0.42	0.43

tives may lead them to seek labor market opportunities to supplement household incomes particularly during the pre-retirement ages.

Table 4.2, which reports employment as fraction of year worked by age at arrival for Mexico-born men, shows that the employment rate of Mexico-born men who arrived in the United States at ages 50 and over fell from 48 to 43 % between 1980 and 2010. In fact, the largest increase in employment rates of Mexican origin seniors correspond to men who arrived in the United States between ages 20 and 49, which gradually converged with their compatriots who arrived as adolescents or children. On balance, arrival-age-specific employment rates are not consistent with claims that late-life Mexican immigrants prolong labor market activity in order to qualify for Social Security or SSI benefits. Partly this may reflect the disproportionate representation of Mexican immigrants in unstable low-wage jobs, including many that are not covered by the Social security system. For example, in seasonal agricultural jobs, where piece rates are often used as a form of compensation, employers can easily skirt obligations to make Social Security contributions for workers[8].

Even as the agricultural sector contracted to less than 1 % of total employment, Mexican origin workers remained overrepresented in farming occupations, as well as service and construction jobs that attract unskilled workers willing to accept unstable and dangerous working conditions for low pay. Table 4.3 reveals that Mexico-born seniors are even more highly concentrated in agriculture than their U.S.-born counterparts, and the disparity increased over time. In 1980, for example, the share of foreign-born Mexican seniors engaged in agriculture was double that of natives, but in 2010, 12 % of the foreign-born worked in agriculture compared with 3 % of native-born Mexican origin seniors. During this period Mexican immigrants also overtook native seniors in construction and manufacturing industries, such as food processing and textiles. The significance of the changing industrial representation of native and foreign-born seniors is that it is related to acquisition of requisite credits to qualify Social Security, as we illustrate below.

---

[8] Regularly employed farm workers have been covered by Social Security since 1950, but those paid cash or on a piece-rate basis were largely excluded from the system. All workers should be covered by Social Security whenever they earn more than $ 150 from a single employer, but income from such casual work tends to go unreported and thus is not counted towards the 40 quarters of covered employment required to qualify for benefits (Social Security Administration 2005).

Table 4.3 Industrial sector distribution of Mexican origin men ages 50–80 by nativity. (Source: Authors' tabulations of IPUMS USA 5% census samples from 1980, 1990, and 2000 and ACS 2009–2011 3% sample, person weights applied)

	1980		1990		2000		2010	
	Foreign	Native	Foreign	Native	Foreign	Native	Foreign	Native
Agriculture	0.15	0.08	0.17	0.06	0.12	0.03	0.12	0.03
Construction	0.08	0.10	0.10	0.09	0.11	0.08	0.14	0.09
Manufacturing	0.19	0.18	0.18	0.15	0.18	0.13	0.15	0.11
Retail	0.08	0.08	0.09	0.08	0.09	0.07	0.10	0.08
All others	0.51	0.56	0.46	0.63	0.49	0.69	0.49	0.70

Table 4.4 Proportions of Mexican origin men ages 50–80 receiving Social Security pensions. (Source: Authors' tabulations of IPUMS USA 5% census samples from 1980, 1990, and 2000 and ACS 2009–2011 3% sample, person weights applied)

	1980		1990		2000		2010	
Ages	Foreign	Native	Foreign	Native	Foreign	Native	Foreign	Native
50–54	0.04	0.06	0.04	0.04	0.03	0.05	0.03	0.06
55–59	0.07	0.10	0.06	0.07	0.06	0.08	0.06	0.09
60–64	0.22	0.27	0.22	0.30	0.20	0.31	0.22	0.30
65–69	0.65	0.70	0.64	0.75	0.59	0.77	0.65	0.81
70–74	0.77	0.79	0.72	0.83	0.67	0.83	0.76	0.92
75+	0.78	0.79	0.76	0.83	0.68	0.82	0.76	0.90

Social Security Participation Among Mexican Origin Seniors

Kim and Torres-Gil (2011) explain that more than half of Hispanic retirees would live in poverty without Social Security income. Tables 4.4 and 4.5 illustrate nativity differentials in Social Security participation by nativity and age-at-arrival for Mexico-born seniors. Provided the minimum threshold of 40 quarters of covered employment is reached, retirees can begin receiving benefits as early as age 62, albeit at a reduced amount. For younger workers, benefits may be accrued through a spouse who earned eligibility. As Table 4.4 shows, Mexico-born seniors have substantially lower rates of Social Security participation than their native-born compatriots. In 2010, over 90% of U.S.-born Mexican men ages 70 and over received Social Security income, compared with about 80% in 1980. Because eligibility rises with age, the proportions receiving the benefit rise with age, especially after age 65. Furthermore, for every age group and every year, higher shares of native-born men receive Social Security income.

Table 4.5 Proportions of Mexican immigrants receiving Social Security pensions by age at arrival. (Source: Authors' tabulations of IPUMS USA 5% census samples from 1980, 1990, and 2000 and ACS 2009–2011 3% sample, person weights applied)

A. Receipt at ages 62+				
Age at arrival	1980	1990	2000	2010
<20	–	–	0.69	0.72
20–49	0.70	0.71	0.62	0.67
50+	0.59	0.36	0.35	0.38
B. Receipt at ages 65+				
<20	–	–	0.78	0.83
20–49	0.79	0.79	0.73	0.78
50+	0.64	0.42	0.40	0.43

Nativity differentials in Social Security participation widened over time among age-eligible seniors. Among 60 to 64-year olds, for example, nativity differentials in rates of Social Security receipt rise from 5 percentage points in 1980 to between 8 and 11 percentage points thereafter, with the largest gap in 2000—the year that the eligibility age for full benefits was raised from 65 to 67 years. Among the oldest age groups, the nativity gap is particularly striking. In 1980, Mexican origin men's claims on Social Security benefits were virtually identical between natives and immigrants ages 70 and over, differing by about 1 or 2 percentage points. However, beginning in 1990 and continuing through 2010, the nativity gap in benefit claims widened for them, as well as for men ages 65 and over. Both in 2000 and 2010, nativity differentials in Social Security participation rates ranged between 16 and 18 percentage points for men ages 65–74, and 14 percentage points for men ages 75–80.

Given that Social Security is an earned benefit, nonparticipation among age-eligible seniors indicates failure to fulfill the 10-year minimum threshold of work in covered employment. Late-life migrants face formidable challenges satisfying the 40-quarter contribution to the Trust Fund in order to qualify for benefits, as Table 4.5 reveals. The top panel reports changes in participation rates by arrival age for all Mexico-born men ages 62 and over and the lower panel reports participation rates at ages 65 and over. That participation rates are systematically higher among seniors ages 65 and over (lower panel) reflects both normative expectations about retirement age and administrative guidelines that stipulated age 65 as the threshold for receiving full benefits prior to 2000.

Lower shares of late-life immigrants are likely to receive benefits than their age counterparts who arrived during prime working ages or adolescence. Nearly three quarters of Mexican immigrants ages 62 and over who arrived in the United States before age 20 claimed Social Security benefits in 2010 compared with less than 40% of their age counterparts who arrived at ages 50 and over. Restricting the samples to ages 65 and above (panel B) shows higher participation rates for all arrival ages compared with Panel A. However, pronounced differences by arrival age persist for all years, but especially since 1990.

The descriptive tabulations, while suggestive, do not reveal why nativity differentials in labor force and Social Security participation of Mexican origin seniors widened between 1980 and 2010. We have offered several plausible explanations— the rise of late-life immigration; the changing policy context; unequal participation in jobs covered by Social Security; however, to disentangle the relative contributions of each we turn to a multivariate analysis that considers period and age variations in employment and Social Security participation of Mexican origin seniors since 1980.

Multivariate Analysis

Our analysis aims to explain widening nativity differentials in labor market activity and Social Security participation among Mexican origin seniors during a period of mass migration and changing requirements for work-linked entitlements. Prior research indicates that the 1996 PRWORA legislation altered work and welfare participation of immigrants by imposing restrictions on access to SSI; therefore, to understand whether and how changes in employment behavior are related to changes in Social Security participation, it is necessary to consider whether respondents reach retirement ages before or after welfare reforms went into effect (Angel 2003; Van Hook 2000, 2003; Kaushal 2010; O'Neil and Tienda forthcoming).

Model 1 addresses whether potential eligibility for insurance benefits predicts late-life labor market activity and Social Security receipt among Mexican immigrants and allows us to ascertain whether results for Mexican origin men are consistent with those of Borjas (2011), who compared the labor supply of all immigrant seniors to comparably aged natives.

$$Y_{it} = \alpha + T\beta_t + \gamma A_i + \delta_n R_i + \eta_n I_i + \theta E_i + \lambda_n (I_i * E_i) + \sigma G_i + \Pi C_i + \varepsilon_{it} \qquad (4.1)$$

In this specification, Y_{it} is the outcome (number of weeks worked in past year or receipt of Social Security benefits) of individual i observed in period t. T includes three indicator variables for census years 1980, 1990, and 2000 with 2010 omitted as the referent period. A is a linear term for age, which is centered at age 50. The vector R includes two indicators for ages of retirement eligibility 62–64 and 65 and above. I is a vector of indicators for whether migration occurred prior to age 20, between ages 21 and 49, or at ages 50 and above. E indicates potential ineligibility for Social Security benefits, indexed as fewer than 15 years of U.S. residence. G is an indicator for those who report working or having worked in the agricultural sector. Finally, C is a vector of control variables, which includes indicators of marital status with the modal category of married with spouse present omitted as the referent; indicators of educational attainment with less than a high-school education, the modal category among Mexican immigrant seniors, omitted as the referent; and a linear term for the number of own relatives in one's household centered at the median family size of 3[9].

Length of U.S. residence serves as a proxy for Social Security eligibility, which is not directly observed. Because immigrants, particularly those with low skill levels, experience gaps in covered employment, we use 15 years of U.S. residence as the bar indicating potential ineligibility for Social Security retirement benefits. Individuals who have not yet completed the required 40 quarters of covered work presumably have a strong incentive to stay in the labor market in order to later qualify for Social Security. The λ_n coefficients, represent interactions with fewer than 15-years of U.S. residence, respectively, for arrival ages 20–49 (λ_1); and 50 and over (λ_2)[10]. Thus, Eq. 4.1 addresses whether potential ineligibility for Social Security benefits, indexed by having spent fewer than 15 years as a U.S. resident, predicts the labor supply and Social Security receipt of older Mexican immigrant men who arrived at similar ages.

The vast majority of Mexican immigrants who arrived before age 50 have at least 15 years of U.S. residence (see Table 4.6), and hence should qualify for benefits. If Mexican immigrants' late-age labor force participation is driven by a desire to earn Social Security benefits, as Borjas (2011) concluded based on all immigrants, the interaction term (λ_3) between late-age migration (ages 50+) and potential ineligibility (<15 years of U.S. residence) should be positive.

Model 1 assumed that the work behavior of Mexican origin seniors was unaffected by PRWORA, which Model 2 relaxes. To test whether changes in program eligibility that rendered noncitizens eligibility for SSI and Social Security relatively similar are responsible for nativity differentials in Mexican seniors' employment behavior and Social Security participation, Model 2 evaluates whether the associations differ before and after the eligibility requirements were changed. Specifically, Model 2 adds an indicator variable, P, to designate the period before and after PRWORA ($P=1$ if observation i occurs in period $\geq$ year 2000), which is interacted with the vector of immigrants' arrival ages I[11].

$$Y_{it} = \alpha + T\beta_t + \gamma A_i + \delta_n R_i + \eta_n I_i + \theta E_i + \lambda_n (I_i * E_i) + \rho_n (P_i * I_i) + \sigma G_i + \Pi C_i + \varepsilon_{it} \quad (4.2)$$

[9] In an alternative specification, we also included state fixed-effects to account for potential changes in the work and retirement behaviors of Mexican origin seniors as the Mexican origin population moved into new destination states. We present the more parsimonious model, which provides national mean estimates of number of weeks worked and probability of Social Security receipt for Mexican origin seniors, because the more complex specification did not improve the model's fit or alter the substantive interpretation of the model coefficients.

[10] The parameter λ is not estimated for the interaction of fewer than 15 years of U.S. residence and ages of arrival before age 20 because early arrivals (by age 20) had lived in the U.S. for at least 30 years when they reached age 50.

[11] Instead of estimating a marginal effect for the post-PRWORA periods indexed by P, we estimate marginal effects for years 1980, 1990, and 2000 relative to the reference year of 2010, which allows post-PRWORA estimates of number of weeks worked and probability of Social Security receipt for years 2000 and 2010 to differ.

Table 4.6 Characteristics of Mexican origin men ages 50 and above by nativity and ages at arrival, weighted means or proportions. (Data source: Authors' tabulations of IPUMS USA 5% census samples from 1980, 1990, and 2000 and ACS 2009–2011 3% sample, person weights applied)

	Natives[a]	Immigrants' ages at arrival		
		<20	20–49	50+
Employed	0.53	0.63	0.59	0.44
Weeks worked	28.65	34.03	32.10	22.58
Fraction of year worked	0.55	0.65	0.62	0.43
Receives social security	0.32	0.19	0.24	0.24
Below poverty	0.14	0.14	0.19	0.28
Age (years)	60.18	57.13	59.23	65.04
In U.S. ≥ 15 years	1.00	1.00	0.94	0.56
In U.S. ≥ 10 years	1.00	1.00	0.98	0.78
Agricultural worker	0.04	0.11	0.14	0.15
Census year (proportions by row)				
1980	0.60	<0.01	0.33	0.07
1990	0.58	0.02	0.34	0.06
2000	0.45	0.10	0.39	0.07
2010	0.40	0.16	0.40	0.05
Education (proportions by column)				
<High school	0.41	0.59	0.75	0.81
High school graduate	0.31	0.25	0.16	0.11
College (non-grads and grads)	0.28	0.16	0.09	0.08
Marital status (proportions by column)				
Married, spouse present	0.69	0.72	0.70	0.52
Married, spouse absent	0.02	0.05	0.09	0.22
Widowed	0.05	0.03	0.04	0.10
Single, divorced, or separated	0.24	0.20	0.16	0.16
Family size (count)	2.90	3.61	3.94	4.35
Number of observations	78,902	15,013	59,751	9,449

[a]Natives are U.S.-born men of Mexican ancestry

Because there are many potential time-varying confounders that we do not account for in these models, our approach is descriptive. Thus, the post-PRWORA coefficients (ρ_n) in Model 2 are not presented as an estimate of the causal effect of the policy on the labor force activity and Social Security receipt of Mexican immigrant men. Rather, these point estimates describe differences in labor force activity and benefit receipt before and after the policy by arrival age after controlling for age, secular trend (period), length of U.S. residence, agricultural sector employment, educational attainment, marital status, and family size.

Table 4.6 summarizes the key variables by nativity and arrival ages for the pooled analytic sample. Approximately 71 % of foreign-born seniors arrived in their prime working ages, but over one-in-ten qualify as late-life migrants. On average, late-age Mexican immigrants worked 10 fewer weeks than their foreign-born counterparts who arrived during their prime working ages. U.S.-born Mexican seniors have lower employment rates than Mexican immigrants who arrived as prime-age workers or adolescents, but higher rates than those who immigrated at age 50 or higher. Immigrant poverty rates are also notably higher compared with that of natives, particularly among late-age arrivals; this suggests that economic need may drive Mexico-born men to postpone retirement. Among the foreign-born, potential non-eligibility for retirement benefits is concentrated among late-age migrants, who also have lower levels of labor force activity.

Nativity Differentials in Labor Force Activity

As shown in Table 4.1, nativity differentials in labor force activity at older ages tend to favor Mexico-born men. Table 4.7 reports results for models (1) and (2) for number of weeks worked in the previous year. In both models the intercept gives the predicted number of weeks worked in a year for a 50-year old native with less than a high school education, married, and living with a spouse and two more relatives in 2010. In addition to a linear effect of age, which is centered at age 50, the empirical specification includes indicator variables designating age groups 62–64 and age 65 + to account for the observed drops in employment at these key ages of eligibility for Social Security retirement with reduced benefits. To facilitate interpretation, Table 4.7 reports the net effects of being exactly age 62 and exactly age 65 relative to the intercept (age 50)[12][13].

At age 62 Mexican origin men work about 3.6 fewer months than 50-year olds and at 65, when many reach full retirement age, seniors have reduced their labor force activity by about two thirds relative to 50 year-olds. The labor force participation of Mexican immigrant seniors is modestly conditioned by their ages-at-arrival, even after controlling for aging, period, engagement in agricultural jobs, education and marital status. Whereas immigrants work 1–3 weeks more than natives, immigrants who entered the country before age 20 work only 1.5 weeks more than

---

[12] A table reporting all coefficients for these models as well as two more models with an alternative specification of Social Security ineligibility (fewer than 10 years of U.S. residence), is included in the appendix (Table 4.10). This additional specification is how Borjas (2011) operationalized Social Security ineligibility. The key difference between our operationalization of fewer than 15 years of U.S. residence and Borjas' is that none of the ineligibility coefficients is statistically significant under the 10-year specification.

[13] We also considered using polynomials of age; however, using a procedure suggested by Box and Tidwell (1962), we established that these more complex specifications did not improve the statistical fit of the model. Another specification tested whether the slope shifts again at age 67, the new age of full retirement, but we find no evidence that this had occurred.

Table 4.7 OLS models of number of weeks worked by Mexican origin men ages 50 and above. (Data source: IPUMS USA 5% census samples from 1980, 1990, and 2000 and ACS 2009–2011 3% sample)

	Model 1[a]	Model 2[a]
Intercept	39.720***	39.064***
	(0.168)	(0.176)
Retirement ages		
Age 62	− 15.613***	− 15.613***
	(0.183)	(0.183)
Age 65	− 24.990***	− 24.905***
	(0.123)	(0.123)
Immigrant age at arrival		
< 20 at arrival	1.517***	− 0.800
	(0.188)	(0.769)
20–49 at arrival	2.797***	1.038***
	(0.120)	(0.193)
50+ at arrival	3.395***	1.204***
	(0.276)	(0.639)
Immigrant time in U.S.		
50+ at arrival and < 15 years in U.S.[b]	− 1.609*	1.372*
	(0.442)	(0.524)
Agricultural worker	6.382***	6.442***
	(0.172)	(0.172)
Census year		
1980	2.211***	3.532***
	(0.172)	(0.200)
1990	0.123	1.457***
	(0.148)	(0.180)
2000	0.051	0.008
	(0.123)	(0.123)
Post-PRWORA (2010)		
if < 20 at arrival		2.955***
		(0.791)
if 20–49 at arrival		2.793***
		(0.237)
if 50+ at arrival		4.564***
		(0.534)
R^2	0.299	0.300
Adj.R^2	0.299	0.300
Num. obs	163,115	163,115

***$p < 0.001$; **$p < 0.01$; *$p < 0.05$
[a]Both models include controls for educational attainment, marital status and family size
[b]Sum of the coefficient for "< 15 years in U.S." and its interaction with "arrival age 50+"

natives, on average, which reflects their early socialization into the host country institutions. Mexican seniors who arrived at ages 20–49 and after age 50 work about 3 weeks more than natives, *provided they have lived in the United States for 15 or more years*. This is an important caveat, as discussed below.

Model 1 tests the hypothesis that immigrants' length of U.S. residence, our proxy for retirement benefit eligibility, accounts for nativity differentials in labor force activity. Because only late-life immigrants are disproportionately represented among recent arrivals (see Table 4.6), we focus on the interaction between arrival after age 50 and having 15 or fewer years of U.S. residence. The interaction term is negative and statistically significant (point estimate is − 1.6); substantively this indicates that late-age migrants with fewer than 15 years of U.S. residence work about 1–2 weeks less per year than late-age migrants with more than 15 years of U.S. residence and only about as much as immigrants who arrived by age 20.

Model 2 addresses whether the labor force activity of Mexican origin men differed in the pre- and post-PRWORA periods. As expected, the association between labor force participation and arrival age differ before and after PRWORA went into effect. Even Mexicans who entered the United States as adolescents, for whom PRWORA-imposed restrictions on benefits were expected to be negligible, worked 2 weeks more than natives and 3 weeks more than similar immigrants did pre-PRWORA. Effects for immigrants arriving during prime working ages are larger— about 4 more weeks of employment relative to natives in the same period and 3 weeks more than similar immigrants pre-PRWORA. For late-age Mexican immigrants, PRWORA predicted even more work activity. Specifically, late-life Mexican immigrants with more than 15 years of U.S. residence averaged nearly 6 more weeks at work than natives and 4.5 more weeks than similar immigrants did pre-PRWORA. Furthermore, after accounting for PRWORA, late-age migrants with fewer than 15 years of U.S. residence are found to work marginally more than similar immigrants with more than 15 years of U.S. residence.

To further clarify these relationships, Table 4.8 shows the predicted fraction of the year worked by natives and immigrants in each of the four periods (census years) as well as the associated predicted probabilities of receiving Social Security under model 2. Late-life Mexican immigrants who resided in the United States fewer than 15 years work less than natives and other Mexico-born men pre-PRWORA, in 1980 and 1990 and marginally less than migrants who arrived in their prime working years in 2000 and 2010, post-PRWORA. However, to understand whether their work behavior is motivated by a desire qualify for retirement insurance, the key question is not whether they work more than other Mexican origin seniors, but whether they work more than they otherwise would. Our estimates suggest that late-life migrants' labor supply increased over time, peaking in 2000 and 2010.

Model 2 predicts that in 2010, a 65 year-old late-life immigrant with 15 or more years of U.S. residence was employed 37% of the time, which is only 2 percentage points more than a comparable compatriot worked in 1980, but represents a more substantial 6 percentage-point increase relative to 1990. Similarly, at age 65, late-age immigrants with fewer than 15 years of U.S. residence increased their work

Table 4.8 Mexican origin men's predicted employment and predicted probabilities of Social Security receipt, model 2. (Data source: IPUMS USA 5% census samples from 1980, 1990, and 2000 and ACS 2009–2011 3% sample)

	Fraction of year worked				Prob. of social security receipt			
			Post-PRWORA				Post-PRWORA	
	1980	1990	2000	2010	1980	1990	2000	2010
Native								
Age 62	0.51	0.47	0.44	0.44	0.38	0.39	0.42	0.47
Age 65	0.33	0.29	0.26	0.26	0.63	0.65	0.67	0.72
Immigrant by age at arrival and time in U.S.								
Age 62, arrival age <20	0.49	0.45	0.48	0.48	0.40	0.42	0.36	0.42
Age 65, arrival age <20	0.32	0.28	0.30	0.30	0.66	0.67	0.61	0.67
Age 62, arrival age 20–49	0.53	0.49	0.52	0.52	0.34	0.36	0.31	0.36
Age 65, arrival age 20–49	0.35	0.31	0.34	0.34	0.59	0.61	0.55	0.61
Age 65, arrival age 50+ and ≥15 years in U.S.	0.35	0.31	0.37	0.37	0.25	0.27	0.26	0.30
Age 62, arrival age 50+ and <15 years in U.S.	0.47	0.43	0.49	0.49	0.06	0.07	0.06	0.08
Age 65, arrival age 50+ and <15 years in U.S.	0.29	0.25	0.31	0.31	0.16	0.17	0.16	0.19
Agricultural worker at age 65								
Native	0.45	0.41	0.39	0.39	0.55	0.57	0.59	0.65
Immigrant, arrival age <20	0.44	0.40	0.43	0.43	0.58	0.60	0.54	0.59
Immigrant, arrival age 20–49	0.47	0.43	0.46	0.46	0.51	0.53	0.48	0.54
Immigrant, arrival age 50+ and ≥15 years in U.S.	0.48	0.44	0.50	0.50	0.20	0.21	0.20	0.24
Immigrant, arrival age 50+ and <15 years in U.S.	0.41	0.37	0.43	0.43	0.12	0.13	0.12	0.15

Both models include controls for educational attainment, marital status and family size

effort by 6 percentage points from 25% in 1990 to 31% in 2000 and 2010. By contrast, 65-year old Mexican immigrants who arrived at the modal ages of 20–49 only increased their work effort by 3 percentage points over the same period and actually worked marginally more in 1980 than they did post-PRWORA.

Agricultural employment helps to explain the labor force behavior of Mexican origin seniors. Our models indicate that seniors affiliated with agricultural jobs average six extra weeks of work per year relative to Mexican origin men engaged in other industries; moreover, this result is robust across different model specifications. Estimates from model 2 indicate that in 2010 a 65 year-old Mexican immigrant arriving between ages 20 and 49 worked 46% of the year, which is 12 percentage points more than a Mexican immigrant of the same age who immigrated between ages 20 and 49 but does not work in agriculture (Table 4.8, left panel).

Evidence that Mexican migrants who arrive after age 50 work marginally more at age 65 than natives and other compatriots who arrived at earlier ages (Table 4.8) suggests that late-age migrants may be extending their working lives in order to qualify for earned retirement benefits. If so, then the life-cycle timing of migration should also affect immigrants' probabilities of receiving Social Security. Therefore, in the next section, we address (1) whether late-age Mexican immigrants have a lower probability of receiving benefits upon reaching eligibility ages compared with natives and other younger-age compatriots; (2) whether and by how much years of U.S. residence increase late- age Mexican immigrants' probability of receiving Social Security pensions; (3) whether agricultural work undermines their benefit prospects; and (4) whether rates of Social Security receipt changed post-PRWORA.

Nativity Differentials in Social Security Receipt

Table 4.9 summarizes results of estimating models (1) and (2) for Social Security receipt among elderly Mexican origin men. For ease of interpretation, coefficients are transformed into relative odds along with their 95% confidence intervals. In the empirical specification, we depict aging into Social Security eligibility with indicator terms for ages 62–64 and 65+; however, in Table 4.9, the aging effects are summarized as net effects of being exactly age 62 and exactly age 65 relative to the intercept (age 50). The results reported in Table 4.9 show how arrival age influences the likelihood of receiving Social Security benefits among elderly Mexican immigrant men (model 1, Table 4.9), net of aging into eligibility.

Mexican immigrants are significantly less likely to collect Social Security than their U.S.-born counterparts, and late-age migrants are considerably less likely than earlier arrivals and natives to receive the entitlement. For example, model 1 implies that a 65-year old U.S.-born Mexican origin male has a 70% probability of receiving a Social Security pension in 2010. This probability is lowered by 3 percentage points for a Mexican immigrant who arrived before age 20, by an additional 5 percentage points (to 61%) for a Mexican immigrant who arrived between the ages of 20 and 49, and by yet an additional 32 percentage points for late-age Mexican immigrants who have lived in the United States for 15 or more years. Put differently, the likelihood that late-age Mexican immigrants claim Social Security is only 30%, even after 15 years of U.S. residence.

Net of variations in arrival ages, residing in the United States fewer than 15 years is inconsequential for receipt of Social Security. That is, the net effect of migrating after age 50 and having less than 15 years of U.S. residence is not significantly as-

Table 4.9 Estimated odds ratios from logistic regressions of Social Security receipt by Mexican origin men ages 50 and above. (Data source: IPUMS USA 5% census samples from 1980, 1990, and 2000 and ACS 2009–2011 3% sample)

	Model 1[a]			Model 2[a]		
	Odds	Confidence Interval		Odds	Confidence Interval	
	Ratios	2.5%	97.5%	Ratios	2.5%	97.5%
(Intercept)	0.063	0.060	0.066	0.067	0.064	0.071
Retirement ages						
Age 62	16.207	14.645	17.779	16.169	14.672	17.816
Age 65	45.653	40.211	51.830	45.425	40.008	51.563
Immigrant age at arrival						
<20 at arrival	0.844	0.796	0.895	1.126	0.853	1.461
20–49 at arrival	0.698	0.673	0.723	0.849	0.802	0.898
50+ at arrival	0.182	0.169	0.196	0.198	0.177	0.220
Immigrant time in U.S.						
50+ at arrival and <15 years in U.S.[b]	0.549	0.017	18.398	0.552	0.016	18.472
Agricultural worker	0.736	0.697	0.778	0.734	0.694	0.775
Census year						
1980	0.765	0.727	0.805	0.670	0.631	0.711
1990	0.820	0.785	0.856	0.723	0.687	0.761
2000	0.795	0.766	0.825	0.791	0.762	0.821
Post-PRWORA						
If <20 at arrival				0.704	0.539	0.935
If 20–49 at arrival				0.733	0.683	0.786
if 50+ at arrival				0.864	0.749	0.997
AIC	115379.53			115307.89		
BIC	115579.58			115537.94		
Log Likelihood	−57669.77			−57630.94		
Deviance	115339.53			115261.89		
Num. obs.	163,115			163,115		

[a]Both models include controls for educational attainment, marital status and family size
[b]Exponentiated sum of the log-odds coefficient for "<15 years in U.S." and its interaction with "arrival age 50+"

sociated with receipt of retirement benefits, as indicated by the confidence interval of the odds-ratio (Table 4.9). As expected, agricultural workers are less likely than other workers to receive Social Security pensions once eligibility age, arrival age, observation period, education, and marital status are modeled. Results for Model 2 suggest that immigrants were less likely to claim Social Security after PRWORA than they had been in previous years.

To expand on the results, in the right panel of Table 4.8 we report predicted probabilities of Social Security receipt by Mexican origin men at ages 62 and 65. For 65-year olds, the predicted probabilities are very similar for natives and immigrants who arrived as adolescents; however, before PRWORA (i.e., 1980 and 1990), Mexican immigrants who arrived between the ages of 20 and 49 have a 4 percentage point lower probability of claiming benefits compared with U.S.-born Mexican seniors; after PRWORA, prime-age Mexican immigrants had a 11–12 percentage points lower probability receiving Social Security than natives.

For late-life Mexican immigrants, the nativity differential in Social Security receipt hovers around 40 percentage points for all periods. Thus, age at arrival is highly consequential in determining the likelihood that Mexican immigrant men will retire with Social Security benefits, even when they extend their working life. Mexican origin men who worked in the agricultural sector are especially disadvantaged in this regard; on average, agricultural workers have a 7 percentage point lower probability of claiming Social Security benefits than non-agricultural workers. For a 65-year old native-born Mexican in 2010, agricultural employment implies a 65 % chance of claiming the public retirement benefit. For Mexico-born men who arrived before age 20, between 20 and 49, and ages 50+, working in agriculture lowers their chances of receiving Social Security, respectively, to 59, 54, and 24 %.

Conclusions

We sought to describe and understand nativity differentials in the labor market activity and retirement behavior of Mexican origin seniors since 1980. Age-specific employment rates revealed a widening gap favoring the foreign-born over this period, which was accompanied by larger nativity differentials in Social Security receipt favoring the native-born. As these trends coincided with rising ages at migration, we investigated how these trends varied by age at arrival. Although most studies of immigrants' labor market behavior and economic wellbeing focus on differences associated with length of U.S. residence (Duncan et al. 2006), several recent studies have called attention to the importance of age at arrival, emphasizing that integration prospects differ for youth, prime-age workers, and seniors (Angel et al. 2000; Treas and Batalova 2007; O'Neil and Tienda forthcoming). Whether late-age immigrants achieve economic self-sufficiency depends not only on their age at admission, but also whether they enter the U.S. labor market; whether they work in jobs covered by Social Security; and whether they acquire the requisite 40-quarters of covered employment in order to qualify for the entitlement.

Borjas (2011) claims that immigrant men work longer in order to qualify for Social Security by considering variations by age at arrival and years since immigration in the 1960–2000 period. We consider whether his claim applies to Mexican origin men for the 1980–2010 period and find little support. The net effects of late-age migration and length of U.S. residence on Mexican immigrant men's employment appear to be relatively modest: arriving at age 50 and over and having fewer than 15 years of U.S. residence is associated with lower employment rates relative to both natives and immigrants who arrived in their prime working years. However, we also find that the labor supply of Mexican immigrants differed before and after PRWORA; specifically, Mexican immigrants work longer since the 1996 welfare reforms that restricted SSI to citizens. If, as Borjas claims, immigrants extend their working lives in order to qualify for retirement benefits, our findings suggest that late-age immigrants may need more than 15 years to accumulate 10 years of covered employment. After accounting for Mexican's concentration in agricultural work, the probability that late-life immigrants will claim Social Security benefits at age 65 is about 40 percentage points lower than statistically comparable natives.

Therefore, consistent with Kaushal (2010) we find suggestive evidence that elderly Mexican immigrant men increased their labor supply after the PRWORA reforms relative to how much they would have worked before the policy. Moreover, we show that increases in Mexican immigrants' work effort depend on age at immigration: the largest increases in weeks worked correspond to men who immigrated at ages 50 and beyond. Among late-age immigrants, Mexican seniors with fewer than 15 years of U.S. residence increase their work effort only marginally relative to their counterparts who resided in the United States for more than 15 years. This finding underlines the importance of age at arrival for understanding immigrants' labor market integration.

This study advances our understanding of the labor market and retirement behaviors of Mexican origin seniors but it is limited in several ways. Because we only observe Mexicans north of the border, we cannot speak to the work and retirement behaviors of Mexican immigrants who return home after spending much of their working lives in the United States. We also do not consider how the labor market and retirement behavior of Mexican origin seniors is influenced by their living arrangements or how it affects their economic wellbeing at advanced ages. Thus, future research could integrate Mexico and U.S. data to extend our findings. It would also be possible to build on this research and on the work of Angel et al. (2000) to consider how the labor market and retirement behaviors of Mexican origin seniors depend on whether they reside with spouses, children or other relatives and how their ages at admission influence their chances of entering the labor market and achieve some degree of economic self-sufficiency or to contribute to household incomes.

Appendix

Table 4.10 : OLS models of number of weeks worked by Mexican origin men ages 50 and above (IPUMS USA 5% census samples from 1980, 1990, and 2000 and ACS 2009–2011 3% sample)

	Potential social security ineligibility			
	<15 years in U.S.		<10 years in U.S.	
	Model 1	Model 2	Model 1	Model 2
Intercept	39.720***	39.064***	39.725***	39.043***
	(0.168)	(0.176)	(0.167)	(0.167)
Age (linear)	−1.017***	−1.016***	−1.019***	−1.016***
	(0.013)	(0.013)	(0.013)	(0.013)
Age 62–64	−3.404***	−3.417***	−3.415***	−3.432***
	(0.199)	(0.198)	(0.199)	(0.198)
Age 65+	−9.729***	−9.659***	−9.707***	−9.639***
	(0.243)	(0.243)	(0.243)	(0.243)
1980	2.211***	3.532***	2.233***	3.542***
	(0.172)	(0.200)	(0.171)	(0.200)
1990	0.123	1.457***	0.141	1.467***
	(0.148)	(0.180)	(0.147)	(0.180)
2000	0.051	0.008	0.033	0.026
	(0.123)	(0.123)	(0.124)	(0.124)
Arrival age <20	1.517***	−0.800	1.521***	−0.793
	(0.188)	(0.769)	(0.188)	(0.769)
Arrival age 20–49	2.797***	1.038***	2.845***	1.038***
	(0.120)	(0.193)	(0.120)	(0.193)
Arrival age 50+	3.395***	1.204***	2.897***	1.206***
	(0.276)	(0.366)	(0.245)	(0.366)
Ineligible for social security	7.414	7.945*	5.648	6.196
	(3.791)	(3.790)	(5.016)	(5.013)
Arrival age 20–49* ineligible	−5.567	−6.573	−3.682	−4.686
	(3.817)	(3.815)	(5.080)	(5.078)
Arrival age 50+* ineligible	−9.022*	−11.353**	−5.985	−7.235
	(3.816)	(3.825)	(5.054)	(5.055)
Post-PRWORA* arrival age <20		2.955***		2.954***
		(0.791)		(0.791)
Post-PRWORA* arrival age 20–49		2.793***		2.845***
		(0.237)		(0.236)
Post-PRWORA* arrival age 50+		4.564***		2.942***
		(0.534)		(0.473)
Agricultural worker	6.382***	6.442***	6.390***	6.443***
	(0.172)	(0.172)	(0.172)	(0.172)

Table 4.10 (continued)

	Potential social security ineligibility			
	< 15 years in U.S.		< 10 years in U.S.	
	Model 1	Model 2	Model 1	Model 2
High school grad.	3.997***	4.166***	4.004***	4.172***
	(0.132)	(0.132)	(0.132)	(0.132)
College (non-grads and grads)	7.530***	7.773***	7.537***	7.779***
	(0.146)	(0.148)	(0.146)	(0.148)
Married, spouse absent	−3.929***	−3.896***	−3.931***	−3.934***
	(0.220)	(0.220)	(0.220)	(0.220)
Widowed	−4.890***	−4.858***	−4.882***	−4.852***
	(0.242)	(0.242)	(0.242)	(0.242)
Single, divorced or separated	−6.128***	−6.088***	−6.122***	−6.084***
	(0.138)	(0.138)	(0.138)	(0.138)
Family size	−0.167***	−0.174***	−0.166***	−0.173***
	(0.025)	(0.025)	(0.025)	(0.025)
R^2	0.299	0.300	0.299	0.300
Adj.R^2	0.299	0.300	0.299	0.299
Num. obs.	163,115	163,115	163,115	163,115

***$p<0.001$; **$p<0.01$; *$p<0.05$

References

Angel, J. L. (2003). Devolution and the social welfare of elderly immigrants: Who will bear the burden? *Public Administration Quarterly, 63*(1), 79–89.

Angel, J. L., Angel, R. J., & Markides, K. S. (2000). Late-life immigration, changes in living arrangements, and headship status among older Mexican origin individuals. *Social Science Quarterly, 81*(1), 389–403.

Batalova, J. (2012, May 30). Senior immigrants in the United States. Migration Information Source. http://www.migrationinformation.org/usfocus/display.cfm?ID=894. Accessed 1 Nov 2014.

Bean, F. D., & Tienda, M. (1987). The Hispanic population of the United States. New York: Russell Sage Foundation.

Binstock, R. H., & Jean-Baptiste, R. (1999). Elderly immigrants and the saga of welfare reform. *Journal of immigrant health, 1*(1), 31–40.

Borjas, G. J. (2003). Welfare reform, labor supply, and health insurance in the immigrant population. *Journal of Health Economics, 22*(6), 933–958.

Borjas, G. J. (2011). Social Security eligibility and the labor supply of older immigrants. *Industrial and Labor Relations Review, 64*(3), 485–501.

Box, G. E., & Tidwell, P. W. (1962). Transformation of the independent variables. *Technometrics, 4*(4), 531–550.

Broder, T., Wheeler, C., & Bernstein, J. (2005). Immigration nationality law handbook. In American Immigration Lawyers Association (pp. 759–781).

Bureau of Labor Statistics. (2008). Labor force participation of seniors, 1948–2007. http://www.bls.gov/opub/ted/2008/jul/wk4/art02.htm. http://www.bls.gov/opub/ted/2008/jul/wk4/art02.htm. Accessed 1 Nov 2013.

Carr, S., & Tienda, M. (2013). Family sponsorship and late-age migration in aging America: Revised and expanded estimates of chained migration. *Population Research and Policy Review, 32*(6), 825–849.

Daly, M. C., & Burkhauser, R. V. (2003). The supplemental security income program. In R. A. Moffitt (Ed.), *Means-tested income programs in the United States* (pp. 79–139). Chicago: University of Chicago Press.

Duncan, B., Hotz, V. J., & Trejo, S. J. (2006). Hispanics in the U.S. labor market. In M. Tienda & F. Mitchell (Eds.), *Hispanics and the future of America*. Washington, D.C.: National Academies Press.

Dunn, A. (1995, April 16). For elderly immigrants, a retirement plan in the U.S. New York Times.

Fullerton, H. (1999). Labor force participation: 75 years of change, 1950–1998 and 1998–2025. *Monthly Labor Review, 122,* 3–12.

Gendell, M. (2006). 2013, Full time work among elderly increases. http://www.prb.org/Publications/Articles/2006/FullTimeWorkAmongElderlyIncreases.aspx. Accessed 1 Nov 2013.

Gonzalez-Barrera, A., & Lopez, M. H. (2013). A demographic portrait of Mexican origin Hispanics in the United States. Washington, DC: Pew Hispanic Center.

Gustman, A. L., & Steinmeier, T. L. (2000). Social Security benefits of immigrants and U.S. born. In G. J. Borjas (Ed.), *Issues in The Economics Immigration* (pp. 309–350). Chicago: University of Chicago Press.

Kaushal, N. (2010). Elderly immigrants' labor supply response to supplemental security income. *Journal of Policy Analysis and Management, 29,* 137–162.

Kim, B. J., & Torres-Gil, F. (2011). Social Security and its impact on older latinos. *Journal of Applied Gerontology, 30,* 85–103.

Leach, M. A. (2008). America's older immigrants: A profile. *Generations, 32,* 34–39.

Monger, R., & Yankay, J. (2012). U.S. legal permanent residents: 2012 (Tech. Rep.). Washington, DC: U.S. Department of Homeland Security.

Mosisa, A. T. (2013). Foreign-born workers in the U.S. labor force. http://www.bls.gov/spotlight/2013/foreign-born/pdf/foreign-born.pdf. Accessed 1 Nov 2013.

Nam, Y. (2011). Welfare reform and immigrants: Noncitizen eligibility restrictions, vulnerable immigrants, and the social service providers. *Journal of Immigrant & Refugee Studies, 9,* 5–19.

Nam, Y., & Jung, H. J. (2008). Welfare reform and older immigrants: Food stamp program participation and food insecurity. *The Gerontologist, 48*(1), 42–50.

Nam, Y., & Kim, W. (2012). Welfare reform and elderly immigrants' naturalization: Access to public benefits as an incentive for naturalization in the United States. *International Migration Review, 46*(3), 656–679.

Nuschler, D., & Siskin, A. (2010). Social Security benefits for noncitizens (Tech. Rep. No. RL32004). Congressional Research Service.

O'Neil, K., & Tienda, M. (forthcoming). Welfare consequences of late-age migration. Gerontology.

Passel, J., & Cohn, D. (2009). A portrait of unauthorized immigrants in the United States. Washington, DC: Pew Hispanic Center. http://www.pewhispanic.org/files/reports/107.pdf. Accessed 4 May 2014.

Passel, J., Cohn, D., & Gonzalez-Barrera, A. (2013). Population decline of unauthorized immigrants stalls, may have reversed. Washington, DC: Pew Hispanic Center.

Population Reference Bureau. (2013, October). Elderly immigrants in the United States (Today's Research on Aging No. 29). Washington, DC. http://www.prb.org/Publications/Reports/2013/us-elderly-immigrants.aspx. Accessed 14 Nov 2013.

Ruggles, S., Alexander, J. T., Genadeck, K., Goeken, R., Shroeder, M. B., & Sobek, M. (2010). Integrated public use microdata series: Version 5.0. Machine-readable database.

Social Security Administration. (2005). A guide for farmers, growers, and crew 15, 2014, leaders. Publication number 05–10025, ICN 455350. http://www.ssa.gov/pubs/EN-05-10025.pdf. Accessed 15 Jan 2014.

Terrazas, A. (2009, May). Older immigrants in the United States. 2014, Migration Policy Institute Report. http://globalaging.org/elderrights/us/2009/immigrants.pdf. Accessed 1 Feb 2014.

Tienda, M., & Mitchell, F. (2006). Multiple origins, uncertain destinies: Hispanics and the American future. Washington, DC: National Academies Press.

Toossi, M. (2012). Labor force projections to 2020: A more slowly growing workforce. *Monthly Laboratory Review, 135,* 43–64.

Treas, J., & Batalova, J. (2007). Older immigrants. In K. W. Schaie & P. Uhlenberg (Eds.), *Social structures: The impact of demographic changes on the well-being of older persons.* (pp. 1–24). New York: Springer.

U.S. Department of Homeland Security. (2006). Yearbook of immigration statistics: 2005. Washington, DC: Department of Homeland Security, Office of Immigration Statistics.

Van Hook, J. (2000). SSI eligibility and participation among elderly naturalized citizens and noncitizens. *Social Science Research, 29*(1), 51–69.

Van Hook, J. (2003). Welfare reform's chilling effects on noncitizens: Changes in noncitizen welfare recipiency or shifts in citizenship status? *Social Science Quarterly, 84*(3), 613–631. doi:10.11111540-6237.8403008.

Wilmoth, J. M. (2012). A demographic profile of older immigrants in the United States. *Public Policy & Aging Report, 22*(2), 8–11. doi:10.1093/ppar/22.2.8.

Chapter 5
Ageing and Retirement Security: United States of America and Mexico

Jorge Bravo, Nicole Mun Sim Lai, Gretchen Donehower and Ivan
Mejia-Guevara

Introduction

As populations age in both industrialized and developing countries, the adequacy
of different mechanisms to provide income security in old age receives increasing
attention. A large literature has examined public and private old-age pension systems in the United States of America, Mexico and other countries, as well as rising
public health care costs, especially in countries at the more advanced stages of ageing (Holzman and Hinz 2005; Barrientos 2008; OECD 1996; Alonso-Ortiz 2010).
Pensions and other public transfers are certainly important elements of retirement
security, but as will be seen in what follows, they are not the only or always the most
significant source of income in old age in the Americas.

Before presenting the specific analytical approach and main results of the paper,
it will be useful to briefly review some key general features and aggregate indict-
ors of ageing and retirement security in the three populations of older persons on
which this paper focuses: Mexicans residing in Mexico, Mexican Americans and
non-Mexican Americans in the United States of America.

Annex Table A.1 contains several indicators related to population ageing and
economic security in old age for selected countries of Latin America and Northern

---

The opinions expressed in this paper are those of the authors and do not necessarily reflect those
of the United Nations or its Member States.

J. Bravo (✉) · N. M. S. Lai
Population Division, United Nations, 2 United Nations Plaza, New York, NY, 10017, USA
e-mail: bravo1@un.org

G. Donehower
Center for the Economics and Demography of Aging of the Department of Demography,
University of California, Berkeley, USA

I. Mejia-Guevara
Harvard Center for Population and Development Studies,
Harvard University, Cambridge, MA, USA

© Springer International Publishing Switzerland 2015
W. A. Vega et al. (eds.), *Challenges of Latino Aging in the Americas,*
DOI 10.1007/978-3-319-12598-5_5

Table 5.1 Selected socio-demographic indicators for Mexico, Hispanic and non-Hispanic White populations in the United States of America

	Mexico	Mexican Americans or Hispanics in the U.S.	Non-Hispanic White, USA
Percentage of population aged 65 years and over	5.4%[a] (2010)	4.3%[b] (Mexican Americans) 5.5%[b] (Hispanic)	15.5[b]%
Total fertility rate	2.4[a] (2005–2010)	3.1[b] (Mexican Americans, 2009) 3.0[b] (Hispanic, 2009)	1.9[b] (2009)
Life expectancy at birth	76.3[a] (2005–2010)	83.1[b] (female, 2006) 77.9[b] (male, 2006)	80.5[b] (female, 2006) 75.6[b] (male, 2006)
Poverty rate among older persons	29%	20.5%[b] (Mexican Americans)	6.1%[b]

[a] World Population Prospects: The Revision 2012 (United Nations 2013c)
[b] Arias (2010), Centers for Disease Control and Prevention

America, including Mexico and the United States of America (also referred as "United States" from now on). The data shows that Mexico and the United States have similar overall and old-age mortality levels, with Mexico trailing the United States by only 1 year in female life expectancy at birth and at age 60 years for the period 2010–2015. Mexico's 9% of the population aged 60 years or over in 2013, denotes a slightly younger age structure than the Latin American average (11%), but a much younger one than the United States of America (20%). Also, older people in Mexico have very different living arrangements than in the United States of America: three quarters of older Mexicans live in multi-generational households, versus one quarter of older people in the United States of America. This is a significant fact given that co-residence tends to facilitate familial financial and other kinds of intergenerational support. Finally, older Mexicans have higher labour force participation rates (nearly one half) than older people in the United States of America (almost one third); and much lower social security coverage: one third in Mexico relative to 90% in the United States of America.

Table 5.1 reports on a smaller set of indicators for Mexicans and people living in the United States of America, including Mexican Americans or more generally, "Hispanics", as noted below. The proportion of older persons among Mexican Americans is much lower than other Americans and Mexicans, partly because of the high fertility of Mexican Americans, but also because of the continued inflow of young Mexican migrants to the United States of America.

These data also confirm the "Hispanic paradox" of higher life expectancy of Mexican Americans as compared to non-Hispanic whites. Older Mexican Americans are subject to much higher poverty rates than non-Hispanic whites, but significantly lower incidence of poverty than older people in Mexico. Note, however, that these last two figures are not strictly comparable, given the very different levels of income and of the poverty line in the two countries.

Table 5.2 National transfer accounts (NTA). (Source: United Nations 2013; Mason and Lee 2011)

A classification and examples of national transfer account age reallocations

	Asset-based reallocations		
	Capital income	Property income	Transfers
Public	Negligible	Public debt Student loan programmes Sovereign wealth funds	Public education Public health care Unfunded pension plans
Private	Housing Consumer durables Structures, production facilities, vehicles, other machinery	Consumer debt Land Subsoil minerals	Familial support of children and parents Charitable contributions Remittances

The aggregate figures referred to in the previous paragraphs tend to confirm the a priori hypothesis that Mexicans are less economically secure in old age, and less well protected against economic hardship than older persons in the United States of America. It may also seem natural to presume that Mexican Americans are likely to be somewhere in between the situation of Mexico and of the non-Mexican American population of the United States in this regard. The more detailed, though still preliminary evidence reviewed next only partially supports these hypotheses. The role of the various sources of economic sustenance in the three populations under study was examined and reflected on what that entails for their economic security in old age.

Analytical Approach

For the main part of this paper, we adopt the National Transfer Accounts framework, that considers the economic flows between nationals of a given country over the life course. The framework is based on the following classification (Table 5.2) of economic flows or "reallocations". They include pensions and public health care, which are very significant in more developed countries, especially in the United States. The framework also includes other important sources of intergenerational support, namely familial transfers and remittances, very important for Mexicans with relatives living in the United States, as well as income from financial and real assets (e.g., stocks, bonds, savings, as well as physical capital, land and real estate). A comprehensive exposition and numerous applications of the framework, including to the United States of America and Mexico, is available in Lee and Mason (2011).[1]

Two additional variables need to be introduced: total final consumption, which includes both private consumption expenditures and government consumption spending, and labour income, a major source of economic support over the life-

---

[1] See Lee et al. (2011), Chap. 15, and Mejía-Guevara (2011), Chap. 13.

cycle. Thus the complete accounting of life-cycle consumption and its sources of finance can be summarized as:

$$Consumption = Labour\ income + Asset\text{-}based\ reallocations +$$
$$Net\ Transfers\,(public\ and\ private)$$

In other words, consumption at any given age can be financed by working (thus perceiving labour income), drawing income from own assets, including dis-saving, and by receiving (net) transfers from the government or private individuals, most commonly, from family members. Details on the concepts, measures and the estimation procedures of the various NTA components are presented in the newly released *National Transfer Accounts Manual* (United Nations 2013a).

Data

In the case of Mexico, the NTA estimates are for 2004 and are based on micro-data from the Household Income and Expenditure Survey for 2004 (ENIGH-2004, see INEGI 2008b), National Accounts of Mexico (INEGI 2006), as well as administrative records from the Ministry of Finance (SHCP 2008) and the National Statistical Institute (INEGI 2008a).

In the case of the United States of America, the estimates used in this paper are for 2007, based on the Consumer Expenditure Survey (CEX), the Current Population Survey (CPS), and National Income and Product Accounts (NIPA). Micro survey data, sometimes supplemented with administrative records are used to estimate the age profiles of consumption and all types of income, while the national accounts are used as macro controls for the component elements of the accounts.

In this paper, Mexican Americans are identified through the self-reported questionnaires in the CEX and CPS as "Mexican, Mexican American or Chicano". Therefore, Mexican Americans are self-identified persons born in the United States of America or in Mexico that are currently living in the United States of America. Non-Mexican Americans are all other respondents not self-reported in the aforementioned category.

Results

Consumption

Mexico's overall per capita consumption age profile (Fig. 5.1) is characterized by relatively flat consumption through most of the adult ages, dropping moderately later in the life-cycle, after age of 60. This age pattern is not typical of Latin America; other countries like Chile have a flatter per capita consumption curve across all of the adult ages, including after age 60, while in others like Brazil, Costa Rica

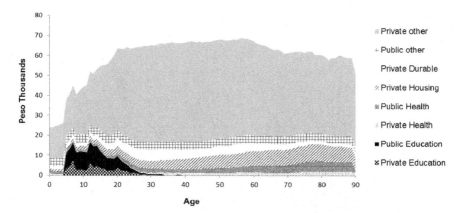

Fig. 5.1 Per capita consumption, Mexico, 2004. (Source: Calculations based on NTA methodology)

and Uruguay, the average level of consumption among adults increases with age. The overall age profile of consumption in Mexico is in fact more like that of certain developing countries outside of Latin America, such as Indonesia and Thailand.

Mexico's decline in consumption at older ages is not consistent with consumption-smoothing models, which predict that individuals would maintain an even level of consumption over their lifetime, through borrowing, saving and dis-saving.[2] The data shows that most of the components of consumption in Mexico drop at the older ages, with two exceptions: consumption of housing is stable through the older ages, and the consumption of health care increases. However, the impact of this increase is fairly modest, as public and private health together account for 10 % of the total consumption for Mexicans aged 65 years or more (and 5 % for younger adults), a relatively low figure even by developing country standards.

Mexican Americans, on the other hand (Fig. 5.2a), have a consumption profile that is increasing with age, similar to that of other Americans (Fig. 5.2b), a pattern which is typical of more developed countries. In virtually all industrialized countries, especially in the United States of America, the upward trend of consumption by age is driven by a sharp increase of health care expenditures, and also higher housing per capita consumption of housing of older adults,[3] while most of the other consumption components stay relatively constant. Persons aged 65 years or older in the United States of America dedicate 37 % of their total consumption to health services, as compared to younger adults who on average dedicate 18 % of their total consumption to health care.

---

[2] The fact that Mexicans do not appear to smooth out consumption over the adult ages could be a reflection of insufficiently developed financial markets and high income inequality, which prevents large segments of the population from accumulating substantial savings over their life-cycle.

[3] Most older persons, especially in the case of the United States of America, tend to live primarily alone or with a spouse only (see Annex Table A.1, also United Nations 2012), driving up the per capita value of housing consumption.

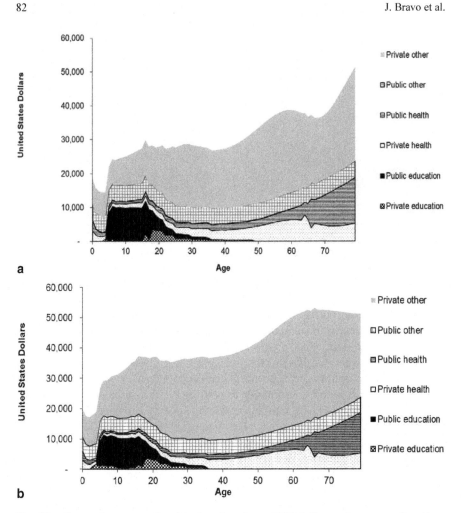

Fig. 5.2 **a** Per capita consumption, Mexican Americans, 2007. **b** Per capita consumption, Non-Mexican Americans, 2007. (Source: Calculations based on NTA methodology)

All together, the average American aged 65 years or over consumes one third more than an average adult aged 30–64 years. Comparing across ethnic groups, the average Mexican American consumes 20% less, in absolute dollar amounts, than an average non-Mexican American, mainly because of their significantly lower income, as discussed in following sections.

Another way to compare the three populations is to look at their per capita consumption normalized by their average labour income[4] (Fig. 5.3). Factors affecting

---

[4] Throughout this paper, the average labour income is restricted by ages 30 and 49, as this abstracts from variations at the very young and older working ages, and thereby facilitates standardized international comparisons (see Lee and Mason 2011).

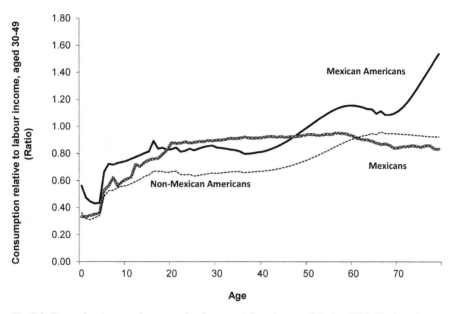

Fig. 5.3 Per capita consumption as a ratio of average labour income, Mexico 2004, Mexican Americans 2007 and Non-Mexican Americans 2007. (Source: Calculations based on NTA methodology)

the age profile of the ratio of per capita consumption to labour income include labour productivity, household composition, and the receipt of other sources of income over the life-cycle. In most of the 23 countries with NTA data (United Nations 2013), adults consume typically between 60 and 70 % of the average labour income. In this regard, Mexico stands out because it has a rather high consumption to labour income ratio, of 80–90 %.

The flip side to this high ratio, as we will see in more detail in the next section, is Mexico's heavy reliance on non-labour income, including remittances, other transfers and asset reallocations to finance their consumption. The age pattern of the consumption to labour income ratio of the United States of America is similar to European countries, slightly over 60 % (Tung 2011). As shown in Fig. 5.3, older Mexican Americans, like older Mexicans, consume a higher proportion of their labour income than older non-Mexican Americans, but the age pattern of consumption of Mexican Americans is more like that of non-Mexican Americans.

Despite this particular similarity of Mexican Americans with the general United States population, it is important to keep in mind their very different levels of income and consumption, as this has an important bearing for the interpretation of the results of sources of retirement security, presented next.

Finance of Consumption

Coming to the central question of this chapter: what are the sources of old-age economic support in these three populations? Figure 5.4 presents the results on the major sources of income that support older persons' consumption in Mexico and the United States of America, including labour income, public transfers, private transfers subdivided in intra and inter-household transfers, as well as asset-based reallocations.

Figure 5.4 shows, first, that older people in Mexico and older non-Mexican Americans finance their consumption in a roughly similar manner. Both labour income and public transfers are quite significant for them (they each finance one fifth to a quarter of old-age consumption), but asset income is their most important source of retirement sustenance, accounting for one half to two thirds of consumption for those aged 65 years or over.

Older Mexican Americans rely even more heavily on public transfers, which finance almost a half of their consumption. Asset reallocations (asset income and dis-savings) are also significant, but come in second place, accounting for one third of their consumption.

Second, familial transfers are positive and important for the Mexican American elderly, but not for older persons in Mexico or for non-Mexican Americans. Older Mexican Americans receive net familial transfers that represent an average of 4% of their consumption, and this share increases as they age further: those aged 75

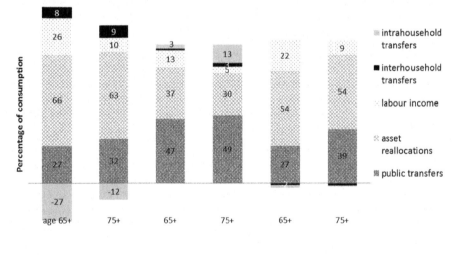

Fig. 5.4 Finance of consumption for persons age 65 years or over, Mexico (2004), United States of America (2007) and Mexican Americans (2007). Labour income, public transfers, private transfers (intra and inter-household) and asset-based reallocations, as a percentage of total final consumption. (Source: Calculations based on NTA methodology)

years or over finance 16% of their consumption with familial transfers, of which 13% is transfers within their household (intra-household transfers) and the remaining 3% is between households (inter-household transfers).

The pattern is the reverse for non-Mexican elders in the United States. Instead of net receivers, they are net givers of familial transfers to their children and grandchildren, in an amount equivalent to 4% of their consumption. These very different forms of intergenerational familial support may be explained partly by the distinctive living arrangements. Approximately 40% of Mexican American people aged 65 years or over live with adult children, an arrangement that is known to facilitate intra-familial transfers, while only 15% of non-Mexican Americans do so. Also, non-Mexican elders are wealthier and they have significantly higher lifetime labour and asset income compared to Mexican American elders. They are therefore better able to self-finance their consumption with asset income and dis-savings and do not need to rely on familial transfers. Cultural factors may also play a role, directly or through the mediating effect of co-residence.

Older people in Mexico are the only group studied here that receives net inter-household transfers, and in quite significant amounts: 8% of the consumption of those aged 65 years or more and 9% of those aged 75 years or more.[5] However, older Mexicans transfer even larger amounts to their younger relatives, in the form of intra-household transfers (27% and 12%, respectively), which makes them net givers of familial transfers, even at the oldest ages (see Fig. 5.4, column labelled 75+).

Third, and not surprisingly, public transfers are an essential source of old-age support both in Mexico and the United States of America, accounting for 27 to 32% of the consumption of people aged 65 years or over. A somewhat unexpected result, however, is that public transfers finance an even larger share (about half) of the old-age consumption of Mexican Americans. A small proportion of older Mexican Americans may receive a pension from Mexico in addition to the United States Social Security benefits, but we believe that there are two more important factors: (a) the lower overall level of income and consumption of Mexican Americans keeps the denominator down and drives up the per capita value of whatever public transfers they receive, and (b) public transfers include means-tested welfare benefits and all the other components of government final consumption, which is assumed to benefit all residents of the United States equally, but that represent a higher proportion of the income and consumption of Mexican Americans

Fourth, income generated from accumulated assets is the primary source for old-age financing in Mexico and the United States. This result does not surprise us for the United States, where financial and capital markets are highly developed, but it is a bit more unexpected for Mexico, where private pension funds are smaller and cover a much lower proportion of the population. Perhaps the accumulation and dis-accumulation of physical assets, including land and housing, are providing more of the asset reallocations of older Mexicans. Assets contribute a more modest share of financing for old-age consumption among Mexican Americans.

---

[5] A large share of this flow is likely to come from remittances from migrant children living in the United States of America.

Conclusions and Discussion

The examination of the major sources of income security confirms that public transfers are important for older persons in both Mexico and the United States of America, but such transfers are even more significant for Mexican Americans. On average, an older Mexican American finances up to one half of his/her consumption from public transfers, while such transfers support only one quarter of consumption of older persons in Mexico and older non-Mexican Americans in the United States.

The data show that older persons in both Mexico and the United States rely little on family transfers; however, familial support is a significant source of old-age support for older Mexican Americans. Cultural factors, including the tighter communities especially of first to second generation migrants, may play a role in producing this result, either directly or through the mediating effect of the higher incidence of intergenerational co-residence among Mexican Americans.

According to the results, asset income and dis-saving have become the major source of retirement income in Mexico and the United States of America, financing from two thirds to one half of their old-age consumption.

One additional key variable of retirement security that needs to be considered for a fuller understanding of retirement security is the reliability of the different sources of income in old age. In this regard, we conjecture that *labour income* will continue to provide some income security in old age in many countries of the Americas as the new cohorts of older people live longer, especially in the case of the United States of America and Canada, where higher statutory ages at retirement are being implemented. However, labour income may be an unstable source of support especially for youth and older workers because wage employment in the United States of America and Mexico is closely tied to the fluctuations of the business cycle, and the bargaining power of labour unions is diminishing over time In addition, the young and the old often experience higher unemployment than the rest of the population, particularly during economic downturns.

Family has been and will continue to be an important source of emotional and economic intergenerational support, especially in times of economic distress (Donehower 2013). However, family becomes a less significant financial source of support for old age as fertility continues to fall in the Americas and independent living becomes more common in the ageing Latin American societies. Familial transfers can be unreliable because family members are susceptible to the effects of macroeconomic cycles.

Asset reallocations have become a major source of income in old age, both in Mexico and the United States of America, as well as in other countries with NTA data. These results are likely to persist as individuals increasingly save for retirement through pension funds and other financial instruments for retirement in developed and developing countries. However, there are reasons to believe that asset reallocations may not always be reliable, as the recent housing and financial crises have shown.

The importance of old-age security emanating from *public sector* programmes (mainly social security and public health) during the twentieth century was on the whole increasing (Miller 2011), although it is unclear if that trend will continue into the future. In the short term, defined-benefit public sector transfers have proven to be more stable than either employment, private pension funds or other assets, particularly during period of economic and financial instability, as evidenced during the recent crisis and subsequent slow recovery. In a longer term perspective however, this conclusion may be affected by the sustainability of public pension systems, put into question as populations continue to age.

In sum, older persons in the three populations studied have diverse sources of retirement security. In most cases, these sources combined provide protection and sustenance of basic consumption needs in old age. However, Mexican Americans are more vulnerable to live in poverty and therefore rely more on familial and public transfers than other Americans. Over the longer-term, all population groups would benefit from further diversification and from policies to expand life-cycle investment in human capital (Lee and Mason 2014) to supplement financial and physical capital assets, as family sizes continue to decline and therefore provide a narrower base for old-age support (Annex Table A.2).

Appendix

Annex Table A.1 Selected demographic indicators, Mexico and United States of America (compared to selected countries and regional average values for Latin America and the Caribbean and Northern America)

| | Population aged 60 years or over | | | | | | | | | |
| | Proportion of total population 60 years or over (percentage) | | Share of persons aged 80 years or over (percentage) | | Sex ratio, 2013 (men per 100 women) | | Life expectancy at birth 2010–2015 | | Life expectancy at age 60 2010–2015 | |
Country or area	2013	2050	2013	2050	60+	80+	Men	Women	Men	Women
Latin America and the Caribbean	11	25	15	23	81	65	72	78	20	23
Costa Rica	11	30	15	25	91	75	78	82	22	25
Mexico	9	26	14	23	80	66	75	80	22	24
Brazil	11	29	15	24	80	66	70	77	20	23
Chile	14	31	16	30	80	57	77	83	22	25
Uruguay	19	27	21	25	69	48	74	80	19	24
Northern America	20	27	19	30	82	60	77	81	22	25
Canada	21	31	19	32	85	62	79	84	23	26
United States of America	20	27	19	29	82	60	76	81	22	25

Annex Table A.2 Selected economic and social indicators, Mexico and United States of America(compared to selected countries and regional average values for Latin America and the Caribbean and Northern America). (Source: Calculations based on NTA methodology)

	Proportion married 60 years or over[a] (percentage)		Proportion living independently, 60 years or over[a] (percentage)		Old-age support ratio[b]		Proportion in labour force 60 years or over[a] (percentage)		Social security coverage[a]	
									Working-age (percentage)	60+ (percentage)
Country or area	Men	Women	Men	Women	2013	2050	Men	Women		
Latin America and the Caribbean	74	42	30	27	9	3	49	22	–	–
Costa Rica	72	45	29	25	10	3	39	12		
Mexico	76	45	26	24	10	3	53	20	50.2[c]	33[d]
Brazil	78	41	32	29	9	3	44	21		
Chile	72	43	32	27	7	2	46	17		
Uruguay	72	40	–	–	5	3	39	20		
Northern America	75	48	77	74	5	3	34	24		
Canada	76	50	–	–	5	2	30	18		
United States of America	75	48	77	74	5	3	35	25	94[e]	90[e]

[a] Latest available information circa 2010 (United Nations 2012, 2013b)
[b] Source: United Nations (2013b)
[c] OECD, Pension coverage 2009
[d] Latin American Economic Outlook 2011—© OECD 2010
[e] Social Security Facts available from http://www.ssa.gov/pressoffice/basicfact.htm

Reference

Alonso-Ortiz, J. (2010). Social security and retirement across the OECD. Instituto Tecnológico Autónomo de México, Centro de Investigación Económica Working Paper 10-07, México.

Arias, E. (2010). U.S. life table by Hispanic origin. Vital health statistics, Centers for Disease Control and Prevention. DHHS Publication No. (PHS) 2011–1352. Maryland.

Barrientos, A. (2008). New strategies for old-age income security in low income countries. Social security essentials, Technical Report 12. Geneva: International Social Security Association.

Donehower, G. (2013). Children, adults and the elderly in the great recession: An economic atlas by age. Paper presented at the Annual Meeting of the Population Association of America, New Orleans, 11–13 April 2013.

Holzmann, R. & Hinz, R. (2005). Old-age income support in the Twenty-first century: An international perspective on pension systems and reform. Washington, DC: The World Bank.

Instituto Nacional de Estadística y Geografía (INEGI). (2006). Sistema de cuentas nacionales de México: Cuentas por sectores institucionales 1999–2004, Tomo II, Mexico City.

INEGI (2008a). Banco de información económica (BIE). http://www.dgcnesyp.inegi.org.mx/cgi-win/bdieintsi.exe. Accessed 20 June 2008.

INEGI (2008b). Encuesta nacional de ingreso y gasto de los hogares, 2004. http://www.inegi.gob.mx. Accessed 20 June 2008.

Lee, R., & Mason A. (2014). Is fertility too low? Population ageing, dependency and consumption. In Science, American Association for the Advancement of Science.

Lee, R., Donehower, G., & Miller T. (2011). The changing shape of the economic life-cycle in the United States, 1960 to 2003. In R. Lee & A. Mason (Ed.), *Population aging and the generational economy: A global perspective*. Northampton: Edward Elgar Press.

Mejía-Guevara, I. (2011). The economic lifecycle and intergenerational redistribution in Mexico. In R. Lee & A. Mason (Ed.), *Population aging and the generational economy: A global perspective*. Northampton: Edward Elgar Press.

Miller, Tim (2011). The rise of intergenerational state: aging and development. In R. Lee & A. Mason (Ed.), *Population aging and the generational economy: A global perspective*. Northampton: Edward Elgar Press.

Organization for Economic Cooperation and Development (OECD). (1996). Ageing populations, pension systems and government budgets: Simulations for 20 OECD countries. By Deborah R., Willi L., Douglas F., Eckhard W., Economics Department Working Papers No. 168.

SHCP (2008). *Cuenta de la Hacienda Pública Federal 2004*. Secretaría de Hacienda y Crédito Público. www.shcp.gob.mx SHCP.

Tung, A.-C. (2011). Consumption over the life-cycle: An international comparison. In R. Lee & A. Mason (Ed.), *Population aging and the generational economy: A global perspective*. Northampton: Edward Elgar Press.

United Nations (2012). Wall chart on population ageing and development. Sales no. E.12.XII.6.

United Nations (2013a). National transfer accounts manual: Measuring and analysing the generational economy. New York.

United Nations (2013b). World population ageing 2013. Department of Economic and Social Affairs, New York.

United Nations (2013c). World Population Prospects: The 2012 Revision, DVD Edition.

Part II
New Data and Methodological Approaches on Aging Research in Mexico and the United States

Chapter 6
New Data and Methodological Approaches on Aging Research in Mexico and the United States

Flávia Cristina Drumond Andrade

It is well-known that the populations in Mexico and in the United States are getting older (Anderson and Hussey 2000; Palloni et al. 2002). The median age was 25.8 in Mexico in 2010 and it is expected to increase to 32.3 by 2030 (Celade 2013). During the same period, median age in the United States will have increased from 37.2 to 39.5 years (Howden and Meyer 2011; United Nations, Department of Economic and Social Affairs, Population Division 2013). The population of older adults (those aged 60 years and over) in Mexico reached 10.2 million in 2010 and it is expected to more than double by 2030, reaching 23 million (United Nations, Department of Economic and Social Affairs, Population Division 2013). During the same time, the older adult population in the United States will increase from 57.8 to 93 million individuals (United Nations, Department of Economic and Social Affairs, Population Division 2013). The fast growth of the old adult population will challenge the way these societies provide care and assistance.

Besides the increased knowledge about the population size and composition, we also have significant information about the population and health trends in both sides of the border. For example, both countries have high prevalence of obesity, an important risk factor for many chronic conditions such as diabetes and hypertension. In Mexico, estimates based on recent data from the Encuesta Nacional de Salud y Nutrición 2012 (ENSANUT 2012; National Health and Nutrition Survey) show that 32.4% of the adults in Mexico are obese (Barquera et al. 2013), 9.2% have been previously diagnosed with diabetes (Hernández-Ávila et al. 2013) and 31.5% have hypertension (Campos-Nonato et al. 2013). In the United States, estimates based on recent data collected by the National Health and Nutrition Examination Survey (NHANES) show that 35.9% of adults are obese (Flegal et al. 2012), 12.1% have diabetes (Cheng et al. 2013) and 29.5% have hypertension (Guo et al. 2012). Besides obesity, diabetes, and hypertension, many studies have focused on mobility issues, with falls being a major concern among older individuals, and disability. In

F. C. Drumond Andrade (✉)
University of Illinois at Urbana-Champaign, Champaign, IL, USA
e-mail: fandrade@illinois.edu

© Springer International Publishing Switzerland 2015
W. A. Vega et al. (eds.), *Challenges of Latino Aging in the Americas,*
DOI 10.1007/978-3-319-12598-5_6

addition, a growing number of studies have addressed problems related to dementia and mental health problems, such as depression.

Health care delivery and access are critical issues to prevent the development of chronic conditions, which are responsible for a large share of public expenditure in both countries. In this aspect, both countries are making efforts to expand access and improve health care delivery. Mexico, for example, has expanded the coverage of health care in recent years based on the increased coverage provided by Seguro Popular. From 2000 to 2012, the proportion of the Mexican population without health care coverage declined from 57.6 to 21.4 % (Gutiérrez and Hernández-Ávila 2013). The United States has also embarked on an ambitious plan to extend health care coverage with the Affordable Care Act, which has the potential to improve the health status of the American population (Corbett and Kappagoda 2013).

The populations in Mexico and the United States are also strongly connected. Currently 51 million Latinos live in the United States. Among those, 33 million (65 %) have Mexican ancestry and other 11.7 million were born in Mexico (Pew Hispanic Center 2012). Latinos play an important and growing role in the U.S. economy, but the Latino population is poorer, have lower educational levels, and has less access to health care than the general population (Andrade and Viruell-Fuentes 2011). Therefore, there is an urgent need to reexamine the current health status of these populations as well as organizations and infrastructures that provide vocational, educational and health services for them (Andrade and Viruell-Fuentes 2011).

In recent decades, Mexico and the United States have increased the availability of health and social datasets that can be used to evaluate the health and socioeconomic wellbeing in Mexico and in the United States. Many of these datasets have good representation of older adults or are focused on this population segment. First, I discuss some of these datasets and the methodological advances to analyze them. Next, I present the main contributions of three papers in this section which highlight new data and pertinent methodologies to address important issues to the health and wellbeing of populations in Mexico and the United States. After, I discuss data limitations and further data needs. Final remarks are presented at the end.

Availability of Data for Research on Aging in Mexico and the United States

Mexico and the United States have a long history of collecting data on their population and households. In Mexico, the first general census (Censo General de la República Mexicana) was conducted in 1895 and the most recent national census (Censo de Población y Vivienda) in 2010. Census data collection in the United States dates back to 1790, with its most recent data collection in 2010. Currently both countries have high quality vital statistics systems which collect data on more than 90 % of deaths (Mathers et al. 2005). Both countries have also increased the availability of population and health related data in recent decades.

In Mexico, two nationally representative data sources are notable for their contribution to the understanding of health and aging—the Encuesta Nacional de Salud y Nutrición (ENSANUT) and the Mexican Health and Aging Study (MHAS). ENSANUT is one of the main sources of health data in Mexico. ENSANUT is a nationally representative study which collects information on a range of health issues and health care utilization (Romero-Martínez et al. 2013). ENSANUT builds upon previous studies conducted in Mexico such as the Mexican National Nutrition Survey conducted in 1988 and Mexican National and Health Survey in 2000. ENSANUT has been carried out in 2006 and 2012. In 2012, ENSANUT interviewed 96,031 individuals and conducted 14,104 of ambulatory health services (Romero-Martínez et al. 2013). MHAS is a longitudinal study of a nationally representative cohort of Mexicans. The study was designed with a field protocol and content similar to the Health and Retirement Survey conducted in the United States. The baseline was collected in 2001, followed by a second study in 2003 and a follow-up in 2012. In the first wave, a total of 15,186 complete interviews were obtained (Wong et al. 2006).

In the United States, several data sources provide relevant health data, with some of them providing nationally representative estimates. Many of these sources, such as National Health and Nutrition Examination Survey (NHANES), National Health and Interview Survey (NHIS), Behavioral Risk Factor Surveillance System (BRFSS) and Health and Retirement Study (HRS), sample large numbers of Latinos. Other data sources such as the Hispanic Established Populations for Epidemiologic Study of the Elderly (HEPESE) have been providing a long follow-up for a representative sample of older Mexican Americans residing in the Southwestern United States. However, one common problem among datasets in the United States is that few allow exploring the diversity of Latino subgroups. Many only allow for the disaggregation of few subgroups, for example, Mexicans vs. non-Mexicans. Others have been collected many years ago, such as National Latino and Asian American Study (NLAAS)—the most comprehensive study of Latino mental health—which was conducted in 2002–2003. Nonetheless, as a group, these sources of data help us better understand some of the opportunities and challenges that older adults and their families will be facing in the coming years and decades.

Some innovative and recent studies such as the National Health and Aging Trends Study (NHATS) and Hispanic Community Health Survey/Study of Latinos (HCHS/SOL) are collecting important social and health data in the United States. The NHATS is a nationally representative panel study of Medicare beneficiaries ages 65 and older. NHATS has new disability protocol (with self-reported and performance assessments) and detailed information on use of assisted devices, mobility accommodations, and technological environment of the home (Freedman et al. 2011). Information on sensory, cognition and other aspects of health and quality of life are also available. The first wave conducted in 2011 collected data on over 8000 individuals (Freedman et al. 2014). The SOL Study is a multi-center epidemiologic study in Hispanic/Latino populations (Sorlie et al. 2010). The project aims to identify risk and protective factors associated with the health of Hispanics/Latinos. The study focused on Latinos aged 18–74 years in the Bronx, Chicago, Miami and

San Diego. These sites were selected sufficient sample sizes of Mexican, Puerto Rican and Dominican, Cuban, and Central and South Americans needed to stratify analyses by place of origin. Extensive clinic exams and assessments were used to determine baseline risk factors. The initial data collection conducted between March 2008 and June 2011 examined 16,415 individuals who self-identified Hispanic/Latino (Daviglus et al. 2012). Follow-up studies are planned to be conducted in 2–4 years to determine health transitions. In Mexico is noteworthy the efforts to conduct the follow up waves of the MHAS.

There is certainly more available data in both countries and some datasets provide a similar set of variables that can be used for comparative studies. There is also progress in terms of having data sources that allow the analyses of trends and trajectories. More recently, some efforts have been made to start collecting experimental design data that are often used to evaluate the impact of programs, policies and interventions. In addition, the availability of qualitative data has increased in both countries. Qualitative and mixed-methods studies have been increasingly important providing in depth information about health, health care access, socioeconomic wellbeing and social support. They have also provided culturally-rich data that are critical for guiding decisions about initiatives, policy and practice.

Along with improvements in data collection, major methodological advances have also happened in the last decades. Both quantitative and qualitative methods have allowed for more complex data analyses. Available software programs allow the analyses of large amounts of complex data. As a result, our knowledge about the health and wellbeing of populations in Mexico and the United States has grown. Together, these efforts in data collection and methodological approaches will provide innovative and up to date information on the health and aging process in these countries.

Contributions for the Study of Aging in Mexico and the United States

In this section, I discuss the contributions of four papers which all use recently collected data that are used to better understand factors associated with successful aging in both Mexico and the U.S. Some of them also highlight the interconnectedness of these two countries and how creative study designs can better help us understand the complexities related to aging, health, health care access, retirement decisions and migration.

Maria Aranda et al. used data from 2012 ENSANUT (Mexican National Health and Nutrition Survey) to analyze the prevalence and determinants of falls among older Mexicans. Falls are very prevalent at older ages and they can result in injuries, disability and even death. Several factors have been shown to be associated with falls in the previous literature and the authors used the World Health Organization risk factor model for variable selection. ENSANUT is a very rich data source and the authors were able to include socioeconomic, demographic, and several physical and mental health variables in their analyses.

Aranda et al. also tested alternative models (Poisson regression and negative binomial regression models) to take into account the fact that the data on falls were skewed. The authors found that 34% of older adults in the sample reported having had at least one fall in the last 12 months, with most of those having had more than one fall during the last year. The authors identified several factors associated with falls. Older Mexicans who reported having difficulties in performing basic daily activities, depressive symptoms, vision and hearing impairments, as well as memory problems were more likely to fall. As found in previous studies, older age and female gender were also associated with higher risk of having falls.

Ester Apesoa-Varano and colleagues conducted an extensive systematic review of qualitative and mixed-methods studies that addressed issues related to caregiving of Latinos with Alzheimer's disease and related degenerative brain diseases (ADRD). The authors identified over 500 articles and 24 met full criteria for review. Based on these 24 studies, the authors identified three main themes in the current literature: the caregiving experience, caregivers' knowledge of dementia, and the caregiving division of labor.

The review study highlighted the normative expectations and cultural values shared by Latinos influences the caregiving division of labor in which most caregivers are women. Many of these women report being discontent in their role of caregivers. This is not surprising given that they often endure financial losses related to leaving the labor force or reducing hours of work, have their leisure activities modified and less time to care for their own health and needs. Even more, these women also suffer from ambivalent feelings about their role and the social expectations of them as caregivers.

The review also found that most of Latino women caring for individuals with ADRD not only lack biomedical knowledge related to dementia, but they are also unaware of available social services guided to caregivers' support. Latino women caring for those   with ADRD expressed frustration dealing with the health care system and raised issues related to having language barriers.

Certainly, there are many caregiving responsibilities' for these Latino women carrying from someone with ADRD. However, as pointed out by Apesoa-Varano and colleagues, these responsibilities' are not well shared within their larger networks, which are often described as unreliable and with loose ties. Fact that contrasts with the 'myth of familism' among Latino families.

Chenoa Flippen used data from survey and in-depth interview with low-income immigrants collected in 2001–2002 and 2006–2007 in Durham/Chapel Hill, NC. Her study pointed out to the need to adopt sensitive methods for data collection that take into account local conditions. In particular, it highlighted the importance of involving community members when working with immigrant communities. When collecting the data, the investigators used Community Based Participatory Research (CBPR) and targeted random sampling to overcome the difficulties related to obtaining a representative sample. The CBPR members were involved in data planning, data collection, questionnaire development and interpretation of survey results. The integration of community members allowed for very high response rates and high data quality.

Flippen's study addressed how remittances and transnational elder care impact the financial security of low-income Latino immigrants in the United States. This is an important gap in the literature as many of the previous studies have focused on the impact of remittances on the finances and wellbeing in migrant-sending regions in Mexico, but less attention has been given to the consequences of those in the host communities in the United States. She found that a large share of immigrants (over 50% of women and 80% of men) sends remittances to family abroad. The average values of annual remittances were significant for this population—$ 2,459 for women and $ 5,382 for men. Most (over 80%) of the remittances were given to older adult family members abroad. However, there were important gender differences. Women who work were more likely to remit higher values and, among married women, those who have a larger share of the household income were more likely to remit high values as well. Women who have been separated from their children, in many cases due to migration, were also less likely to remit to their parents. Among men, those who are unaccompanied by their wives and children were the least likely to remit to older family members abroad. These men often struggle to remit to their wives and children while supporting themselves. In sum, her study found that depending on how families have been separated across borders inter and intragenerational support are affected, which impacts the health and wellbeing of generations differently.

The study conducted by Emma Aguila et al. used data from MHAS and focused on older adult men aged 50-79 years residing in Mexico. The analyses were restricted to older men who worked full-time in the baseline survey and who were interviewed in the two follow-up surveys in 2003 and 2012. The paper described the characteristics of older men in Mexico who had a salaried job and those who were self-employed. Older men who were self-employed were more likely to live in rural areas, where there are fewer opportunities for formal employment, and had lower income than those who held a salaried job. In addition, those who were self-employed were less likely to have health insurance. These findings confirmed previous studies conducted in Mexico who have shown that self-employed workers are more likely to be older, with lower education and with lower income (Poplin 2010). In sum, the lives of many of these older man have been marked by poverty and economic instability, which certainly impact their health as they age.

These findings are important given that Mexico does not provide universal social security benefits, which creates large inequalities on the social well-being at older ages. Older adults who worked in the informal sector and those who were self-employed are less likely to be eligible for benefits. Aguila and colleagues also show that return migrants who remained between one and nine years in the U.S. are more likely to be self-employed, which also reduces their prospects of being eligible for benefits. Nonetheless, there is some evidence that Seguro Popular has expanded access to health care to older adults in Mexico (Vargas 2011), including those who were self-employed as shown by the authors based on data from MHAS 2012.

Aguila and colleagues limited their analyses to older men, but women's participation in the labor force is increasing in Mexico and a larger share of self-employed are women (Poplin 2010). Therefore, limiting the analyses to men is problematic as a larger share of older adults is composed by women. The study also excluded those who died and this may bias the estimates. Finally, the exclusion of those who work part-time can also impact the findings. Future analyses should then expand the analyses to other groups of older adults in Mexico. The trajectories in the labor market in Mexico are diverse and with distinct implications for access to health care and social security benefits.

Limitations and Further Data Needs for the Study of Aging in Mexico and in the United States

Significant advances have been happening in terms of data availability, but there are still some limitations, such as limited information on smaller geographic units, interventions and their cost-effectiveness. In addition, few data sources have the ability to be linked to additional socioeconomic and environmental sources, and most data are based on self-reports. Finally, even though quantitative longitudinal data are increasingly available, the majority of qualitative studies are crossectional, which limit the analyses of trends.

The availability of data at smaller geographic units is particularly important as many programs are needed at the community level. For example, even though most Latino immigrants continue to be geographically concentrated in a few states and cities, they are increasingly relocating to new destination areas. Many of these new host communities lack the services and community infrastructures that promote health for these migrant groups. Programs and policies at these new destinations will be needed to provide adequate health care to Latinos, particularly to those with limited English proficiency. However, little is known about the needs and characteristics of these communities, which limits the ability to plan.

There are also very few datasets that have the ability to be linked to additional socioeconomic and environmental data or that contain valuable socioeconomic information that could be used for planning and implementation of programs. This limitation affects, for example, the ability to monitor changes on health indicators by socioeconomic groups and taking into account environment conditions. These types of information are critical for understanding socioeconomic, ethnic and racial disparities, an understanding that is needed to mobilize resources and create programs to combat those disparities. The lack of environmental data also limits the ability to evaluate geographical distributions of risk factors, populations at risk, health indicators, and to address how the environment influences health outcomes and changes. Major advances have been done in recent decades to software and computation speed to advance the use of geographic information systems, but there is still limited geocoded data. This is particularly important as many health issues, such as the geographical distribution of health care services, are unevenly distributed within and across communities, which impacts access to care and health outcomes (Cromley and McLafferty 2012).

A large number of datasets are based on self-reports. However, there has been great progress in terms of including biological data, measures of performance and innovative measures to capture advances in several study fields. Also, even though there are some recent attempts to design and test interventions, very little is known about the cost effectiveness of many programs and interventions. Finally, studies using qualitative or mixed-methods for data collection should include, more often, a longitudinal design.

Final Remarks

The last decades have been marked by a considerable increase in the availability of population-based and health data in Mexico and the United States. In addition, data from experimental design and studies using qualitative and mixed-methods approaches have become more available. In sum, datasets have provided a wealth of data and methodological advances have been made to better analyze the available data sources. Additional improvements could further advance our understanding of the health and socioeconomic wellbeing of the older adults in those countries, but certainly major progress has been done in recent decades.

The papers in this section highlight the similarities and interconnectedness of the Mexican and the U.S. populations. Findings from Aranda et al. showed that, even though, older adults in Mexico are more exposed to falls than those of Mexican ancestry living in the U.S., the main risk factors are similar for those two populations (Reyes-Ortiz et al. 2005). Apesoa-Varano and colleagues showed that Latino women caring for individuals with Alzheimer's or other types of dementia are often frustrated on their caregiving roles. This frustration is in part a result of the conflict between their own needs and filial piety. These Latino women also lack the knowledge related to dementia and express frustration dealing with the health care system. There is also evidence that the limited availability of linguistically and culturally competent providers restricts the access to many services among Latinos. Flippen's work adds to the previous works in the literature that have mostly focused on the impact of remittances on the finances and wellbeing in migrant-sending regions in Mexico by focusing on the impact on the other side of the border. Her study highlights that not only most of the immigrants send remittances to older family members abroad, but also that these are significant for the populations that are many times strained financially. The impacts of migration to the U.S. are also explored by Aguila and colleagues. The authors found that older adult Mexican men who had lived in the U.S. for less than 10 years were more likely to be self-employed in Mexico, which is associated with lower earnings, less access to health insurance and older ages at retirement. As both populations age, several opportunities will be shared across the borders. High quality and timely data will be important for providing the basis needed for educated decisions in decades ahead.

References

Anderson, G. F., & Hussey, P. S. (2000). Population aging: A comparison among industrialized countries. *Health Affairs, 19*(3), 191–203. doi:10.1377/hlthaff.19.3.191.

Andrade, F. C., & Viruell-Fuentes, E. A. (2011). Latinos and the changing demographic landscape: Key dimensions for infrastructure building. In L. P. Buki & L. M. Piedra (Eds.), *Creating infrastructures for Latino mental health* (pp. 3–30). New York: Springer.

Barquera, S., Campos-Nonato, I., Hernández-Barrera, L., Pedroza, A., & Rivera-Dommarco, J. A. (2013). Prevalencia de obesidad en adultos mexicanos, 2000–2012. *Salud Pública de México, 55*(S2), S151–S160.

Campos-Nonato, I., Hernández-Barrera, L., Rojas-Martínez, R., Pedroza-Tobías, A., Medina-García, C., & Barquera-Cenera, S. (2013). Hipertensión arterial: Prevalencia, diagnóstico oportuno, control y tendencias en adultos mexicanos. *Salud Pública de México, 55*(S2), 144–150.

Celade. (2013). *Observatorio demográfico: Proyecciones de población, 2012*. Santiago de Chile: Centro Latinoamericano y Caribeño de Demografía.

Cheng, Y. J., Imperatore, G., Geiss, L. S., Wang, J., Saydah, S. H., Cowie, C. C., et al. (2013). Secular changes in the age-specific prevalence of diabetes among US adults: 1988–2010. *Diabetes Care, 36*(9), 2690–2696. doi:10.2337/dc12-2074.

Corbett, J., & Kappagoda, M. (2013). Doing good and doing well: Corporate social responsibility in post Obamacare America. *The Journal of Law, Medicine & Ethics, 41*, 17–21. doi:10.1111/jlme.12032.

Cromley, E. K., & McLafferty, S. (2012). *GIS and public health*. New York: Guilford Press.

Daviglus, M. L., Talavera, G. A., Avilés-Santa, M. L., Allison, M., Cai, J., Criqui, M. H., Gellman, M., et al. (2012). Prevalence of major cardiovascular risk factors and cardiovascular diseases among Hispanic/Latino individuals of diverse backgrounds in the United States. *JAMA: The Journal of the American Medical Association, 308*(17), 1775–1784.

Flegal, K. M., Carroll, M. D., Kit, B. K., & Ogden, C. L. (2012). Prevalence of obesity and trends in the distribution of body mass index among US adults, 1999–2010. *JAMA: The Journal of the American Medical Association, 307*(5), 491–497. doi:10.1001/jama.2012.39.

Freedman, V. A., Kasper, J. D., Cornman, J. C., Agree, E. M., Bandeen-Roche, K., Mor, V., et al. (2011). Validation of new measures of disability and functioning in the national health and aging trends study. *The Journals of Gerontology Series A: Biological Sciences and Medical Sciences, 66A*(9), 1013–1021. doi:10.1093/gerona/glr087.

Freedman, V. A., Kasper, J. D., Spillman, B. C., Agree, E. M., Mor, V., Wallace, R. B., et al. (2014). Behavioral adaptation and late-life disability: A new spectrum for assessing public health impacts. *American Journal of Public Health, 104*(2), e88–e94. doi:10.2105/AJPH.2013.301687.

Guo, F., He, D., Zhang, W., & Walton, R. G. (2012). Trends in prevalence, awareness, management, and control of hypertension among United States adults, 1999 to 2010. *Journal of the American College of Cardiology, 60*(7), 599–606. doi:10.1016/j.jacc.2012.04.026.

Gutiérrez, J. P., & Hernández-Ávila, M. (2013). Cobertura de protección en salud y perfil de la población sin protección en México, 2000–2012. *Salud Pública de México, 55*(S2), S83–S90.

Hernández-Ávila, M., Gutiérrez, J. P., & Reynoso-Noverón, N. (2013). Diabetes mellitus en México. El estado de la epidemia. *Salud Pública de México, 55*(S2), S129–S136.

Howden, L. M., & Meyer, J. A. (2011). Age and sex composition: 2010. 2010 Census Briefs. http://www.census.gov/prod/cen2010/briefs/c2010br-03.pdf. Accessed 27 Jan 2014.

http://ensanut.insp.mx/

Mathers, C., Ma Fat, D., Inoue, M., Rao, C., & Lopez, A. D. (2005). Counting the dead and what they died of: An assessment of the global status of cause of death. *Bulletin of the World Health Organization, 83*, 171–177.

Palloni, A., Pinto-Aguirre, G., & Pelaez, M. (2002). Demographic and health conditions of ageing in Latin America and the Caribbean. *International Journal of Epidemiology, 31*(4), 762–771.

Pew Hispanic Center. (2012). Statistical portrait of Hispanics in the United States, 2010. http://www.pewhispanic.org/files/2012/02/Statistical-Portrait-of-Hispanics-in-the-United-States-2010_Apr-3.pdf. Accessed 18 Nov 2014.

Popli, Gurleen K. (2010). "Trade Liberalization and the Self-employed in Mexico." World Development *38*(6), 803-813.

Reyes-Ortiz, C. A., Al Snih, S., & Markides, K. S. (2005). Falls among elderly persons in Latin America and the Caribbean and among elderly Mexican-Americans. *Revista Panamericana de Salud Pública, 17*(5/6), 362–369.

Romero-Martínez, M., Shamah-Levy, T., Franco-Núñez, A., Villalpando, S., Cuevas-Nasu, L., Gutiérrez, J. P., et al. (2013). Encuesta nacional de salud y nutrición 2012: Diseño y cobertura. *Salud Pública de México, 55*(S2), S332–S340.

Sorlie, P. D., Avilés-Santa, L. M., Wassertheil-Smoller, S., Kaplan, R. C., Daviglus, M. L., Giachello, A. L., et al. (2010). Design and implementation of the Hispanic community health study/study of Latinos. *Annals of Epidemiology, 20*(8), 629–641.

United Nations, Department of Economic and Social Affairs, Population Division. (2013). *World population prospects: The 2012 revision*. New York: United Nations.

Vargas P. (2011). La cobertura en salud y el Seguro Popular. Coyuntura Demográfica. 1:52–56.

Wong, R., Pelaez, M., Palloni, A., & Markides, K. (2006). Survey data for the study of aging in Latin America and the Caribbean: Selected studies. *Journal of Aging and Health, 18*(2), 157–179. doi:10.1177/0898264305285655.

Chapter 7
Self-Employment, Health Insurance, and Return Migration of Middle-Aged and Elderly Mexican Males

Emma Aguila, Raquel Fonseca and Alma Vega

Introduction

The number of Mexicans aged 60 or older is expected to increase from about 8 million in 2000 to more than 36 million in 2050. Their share of the total population is predicted to grow from 7.3 % in 2000 to 17.5 % in 2030 and 28.0 % in 2050 (Zúñiga Herrera 2004; Van Gameren 2008, 2010). This growth will put considerable pressure on Mexico's public health-care and pension plans. Therefore, understanding the determinants of labor-force participation among older adults is an important policy issue.

In this study, we analyze the labor-market dynamics of middle-aged and older men in Mexico. Specifically, we examine their transitions between self-employment and salaried work, as well as between work and retirement, and analyze the role of health insurance and U.S. migration experience in determining self-employment.

Self-employment is characterized by the ability to identify economic opportunities and generate work for one's self. It is perceived as an engine of entrepreneurial activity that has the potential to deliver more jobs in the future or, at the very least, a means of earning income in the absence of other opportunities. Mexico has one of the highest proportions of self-employed in its workforce. One in every three workers reported self-employment status in 2009 (OECD 2011a), with self-employment

E. Aguila (✉)
Sol Price School of Public Policy, University of Southern California (USC) and RAND, 650 Childs Way, Los Angeles, CA 90089, USA
e-mail: eaguilav@usc.edu

R. Fonseca
Département des sciences économiques, Université du Québec a Montréal and RAND, 315 rue Ste-Catherine Est, Montréal, QC H2X 3X2, Canada

A. Vega
RAND, 1776 Main Street, Santa Monica, CA 90407, USA

© Springer International Publishing Switzerland 2015
W. A. Vega et al. (eds.), *Challenges of Latino Aging in the Americas*,
DOI 10.1007/978-3-319-12598-5_7

103

rates being higher for middle-aged and older workers (Fairlie and Woodruff 2007; Duval-Hernández and Orraca 2009). Possible reasons for the high prevalence of self-employment in Mexico include relatively low income levels, with lower income countries displaying higher rates of self-employment in one cross-country comparison (Gollin 2002), numerous tax evasion opportunities (Torrini 2005), a lack of formal employment opportunities in Mexico's rural areas (e.g., Perry et al. 2007; Aguila et al. 2011), and migrant networks which help alleviate capital constraints for entrepreneurial activities (Woodruff and Zenteno 2007).

In Mexico, social-security and health-care services contributions are not mandatory for self-employed workers. This may provide an incentive for the self-employed to work in the informal sector without public health insurance and eligibility for social-security benefits during retirement. While formal-sector workers are entitled to social-security benefits and health-care insurance provided mainly by the Mexican Social Security Institute (Instituto Mexicano del Seguro Social, or IMSS) in the private sector and by the Social Security Institute for Government Workers (Instituto de Seguridad y Servicios Sociales de los Trabajadores del Estado, or ISSSTE) in the public sector, informal-sector workers who do not pay into these systems are only entitled to noncontributory pensions and health care services such as "70 y más" and "Seguro Popular". This has significant implications, given that 58 % of the labor force is in the informal sector (Perry et al. 2007).

There is a considerable amount of empirical and theoretical research on self-employment and entrepreneurship (Blanchflower and Oswald 1998; Guiso et al. 2004; Hurst and Lusardi 2004; Evans and Jovanovic 1989; Evans and Leighton 1989; Holtz-Eakin et al. 1994; Gentry and Hubbard 2000; among others), particularly among older persons (Hochguertel 2004; Fonseca et al. 2007; Zissimopoulos and Karoly 2007; among others). Previous studies have used U.S. and Mexican Census data to analyze entrepreneurship among Mexicans living in the United States and Mexico (Mora 2006; Mora and Davila 2006; Fairlie and Woodruff 2007, 2010). Other studies have analyzed the characteristics of self-employed using the Mexican employment survey (Samaniego 1998; Wong and Parker 1999) and the Mexican income and expenditure survey after the 1994 introduction of NAFTA to analyze poverty and inequality of self-employed (Popli 2010) and self-employment in rural areas (González and Villarreal 2006).

The importance of health-care coverage as an explanation for self-employment at elderly ages has been previously documented for the United States (Parker and Rougier 2007; Zissimopoulos and Karoly 2007; Fairlie et al. 2011). Other studies have analyzed the occupational choices and characteristics of return migrants in Mexico (Papail 2002, 2003). This is the first study to analyze the interactions between self-employment, return migration, and health insurance for middle-aged and older persons in Mexico.

We use a unique source of information, the Mexican Health and Aging Study (MHAS) to examine this issue. The MHAS is similar to the U.S. Health and Retirement Study (HRS) in being a nationally-representative panel survey of individuals at least 50 years of age. We use cross-tabulations to describe the characteristics of the sample and its labor-market transitions, and a probit model to examine predictors of self-employment. Results suggest that return migrants who worked 1–9 years in the United States are more likely to be self-employed in Mexico than those who never migrated. We also find the self-employed are less likely to have health insurance and more likely to retire at older ages than salaried workers and a large proportion of salaried and self-employed workers transition directly to retirement with few movements between salaried work and self-employment.

In Sect. 2, we describe our data and provide descriptive statistics. In Sect. 3, we examine transitions between self-employment and salaried work, and between work and retirement. In Sect. 4, we explore the likelihood of being self-employed, particularly among individuals who had migrated to the United States and returned. In Sect. 5, we discuss the policy implications of this research.

Data

The MHAS is a three-wave panel survey conducted in 2001, 2003, and 2012 among Mexicans at least 50 years of age. The data set is nationally representative and contains information on demographic and employment characteristics, health status, access to health-care services, family transfers, and wealth for 9,862 households. It includes spouses of eligible individuals regardless of age, although we did not include spouses less than 50 years of age in this analysis.

The first wave interviewed individuals between May and August of 2001. It is worth highlighting that information was not gathered for only 10.3% of the 11,000 selected households. These very high response rates show the survey to be of high quality. The second wave re-interviewed these individuals (including those who had moved) from June to September 2003. The second wave also achieved a very high response rate, 94.22% (MHAS 2004; Wong and Espinoza 2004). The third wave interviewed surviving individuals between October and November 2012 and included new respondents to ensure the sample remained representative of the 50-and-older population in Mexico. Altogether, more than 14,000 individuals in the 2001 and 2003 were reinterviewed in 2012, with the total sample now including 20,927 individuals and having an overall response rate of 88% (INEGI 2013).

Variables Description

Self-Employment Status Self-employment can include a broad spectrum of occupations, ranging from family and unpaid workers to business owners. Because we are interested in studying the role of formal insurance in influencing occupational choices in the Mexican labor market, we do not include family and unpaid workers in our analysis of the self-employed. Rather, we define self-employed workers as those individuals who report being a business owner or self-employed in their main job and also report having been employed the previous week.

Health Insurance We are able to identify the social-security system to which MHAS respondents contribute (IMSS, ISSSTE, or other). These social-security institutes provide health-care services and a pension system. In this study, we classify individuals as insured or uninsured according by whether they have access to health insurance through a social-security institution. The uninsured are in the informal sector and the insured in the formal sector of the economy.

Return Migrants MHAS asks respondents if they have worked or lived in the United States (excluding holidays or short visits). We categorize return migrants by the length of their stay in the U.S.

Health and Life Satisfaction We use three categories of self-reported health status, (1) excellent, very good, or good, (2) fair, and (3) poor. We define respondents expressing satisfaction with life as those who reported having enjoyed life for most of the time in the past week.

Education, Income, Wealth, Number of Years in Main Job, and Demographic Characteristics To measure education, we categorize individuals into those who completed a primary education and those who did not. To measure income, we use the MHAS derived variables of total household income and net worth for 2001 and 2003, which include detailed information on the sources and amounts of income and wealth. Unfortunately, the 2012 wave data does not yet have these derived income measures. Total household income includes earned income, business profits, property rent income and expenses, capital-assets income, pension income, family transfers, and transfers from government programs or individuals who are not family members. Net worth includes, net of debt, the value of real-estate properties, business and capital assets, and vehicles, as well as other assets such as the value of savings and deposit accounts.[1] We categorize these variables into tertiles with respect to the entire MHAS population. Number of years in the main job indicates the number of total years the respondent worked in his life for income or profits. Other characteristics of interest include age, marital status, number of children, and residence in urban or rural areas.

Sample

We limit our analysis to male respondents 50–79 years of age and with full-time employment in 2001 for whom occupational type is available and who were interviewed in all three survey years. We do not include females because they have very different labor dynamics, particularly among the cohorts we analyze. We also do not include part-time workers because their small sample size ($N=853$) prohibits a reliable analysis of transitions, particularly those that occurred between 2003 and 2012, given that most had already retired by 2003.

---

[1] The derived-income measures include imputed values from unfolding brackets for nonresponses. A detailed description of the imputation method can be found in Wong and Espinoza (2004).

An important limitation in our data is that in 2001 the MHAS did not ask respondents the occupational type of their current main job, i.e., whether they were self-employed or had salaried employment for their main job. Rather, it only solicits this information for the main job they held throughout their lives. This problem affects 38 % ($N=910$) of the total sample we analyze ($N=2,370$) since these individuals were not currently employed in a job similar to that which they had throughout their lives. Nevertheless, for most of these respondents, we are able to impute self-employment status using indirect information. Specifically, 708 of the 910 respondents missing information in 2001 indicated in 2003 that they had been at their job for 2 or more years, thus providing information on whether they were self-employed or salaried in 2001. Another six reported in 2003 having worked 2 or more years in their current job but being an unpaid worker. We exclude these individuals from the sample. We use information on health insurance type and salary income for 2001 to classify another 152 individuals. In total, we were not able to account for 50 of the 910 individuals, constituting 2 % of the sample of interest.

Characteristics of Middle-Aged and Elderly Self-Employed and Salaried Workers in Mexico

Self-Employment Rates

Table 7.1 shows rates of self-employment by age in 2001, 2003, and 2012 for males who were 50–79 and working full-time in 2001. This table demonstrates that self-employment rates increase with age, suggesting that self-employed workers retire at older ages possibly due to a lack of social security benefits. (Because the table focuses on those working, in 2003 and 2012 it excludes those who retired since 2001).

Table 7.1 Self-employment rates by groups of age (%). (Source: Authors' calculations using the 2001, 2003, and 2012 MHAS)

	2001	2003	2012
Unweighted N	2273	1828	805
Weighted N	2,578,740	2,010,038	922,170
Age			
50–54	53.32	49.31	
55–59	50.14	49.97	
60–64	54.48	54.04	53.68
65–69	54.47	55.02	57.31
70–74	67.92	66.99	68.60
75–79	77.18	77.99	68.56

Notes: Self-employment rates indicate the self-employed as a proportion of salaried workers and self-employed workers

Socioeconomic Characteristics of the Sample

Tables 7.2 and 7.3 shows summary statistics of males aged 50–79 in 2001 who were working full-time that year. We show characteristics of both self-employed and salaried workers in all 3 years of the survey to better understand the role of selection over time. (Again, the table excludes in later years those who retired after 2001).

The age pattern of both self-employed and salaried workers across years shows a tendency for self-employed workers to remain in their current position longer than salaried workers. In 2001, 28 % of self-employed workers were at least 65 years of age, compared to 21 % of salaried workers. This 8-%-point gap did not change much by 2012, suggesting that more salaried workers exited the labor force, possibly as a result of becoming age-eligible for retirement social security benefits.

A striking difference between self-employed and salaried workers is their marital status in 2012, when 83 % of the self-employed were married compared to 75 % of salaried workers. This suggests that salaried employment provides an easier pathway toward retirement for married couples than self-employment.

Lower levels of education also appear to be associated with self-employment. In every year, the proportion of self-employed workers who had completed a primary education is at least 8 % below that for salaried workers.

In addition to their lower levels of education, self-employed workers are in the lower ends of the income and wealth distributions in the 2 years for which data is available. In both 2001 and 2003, a higher proportion of self-employed workers were in the bottom tertile of income and wealth than salaried workers.

Over time, the self-employed included a higher proportion of U.S. return migrants. Among 2001 respondents, 14 % of the self-employed had lived in the United States, compared to 9 % of salaried workers. In 2012, 17 % of the self-employed respondents we analyze had lived in the United States, compared to 10 % of salaried workers. This suggests a link between U.S. migration experience and self-employment that we explore further with probit models.

Health Insurance, Health Status, and Life Satisfaction

Table 7.3 shows the health insurance coverage, health status, and reported life satisfaction of full-time self-employed and salaried workers aged 50–79 in 2001 and of those in this group remaining employed in subsequent years. In every year, a much lower proportion of self-employed workers than salaried workers have health insurance from social-security institutions. By 2012, however, *Seguro Popular* narrows this gap by providing coverage for 42 % of self-employed workers.

Over time, the health disparities between both groups, as captured by self-reported health, converge. In 2001, 45 % of the self-employed in this group rated their health as "Excellent," "Very good," or "Good," compared to 49 % of salaried workers. By 2012, 40 % of the self-employed in this group still working rated their health at least as "good," compared to 42 % of salaried workers, and the self-employed were slightly less likely to rate their health as "poor." This convergence may be

Table 7.2 Socioeconomic characteristics of males self-employed workers and salaried workers. (Source: Authors' calculations using the 2001, 2003, and 2012 Mexican Health and Aging Study MHAS)

	2001		2003		2012	
	Self-employed	Salaried	Self-employed	Salaried	Self-employed	Salaried
Unweighted N	954	1319	757	1071	419	386
Weighted N	1,415,080	1,163,660	1,088,582	921,456	546,343	375,827
Age group (%)						
50–54	23.70	25.23	13.29	16.14	NA	NA
55–59	23.53	28.45	26.97	31.90	NA	NA
60–64	24.99	25.38	24.38	24.49	27.19	34.52
65–69	13.02	13.23	18.99	18.34	40.75	44.14
70–74	11.13	6.39	14.72	8.57	23.15	15.40
75–79	3.63	1.31	1.65	0.55	8.90	5.94
Total	100	100	100	100	100	100
Married (%)	84.89	84.85	84.22	84.96	82.61	75.28
Years in U.S. (%)						
0	85.85	90.50	82.83	88.61	82.96	89.74
1–9	13.04	7.90	15.32	9.40	15.35	8.40
10–19	0.78	1.32	1.37	1.48	1.36	0.88
20–29	0.33	0.00	0.47	0.00	0.24	0.32
30+	0.00	0.28	0.00	0.51	0.08	0.66
Total	100	100	100	100	100	100
More than primary level education	30.32	38.25	28.96	39.61	36.00	45.29
Income (%)						
1st tertile	48.54	24.95	34.07	14.33	NA	NA
2nd tertile	26.12	37.88	31.87	39.60	NA	NA
3rd tertile	25.34	37.17	34.06	46.07	NA	NA
Total	100	100	100	100	NA	NA
Net wealth (%)						
1st tertile	46.98	38.21	42.27	31.88	NA	NA
2nd tertile	29.05	35.30	32.20	34.43	NA	NA
3rd tertile	23.97	26.49	25.52	33.69	NA	NA
Total	100	100	100	100	NA	NA
Household size	4.36	4.40	3.85	3.40	3.93	3.91
Years in main job	43.07	41.83	42.44	41.15	39.14	37.98
Urban (%)	52.34	78.71	51.13	79.78	66.64	68.91

Notes: Estimates are weighted. We only include individuals interviewed in 2001, 2003, and 2012. In 2012, all respondents have aged out of the 50–59 age categories. Income and net wealth variables are not yet available for 2012

Table 7.3 Health insurance, health status, life satisfaction by self-employment or wage/salary worker. (Source: Authors' calculations using the 2001, 2003, and 2012 Mexican Health and Aging Study MHAS)

	2001		2003		2012	
	Self-employed	Salaried	Self-employed	Salaried	Self-employed	Salaried
Unweighted N	954	1319	757	1071	419	386
Weighted N	1,415,080	1,163,660	1,088,582	921,456	546,343	375,827
Health insurance (%)	28.25	64.34	36.81	59.06	34.68	54.17
Seguro popular (%)	NA	NA	NA	NA	42.05	34.74
Health status (%)						
Excellent/very good/ good	44.72	48.86	41.67	41.78	40.35	41.77
Fair	42.84	44.68	43.47	50.52	52.48	48.33
Poor	12.45	6.46	14.86	7.70	7.17	9.90
Enjoyed life in previous week	79.05	73.66	75.74	71.65	82.96	78.34

Notes: Estimates are weighted. Health insurance includes health care services provided by social security institutions excluding *Seguro Popular*. *Seguro Popular* is a noncontributory health care services introduced in 2004. Therefore, this information does not exist for the years 2001 and 2003

due to many unhealthy individuals in both groups having retired by 2012. Interestingly, despite their worse health, self-employed workers were more likely to report having enjoyed life in the previous week. We later explore these differences with transitional models and control variables.

Self-Employment and Salaried Workers Labor and Retirement Transitions

Tables 7.2 and 7.3 show the changing characteristics of self-employed and salaried workers over time. They do not indicate whether these changes are due to retirement or movements between self-employment and salaried employment. In other words, do self-employed workers retire later or do they simply take up salaried work? What proportion remains self-employed over time?

We use the panel component of the MHAS and describe labor market transitions for males aged 50–79 with full-time employment in 2001. It is worth reiterating that we only capture the labor status of these individuals in 2001, 2003, and 2012 and do not observe transitions between survey years. Rather, we calculate transition rates to self-employment, salaried work or retirement in 2003 and 2012, conditional on being self-employed or salaried workers in the immediately preceding survey year.

Table 7.4 shows the labor-market transitions of males aged 50–79 that were full-time salaried or self-employed workers in 2001. This table shows that, by 2003, a greater proportion of the self-employed (86 %) had remained in their 2001 position

Table 7.4 Transitions to salaried worker or self-employment or retired in 2003 or 2012 conditional on being salaried worker or self-employed in 2001 or 2003 (%). (Source: Authors' calculations using the 2001 and 2003 Mexican Health and Aging Study MHAS)

	Salaried	Self-employed	Retired
2001/2003			
Salaried (*N*=1319)	82.52	4.77	12.06
Self-employed (*N*=954)	2.64	86.39	10.12
2003/2012			
Salaried (*N*=1071)	27.03	22.34	48.84
Self-employed (*N*=757)	7.52	40.49	44.13

Notes: Estimates are weighted. The first panel includes males aged 50–79 in 2001 working full-time. The second panel includes males aged 50–79 in 2001 working full-time with data for 2003 and 2012

than salaried workers (83 %). This table also shows that a slightly higher proportion of salaried workers (12 %) than self-employed workers (10 %) retired during this time.

We next examine whether this pattern persists over time. Every year an individual remains in the labor force has tangible effects on Mexico's budget through social-security contributions, labor-force productivity, and reliance on public pensions. Given Mexico's extremely high effective retirement age compared to other OECD countries (OECD 2011b), it is worth exploring how long self-employed workers remain in the labor force compared to salaried workers.

The second panel of Table 7.4 displays the labor force transitions of self-employed and salaried workers between 2003 and 2012. This table shows that even 9 years later, at which point many respondents had surpassed Mexico's retirement age, a lower proportion of self-employed individuals had retired (44 %) compared to salaried workers (49 %). On one level, this is not surprising. As noted, in Mexico, informal workers are not required to pay into the public pension system and may, therefore, remain in the labor force given they cannot receive a public pension.

A higher proportion of self-employed workers remain in the labor force after the normal retirement age, 65, than salaried workers, and few of them transition to salaried work at older ages. However, the data we have analyzed so far do not show whether these patterns persist throughout the years. In other words, does self-employed lead to later retirement only if individuals have been self-employed most of their working lives? Or do those who transition from salaried to self-employment also remain in the labor force longer than those who are salaried?

Table 7.5 displays multi-year transitions for full-time male workers aged 50–79 who were salaried and self-employed in 2001. Unlike Table 7.4, this table displays not only single-year transitions, but also subsequent transitions between 2003 and 2012 conditional on previous transitions between 2001 and 2003. This table shows that a higher proportion of workers who were salaried in both 2001 and 2003 retired in 2012 (48 %) compared to those who were self-employed in these 2 years (42 %). Moreover, nearly all individuals who transitioned from salaried to self-employed

Table 7.5 Transitions to salaried worker or self-employment or retired in 2012 conditional on being salaried worker or self-employed in 2001 and 2003 (%). (Source: Authors' calculations using the 2001, 2003, and 2012 Mexican Health and Aging Study MHAS)

2001	2003	2012		
		Salaried	Self-employed	Retired
Salaried	Salaried ($N=1051$)	27.86	22.57	48.13
Salaried	Self-employed ($N=32$)	0.17	2.61	97.21
Self-employed	Salaried ($N=20$)	1.82	15.46	70.37
Self-employed	Self-employed ($N=725$)	7.78	41.86	42.21

Notes: Estimates are weighted and do not add up to 100% because a small proportion reported "don't know" or refused to answer the question in 2012 or were unpaid family workers

between 2001 and 2003 had retired in 2012 (97%). Because this sample is small ($N=32$), however, further analysis would be needed to understand whether these individuals use self-employment as a "bridge job" before retirement and what happens to their eligibility for social-security benefits.

Self-Employment, Health Insurance and Return Migration

Access to health insurance offered by social-security institutions may signal whether middle-aged and elderly self-employed are more likely to be in the formal or informal sectors, a distinction with implications for income security in old age. We analyze whether self-employed have such access, as well as whether individuals with U.S. migration experience are more likely to become self-employed or salaried workers. Migrants with truncated labor histories may have different incentives to join the formal labor market when they return to Mexico than their non-migrant counterparts. There may also be market rigidities that do not allow them to transition into formal employment.

Empirical Strategy

We use a probit model to analyze the likelihood of being self-employed while controlling for observed individual characteristics:

$$P(SE_{it} = 1) = f(\alpha_0 + \alpha_1 X_{it} + \alpha_2 HI_{it} + \alpha_3 H_{it} + \alpha_4 M_{it})$$

where SE_{it} takes on the value 1 for an individual, i, self-employed in year t. X_{it} is a vector consisting of socioeconomic characteristics including age, education, number of household residents, marital status, residence in urban or rural areas, number of years in main job, household net worth, and household income. HI_{it} is a vector of indicators for access to health care insurance. H_{it} is self-reported health status and life satisfaction. Previous studies have found that self-employed individuals report higher life satisfaction (Andersson 2008). M_{it} includes number of years in the United States for return migrants. Robust standard errors are clustered at the individual level to adjust for serial correlation.

Results

Table 7.6 shows the effect of several explanatory variables in predicting self-employment. Before interpreting their marginal effects, it is important to note that the sample becomes more selected across years as more individuals retire and exit the

Table 7.6 Probit model describing the characteristics of self-employed (marginal coefficients). (Source: Authors' calculations using the 2001, 2003, and 2012 Mexican Health and Aging Study MHAS)

	(1)	(2)
Age		
55–59	−0.0761	−0.0517
	(0.0494)	(0.0482)
60–64	0.0373	0.0159
	(0.0638)	(0.0565)
65–69	0.0308	−0.0137
	(0.0738)	(0.0674)
70–74	0.1332	0.1549*
	(0.0871)	(0.0821)
75–79	0.1769	0.1063
	(0.1189)	(0.0958)
Enjoyed life in previous week	0.1187***	0.1109***
	(0.0418)	(0.0394)
Married	−0.0672	0.0113
	(0.0651)	(0.0703)
Completed primary education or more	0.0828	0.0676
	(0.0504)	(0.0443)
Household size	0.0045	−0.0028
	(0.0084)	(0.0084)
Lives in urban area	−0.1466***	−0.1577***
	(0.0510)	(0.0453)
Total years worked	0.0015	0.0009
	(0.0024)	(0.0020)
Health status		
Fair	−0.0208	−0.0232
	(0.0369)	(0.0321)

Table 7.6 (continued)

	(1)	(2)
Poor	0.1213**	0.0654
	(0.0550)	(0.0495)
Years in U.S.		
1–9	0.1355**	0.1352***
	(0.0609)	(0.0505)
10–19	−0.2938*	−0.2033
	(0.1685)	(0.1542)
20–29	0.0000	0.4823
	(0.0000)	(0.3211)
30+	0.0000	−0.6673***
	(0.0000)	(0.2144)
Health insurance coverage	−0.2970***	−0.2882***
	(0.0363)	(0.0313)
Seguro popular		−0.1578**
		(0.0679)
Income		
2nd tertile	−0.1647***	NA
	(0.0526)	
3rd tertile	−0.1055*	NA
	(0.0591)	
Net wealth		
2nd tertile	0.0433	NA
	(0.0488)	
3rd tertile	0.0335	NA
	(0.0571)	
2003	0.0389	−0.0214
	(0.0301)	(0.0259)
2012	NA	0.0671
		(0.0536)
N	2123	3391

Notes: Estimates are weighted. The second column includes the pooled data for 2001 and 2003 and the third column includes pooled data of 2001, 2003, and 2012 for observations with 3 years of data. NA refers to non-applicable because the 2012 wave does not provide derived measures of income and net wealth and in model 1 we do include the year 2012. We therefore excluded these measures from this specification. Omitted benchmark categories are: age group 50–54 years old; not having mainly enjoyed life in the previous week; single/divorced/separated union/ widow from union/widow from marriage; incomplete primary education or no schooling; lives in rural area; excellent, very good, and good self-reported health status; no migration to the U.S.; not covered by IMSS, ISSSTE, PEMEX, private insurance, or other health insurance (excluding Seguro Popular); 1st tertile of income and net worth; and the year 2001 $*p<0.1$; $**p<0.05$; $***p<0.01$

labor force. Model 1 only contains observations for the years 2001 and 2003, a time period likely less subject to selection given the temporal proximity of both survey years, while model 2 includes pooled observations with data for all 3 years. We also calculated separate models for both time frames given, unfortunately, the MHAS does not yet contain for 2012 information on income and net worth. A comparison of both models serves as a robustness check for the variables available in all years.

It is important to note that Table 7.6 does not provide information on transitions to self-employment but rather only accounts for contributors to self-employment.

Consistent with the literature, we found the self-employed are less likely to have health-insurance coverage in the model 1 which applies to 2001 and 2003 and model 2, which applies to all 3 years (Zissimopoulos and Karoly 2007; Parker and Rougier 2007). Having health insurance decreases the probability of self-employment by 30% points in model 1 and 29% points in model 2. We found a similar but somewhat weaker pattern for *Seguro Popular* in model 2. These findings support the hypothesis that individuals essentially relinquish access to health-insurance coverage upon becoming self-employed.

Also consistent with the literature, both models suggest that self-employment is associated with increased life satisfaction (Andersson 2008). An important related question is whether self-employment generates greater life satisfaction or whether individuals satisfied with their lives select into self-employment. The models indicate the self-employed are less likely to live in urban areas, which is consistent with earlier research (Amuedo-Dorantes and Pozo 2006).

U.S. migration experience affects the probability of self-employment. In models 1 and 2, individuals who had been in the United States from 1 to 9 years were approximately 14% points more likely to be self-employed than those with less migration experience. Previous theoretical work provides a possible explanation for the significance of this variable. Stark et al. (1997) hypothesize that immigrants remain abroad only enough time to save an amount of money that would maximize utility in the home country. Our results suggest that the optimal length of stay may be 1–9 years for older males in Mexico who wish to become self-employed upon returning to Mexico. It is possible that these individuals accrued the financial capital in the United States with which they later became self-employed in Mexico. Durand et al. (1996) find that Mexican return migrants are more likely to accumulate savings while in the United States, savings they use for productive purposes if having access to capital resources in their home country.

Model 2, which includes data for all three MHAS years, suggests a u-shaped relationship between years in the United States and the probability of self-employment. In this model, individuals still working in 2012 and who had spent 1–9 years in the United States were more likely to be self-employed than those with no U.S. migration experience. At the same time, Model 2 shows those who had lived in the United States for at least 30 years were 67% points *less* likely to be self-employed. This u-shape pattern suggests that only individuals who spend an intermediate number of years in the United States are likely to return to Mexico to become self-employed. It is possible that those who spend this amount of time in the United States migrated for the express purpose of accumulating assets they could use to pursue self-employment in Mexico. In contrast, those with more than 30 years of U.S. migration experience lived all or nearly all their adult working years in the United States. Such individuals may have prioritized using their U.S.-accumulated resources on U.S.-based attachments rather than self-employment prospects in Mexico.

Given the descriptive nature of our model, we can only speculate on these issues. Moreover, because this model is highly subject to selection bias resulting from retirement, our coefficients may indicate that there is a negative relationship between years in the United States and self-employment only for individuals who work until very old ages.

Conclusion and Discussion

Self-employment and entrepreneurship are characterized by the ability to identify economic opportunity and generate one's own work and firms. The businesses self-employed workers create are characterized by growth, competitiveness, and efficiency in the allocation of resources or, at the very least, could represent a source of income in the absence of other opportunities as salaried worker. In this study we used longitudinal MHAS data to examine the characteristics of middle-aged and older self-employed Mexican males who work full-time, and the role of health insurance and U.S. migration spells in determining self-employment. We find that return migrants who have worked 1–9 years in the United States are more likely to be self-employed and that the self-employed are less likely to have access to health insurance. This latter finding may indicate that the self-employed are more likely to be in the informal sector and not eligible for social security benefits in retirement. We also find that self-employed workers are less likely to retire than salaried workers.

In Mexico, self-employed workers are not mandated to contribute to the social security system. The findings of this study highlight the potential lack of income security and vulnerability this population faces during old age. It is out of the scope of this study to analyze the incentives to remain in the informal sector during working age. However, previous studies have found that government social assistance programs such as non-contributory pensions and health care services may incentivize individuals to remain in the informal sector (e.g., Aterido et al. 2011; Bosch and Campos-Vázquez 2010; Perry et al. 2007). Additionally, when a worker is affiliated with IMSS, her or his health care insurance is extended to the spouse, children, and economically dependent parents, disincentivizing these family members from contributing to the social security system. Doing so would entail paying twice to receive the same benefits (e.g., Aguila et al. 2011). Another explanation for the high rates of informality is that rural areas may have less access to formal employment (e.g., Perry et al. 2007; Aguila et al. 2011). Our findings suggest that social policies targeted toward increasing the participation of self-employed in the social security system may improve well-being during retirement in Mexico for a large proportion of the population.

Topics for further research include analyzing how economic cycles affect labor-force participation and type of occupation of middle-aged and older workers. Mexico suffered a mild recession in 2001 and 2008. It would be interesting to understand whether macro shocks affect self-employment decisions.

Acknowledgement This study was supported by a grant from the National Institute of Aging (NIA) funded under program project "International Comparisons of Well-Being, Health and Retirement" (2P01AG022481-06). This study was also supported by a NIH Ruth L. Kirschstein-National Service Research Award (T32AG000244) available to Alma Vega as a postdoctoral fellow at the RAND Corporation. We thank Joanna Carroll for her excellent programming assistance and Sarah Kups for her excellent research assistance, and anonymous referees for their valuable comments.

References

Aguila, E., Diaz, C., Manqing, F. M., Kapteyn, A., & Pierson, A. (2011). Living longer in Mexico: Income security and health. RAND Corporation. http://www.rand.org/pubs/monographs/MG1179. Accessed 15 July 2014.

Amuedo-Dorantes, C., & Pozo, S. (2006). Migration, remittances, and male and female employment patterns. *American Economic Review, 96*(2), 222–226.

Andersson, P. (2008). Happiness and health: Well-being among the self-employed. *Journal of Socio-Economics, 37*(1), 213–236.

Aterido, R., Hallward-Driemeier, M., & Pagés, C. (2011). Does expanding health insurance beyond formal-sector workers encourage informality? Measuring the impact of Mexico's Seguro popular. IZA Discussion papers 5996, Institute for the Study of Labor (IZA). http://www.iza.org/en/webcontent/publications/papers/viewAbstract?dp_id=5996. Accessed 15 July 2014.

Blanchflower, D. G., & Oswald, A. (1998). What makes an entrepreneur? *Journal of Labor Economics, 16*(1), 26–60.

Bosch, M., & Campos-Vázquez, R. M. (2010). *The trade-offs of social assistance programs in the labor market: The case of the "Seguro Popular" program in Mexico* (Working Paper No. 12-2010). Centro de Estudios Económicos, El Colegio de México. http://cee.colmex.mx/documentos/documentos-de-trabajo/2010/dt201012.pdf. Accessed 15 July 2014

Durand, J., Kandel, W., Parrado, E. A., & Massey, D. S. (1996). International migration and development in Mexican communities. *Demography, 33*(2), 249–264.

Duval-Hernández, R., & Orraca, P. (2009). A cohort analysis of labor participation in Mexico, 1987–2009. IZA Discussion Papers 4371, Institute for the Study of Labor (IZA). http://ftp.iza.org/dp4371.pdf. Accessed 15 July 2014.

Evans, D., & Jovanovic, B. (1989). An estimated model of entrepreneurial choice under liquidity constraints. *Journal of Political Economy, 97*(4), 808–827.

Evans, S., & Leighton, L. (1989). Some empirical aspects of entrepreneurship. *American Economic Review, 79*(3), 519–535.

Fairlie, R. W., & Woodruff, C. (2007). Mexican entrepreneurship: A comparison of self-employment in Mexico and the United States. In G. J. Borjas (Ed.), *Mexican immigration to the United States* (pp. 123–158). Chicago: University of Chicago Press.

Fairlie, R. W., & Woodruff, C. (2010). Mexican-American entrepreneurship. *The B.E. Journal of Economic Analysis & Policy, 10*(1), 1–44.

Fairlie, R. W., Kapur, K., & Gates, S. (2011). Is employer-based health insurance a barrier to entrepreneurship? *Journal of Health Economics, Elsevier, 30*(1), 46–162.

Fonseca, R., Michaud, P. C., & Sopraseuth, T. (2007). Entrepreneurship, wealth, liquidity constraints, and start-up costs. *Comparative Labor Law and Policy Journal, 28*(4), 637–674.

Gentry, W., & Hubbard, R. G. (2000). Entrepreneurship and household savings. *American Economic Review, 90*(2), 283–287.

Gollin, D. (2002). Getting income shares right. *Journal of Political Economy, 110*(2), 458–474.

González, S., & Villarreal, H. (2006). *More pushed than pulled: Self-employment in rural Mexico ten years after NAFTA* (Working Papers 20063). Escuela de Graduados en Administracioìn Puìblica y Poliìticas Puìblicas, Campus Monterrey. http://www.mty.itesm.mx/egap/deptos/cee/cieds/2006-3.pdf. Accessed 15 July 2014.

Guiso, L., Sapienza, P., & Zingales, L. (2004). Does local financial development matter? *The Quarterly Journal of Economics, 119*(3), 929–969.

Hochguertel, S. (April 2004). Self-employment around retirement in Europe. *Economy, 112*(2), 319–347.

Holtz-Eakin, D., Joulfaian, D., & Rosen, H. S. (1994). Entrepreneurial decisions and liquidity constraints. *The RAND Journal of Economics, 25*(2), 334–347.

Hurst, E., & Lusardi, A. (2004). Liquidity constraints, household wealth and business ownership. *Journal of Political Economy, 112*(2), 319–347.

INEGI. (2013). Encuesta Nacional Sobre Salud y Envejecimiento 2012 en México. http://www.inegi.org.mx/est/contenidos/proyectos/encuestas/hogares/especiales/enasem/default.aspx. Accessed 15 July 2014.

Mexican Health and Aging Study, MHAS. (2004). *Estudio Nacional de Salud y Envejecimientoen México 2001 (ENASEM)*. Documento Metodológico. http://www.mhasweb.org/DataDocumentationNew.aspx. Accessed 15 July 2014.

Mora, M. T. (2006). Self-employed Mexican immigrants residing along the U.S.-Mexico border: The earnings effect of working in the U.S. versus Mexico. *International Migration Review, 40*, 885–898. doi:10.1111/j.1747-7379.2006.00047.

Mora, M. T., & Davila, A. (2006). Mexican immigrant self-employment along the U.S.-Mexico border: An analysis of 2000 census data. *Social Science Quarterly, 87*(1), 91–109.

OECD. (2011a). Labor force statistics 2010, OECD publishing. http://www.oecd-ilibrary.org/employment/labour-force-statistics_19962045. Accessed 15 July 2014.

OECD. (2011b). Pensions at a glance 2011: Retirement-income systems in OECD and G20 countries. http://www.oecd-ilibrary.org/finance-and-investment/pensions-at-a-glance-2011_pension_glance-2011-en. Accessed 15 July 2014.

Papail, J. (2002). De salariado a empresario: La reinserción laboral de los migrantes internacionales en la región Centro-Occidente de México. *Migraciones Internacionales, 1*(003), 79–102.

Papail, J. (2003). Migraciones internacionales y familias en áreas urbanas del Centro-Occidente de México. *Papeles de Población, 36*, 109–131.

Parker, S. C., & Rougier, J. C. (2007). The retirement behaviour of self-employed in Britain. *Applied Economics, 39*(6), 697–713.

Perry, G. E., Maloney, W. F., Arias, O. S., Fajnzylber, P., Mason, A. D., & Saavedra-Chanduvi, J. (2007). *Informality: Exit and exclusion*. Washington, DC: The World Bank.

Popli, G. K. (2010). Trade liberalization and the self-employed in Mexico. *World Development, Elsevier, 38*(6), 803–813.

Samaniego, N. (1998). *Urban self-employment in Mexico recent trends and policies*. Paper presented at the Canadian International Labor Network Conference, Burlington, ON.

Stark, O., Helmenstein, C., & Yegorov, Y. (1997). Migrants' savings, purchasing power parity, and the optimal duration of migration. *International Tax and Public Finance, 4*(3), 307–324.

Torrini, R. (2005). Cross-country differences in self-employment rates: The role of institutions. *Labor Economics, 2*(5), 661–683.

Van Gameren, E. (2008). Labor force participation of Mexican elderly: The importance of health. *Estudios Económicos, 23*(1), 89–127.

Van Gameren, E. (2010). *Labor force participation by the elderly in Mexico* (Working Paper No. 6-2010). Centro de Estudios Económicos, El Colegio de México. http://cee.colmex.mx/documentos/documentos-de-trabajo/2010/dt20106.pdf. Accessed 15 July 2014.

Wong, R., & Espinoza, M. (2004). Imputation of non-response on economic variables in the Mexican health and aging study (MHAS/ENASEM) 2001: Project report. MHAS. http://mhasweb.

org/Resources/DOCUMENTS/2001/Imputation_of_Non-Reponse_on_Economic_Variables_
in_the_MHAS-ENASEM_2001.pdf. Accessed 15 July 2014.

Wong, R., & Parker, S. W. (1999). Welfare of the elderly in Mexico: A comparative perspective.
Unpublished.

Woodruff, C., & Zenteno, R. (2007). Migration networks and microenterprises in Mexico. *Journal
of Development Economics, 82*(2007), 509–528.

Zissimopoulos, J. M., & Karoly, L. A. (2007). Transitions to self-employment at older ages: The
role of wealth, health, health insurance and other factors. *Labour Economics, 14*(2), 269–295.
doi:10.1016/j.labeco.2005.08.002.

Zúñiga Herrera, E. (2004). Tendencias y características del envejecimiento demográfico en
México. In Consejo Nacional de Población (CONAPO) (Ed.), *La situación demográfica de
México* (pp. 31–42). México: CONAPO.

Chapter 8
Lifelines: The implications of Migrant Remittances and Transnational Elder Care for the Financial Security of Low-Income Hispanic Immigrants in the United States

Chenoa Flippen

Brigida, a 26 year old Honduran immigrant who had been living in Durham, NC for 4 years when we interviewed her was one of four sisters (three in Durham and one in Honduras, where her mother also resided):

> *"When my mother got sick it was a big expense for us. Angela (one of U.S. sisters) had just had a baby and couldn't help out, and Dalia had just come a few months before and didn't have any money. And the hospital bill was so high! She needed the operation, there was no choice about that. But my sister (in Honduras) sent her to the private hospital, not the public one. She said, 'If they're not making money, why would they be there?' But it was really hard. The bill was over $ 3,000 and we only had maybe 8 or 900. We borrowed some from brothers from the church, and ... " made up the rest by missing the rent, car, cell phone, and other payments that month. The ensuing penalties and late feeds added more than $ 500 to the cost of her mother's hospitalization.*

Research on population aging generally focuses on a particular elderly population in a specific location or context. However, a growing literature on transnationalism emphasizes that for immigrant populations, societies of origin and settlement are linked through a dense web of economic, cultural, and political connections (Itzigsohn and Giorguli-Saucedo 2002; Vertovec 1999). Indeed, as Brigida's experience illustrates, these connections hold the potential to powerfully shape aging on both sides of the border; the aging of family members abroad impacts the economic stability of immigrants in the United States, and the experience of aging in Latin America is also influenced by the presence of migrant family abroad. In spite of the importance of these effects, the relationship between migration and aging remains relatively understudied. This paucity of information is particularly problematic in the case of Latin American migration to the United States, due to rapid population aging and high levels of need on both sides of the border.

This research was supported by grant #NR 08052-03 from NINR/NIH.

C. Flippen (✉)
University of Pennsylvania, 3718 Locus Walk, Philadelphia, PA 19104-6298, USA
e-mail: chenoa@sas.upenn.edu

© Springer International Publishing Switzerland 2015
W. A. Vega et al. (eds.), *Challenges of Latino Aging in the Americas*,
DOI 10.1007/978-3-319-12598-5_8

While the continuous inflow of young migrants results in a relatively young profile among U.S. Latinos, the number of Latino seniors is growing rapidly, and they are projected to account for 20% of the population over 65 by 2050 (Vincent and Velkoff 2010). Latino immigrants are also a group that experiences a high risk of poverty in old age. Lifetimes of low earnings and employment instability result in perilously low personal retirement savings (Flippen 2001; Hao 2007; Smith 1995), while immigrant and legal status often limit their access to social safety net provisions. Latinos' concentration in physically demanding and unstable occupations likewise often makes it difficult to prolong employment as a means of coping with low retirement incomes (Flippen and Tienda 2000). The Latino population also suffers from high rates of diabetes, Alzheimer's, and other chronic diseases that further complicate both work and retirement security (Markides et al. 1997).

At the same time, Latin America also faces rapid population aging in coming decades. In Mexico, for instance, the birthrate fell from 7 children per woman of reproductive age in the 1960s to 2.3 in 2013, while life expectancy rose from 30 years in 1910 to 76.9 years in 2013. Projections estimate that the share of Mexicans over 60 will rise from 7% in 2000 to 24% in 2050, placing Mexico among the countries with the largest elderly populations in the world (United Nations 2001). Moreover, it is not clear that additional years will be healthy ones, given the prevalence of early childhood disease, exposure to other risk factors, and social and economic inequities in access to healthcare among older cohorts (Palloni et al. 1999). Combined with the context of economic instability and insecurity and lack of government support for the aged (according to 2000 figures, only half of the population over 60 has health insurance coverage and only 30 and 15% of men and women, respectively, have retirement pensions in Mexico (Parker and Wong 2001), these trends imply a heavy reliance on family members for economic and instrumental support in old age.

Despite the importance of population aging in both Latin America and among U.S. Latinos, and the scope and continuity of migration flows from Latin America to the United States, there is little understanding of the link between immigration and aging in the Americas. While survey data in Mexico and other countries in the region have highlighted the elderly's reliance on family for economic and instrumental support, including reliance on remittances from abroad (Aguila 2013; Bravo et al. 2013; Wong and Higgins 2007), our understanding of the determinants and consequences of transnational elder support from the perspective of the immigrant community remains extremely limited. While there is a vast literature on the extent and determinants of remittances to Latin America (see Ruiz and Vargas-Silva 2009 for a review), they often fail to distinguish between savings and family support, and there is virtually no analysis of how much migrants spend supporting elderly family members in particular. Moreover, aside from the observation that men are more likely to send remittances than women (Amuedo-Dorantes and Mazzolari 2010), there is little understanding of how gender and migration interact to shape support for elder relatives abroad.

This paper draws on survey and qualitative data collected in Durham, NC, a new immigrant receiving area in the American Southeast, to explore the relationship between migration and aging in the Americas, focusing on the U.S. side of the border. First, I assess the extent and determinants of transnational elder support, explicitly investigating whether remittances to elders follow the same pattern as the larger literature on undifferentiated remittances. Second, I examine not only overall gender differences in remittances to elders, but also the interaction between gender, acculturation, and family structure. And finally, I consider not only regular remittances, but also more episodic, emergency spending on elderly family abroad. Results show that the relationship between family structure and transnational elder care differs in important ways from the previous literature on undifferentiated remittances; while family reunification is associated with a sharp drop in remittances overall, the same is not true for remittances to elders in particular. Moreover, while family structure, particularly marital status, is a critical determinant of elder remittances among men, women's elder remittances are insensitive to marital status. Men and women's elder remittance patterns also respond differently to household income, and to the type of support offered. Implications for the old-age security of Latino immigrants are also discussed.

Background

Migration from Latin America to the United States has grown tremendously in recent decades; between 1990 and 2010 the size of the foreign born Latino population more than doubled, from 8.4 to 21.2 million. The sheer scope and magnitude of these population flows has led to a radical transformation of families on both sides of the border. The labor migration of men, particularly between Mexico and the United States, has a long history, as does the separation of husbands from wives and children that it often entails. However, as border security has tightened in the post-IRCA and particularly post 9/11 era, the once largely circular and target-earner migration flows have tended more towards longer stays and settlement (Massey 2013). Moreover, there has been an increase in the migration of both married and unmarried women, which also profoundly shapes families (Donato et al. 2006). This phenomenon has increased the presence of skipped generation households in Latin America, or elderly grandparents raising children left behind by working parents. While the consequences of these new family forms have been studied extensively for their implications for the well-being of women and children, and the change they represent to "motherhood" and "fatherhood" (Hondagneu-Sotelo 1997; Montes de Oca Zavala 2009; Parreñas 2005), there has been little commensurate work on the impact of migration on the elderly.

This is problematic because the elderly in much of Latin America exhibit a high degree of reliance on assistance from family for financial and instrumental support. Mexico is an illustrative case. Mexico is often described as being characterized by strong norms of filial responsibility, with traditional living arrangements that include high levels of co-residence and family caregiving of elderly (Bravo et al. 2013). Wong and Espinoza (2003) show that assistance from kin, primarily adult children, amounts to 35 % of the total income of elderly Mexicans. While an extensive body of research has examined the parental and child characteristics that predict intergenerational transfers and co-residence (Kim et al. 2012; Shuey and Hardy 2003; Van Hook and Glick 2007; Wong 1999), most work tends to focus on one side of the border or the other.

Overall aging in Latin America has been studied as it relates to falling fertility and mortality, but far less so as it relates to migration (Montes de Oca Zavala 2009). More work is needed that recognizes that families are increasingly transnational, involving spatial, economic, and social ties that link residents in two countries. A handful of studies have begun to show how migration is shaping aging patterns in Mexico. For example, Wong and Higgins (2007) document that 22 % of Mexicans over 50 have at least one child currently living in United States. Moreover, those with U.S. resident children are more likely to receive financial support from their descendants, and also benefit from more continuous assistance than their counterparts without migrant children. There are also indications of a negative impact of migration on the elderly in Latin America, however. For instance, Kanaiaupuni (2000) shows that traditional patterns of co-residence of elderly with adult children are disrupted by migration; elders with migrant children are substantially more likely to live alone than their peers without migrant children. In addition, Antman (2009) shows that child migration raises the odds that parents in Mexico will be in poor physical and mental health, even after taking into consideration the potential endogeneity of parental health to child migration decisions.

While these studies, based largely on survey data in Mexico, provide tantalizing evidence of the impact of migration on aging, far fewer examine the issue from the perspective of immigrant communities. There is a large body of work that examines the flow of remittances from the United States to Latin America. Much of this work has sought to assess the impact of remittances on economic development in receiving countries (see Massey and Parrado 1994; Duran et al. 1996) or to illuminate the motivation behind remittances, such as altruism, risk diversification, debt repayment, and so on (Duany 2010; Ruiz and Vargas-Silva 2009; Sana 2005; Sana and Massey 2005). Towards those ends a number of studies have also examined the individual-level predictors of remittance behavior. These studies generally find an inverse relationship between an immigrant's level of acculturation and remittances; those with longer durations in the United Sates, higher levels of English proficiency, and naturalized legal status are less likely to remit, as are those who have more family in the United States and fewer family members abroad (Amuedo-Dorantes et al. 2005; Amuedo-Dorantes and Mazzolari 2010; Desipio 2002; Duran et al. 1996; Marcelli and Lowell 2005; Menjivar et al. 1998; Nziramasanga and Yoder 2013; Sana 2005). The association between economic resources and human capital is mixed; some find that immigrants with more resources, such as higher educational attainment or income, are more likely to remit (Funkhouser 1995; Menjivar et al. 1998) but others find that those with lower education are more likely to send money to family abroad, though the amount may be lower due to the more limited resources they command (Amuedo-Dorantes et al. 2005; DeSipio 2002).

These studies, while informative, tell us very little about the link between aging in Latin America and immigrant adaptation in the United States. First, while a number of studies paint a rich portrait of the varied nature of remittances—which flow to diverse household members and for purposes that range from consumption to savings to home construction and business investments—models of remittance behavior almost always center on total remittances. It is therefore unclear if family maintenance and savings, for instance, are similarly shaped by acculturation. More importantly for our purposes, the previous research has not considered the support of elderly family members in isolation, and it is unclear what factors predict this kind of remittance in particular. This is a serious shortcoming for aging scholarship because there is ample reason to expect that support for aging parents may not share the same underlying determinants as investing in property in Mexico, for instance.

A second major limitation of previous studies is their insufficient attention to gender. Much of the research on remittances to Latin America is based on the Mexican Migration Project (MMP), which is an incredibly rich source of information on migrant-sending regions in Mexico but has far less information about host communities in the United States. Roughly 79 % of the immigrant sample consists of return migrants, interviewed in Mexico, who are likely to have far stronger ties to family in Mexico than those who do not return. Moreover, as the MMP is a survey of household heads, the sample is over 95 % male. This is problematic both because migrant women are seriously under-registered, and because we cannot examine within-household variation in the remittance behavior of men and women. While other studies with U.S. immigrant samples have examined gender differences in remittances, they generally have not gone beyond the observation that men are more likely to remit than women. That is, they add women to the analysis, but generally neglect to consider whether and how gender interacts with and other characteristics to shape transnational elder care.

Gender, Migration, and Elder Support

Gender is potentially a critical dimension of the impact of migration on aging, because migration, remittance behavior, and family (including elder) care all vary tremendously between men and women. Migration from Latin America to the United States has long been disproportionately male. The prevailing image is that most women migrate to rejoin husbands or parents already established in the United States, though there is growing recognition of the importance of migration among unmarried and divorced women as well (Donato et al. 2006; Flippen 2014). Indeed, nearly 45 % of adult immigrant Latinos in the United States are women, and as many as one-third migrated prior to marriage (Parrado and Flippen 2005 Parrado et al. 2005a). Settlement patterns also vary by sex, with men more often favoring return to their countries of origin than women (Hondagneu-Sotelo 1994).

There are also important gender differences in remittance behavior. Overall, Mexican men are both more likely to remit to their families abroad and average higher amounts of support as well (Amuedo-Dorantes and Mazzolari 2010). Conversely, in migrant sending communities families with sons abroad are more likely to receive support than those with daughters abroad (Sana and Massey 2005). While a number of studies point to important gender differences in remittance behavior, few move beyond the simple observation that women remit less. The interaction between marital status and gender is potentially important and yet remains understudied. It has been suggested, for instance, that unmarried women who migrate (either internally or internationally) are expected to remit a greater share of their earnings than their unmarried male peers, because they are not expected to be the main breadwinners responsible for providing for their future families. Likewise, there is evidence that remittances from sons and daughters may be spent differently, suggesting that the type of financial support differs by gender (Blue 2004; de la Briere et al. 2002).

And finally, family care behavior also varies tremendously by gender. Women average greater contact with parents than men, with the inverse true of contact with in-laws; daughters are often described as "kin keepers" charged with maintaining ties with extended families. They are far more likely to engage in care of elders, and in U.S. samples their parents are also favored over their husbands' in terms of financial assistance (Shuey and Hardey 2003). In Mexico, however, the bias towards wife's parents is not evident; among mothers of similar financial need, the husband's mother is twice as likely to receive financial assistance as the wife's, while the opposite is true of help with personal care needs (Noel-Miller and Tfaily 2009).

Taken together these patterns imply that the migration-gender-elder care link is potentially important. Migration removes many women from the close proximity to their parents necessary for personal care. At the same time, migration also pulls more women into the labor force, potentially transforming their control over the allocation of resources within the household. It is essential to move beyond simple comparisons of the remittance behavior of men and women to explore how gender interacts with other characteristics, particularly family structure, and to consider how gender differences may vary across different types of transnational elder support.

Data and Methods

The data used in the analysis come from an original, locally representative survey of Latino immigrants in the Durham/Chapel Hill, NC metropolitan area (for the sake of parsimony, referred to as Durham, where the vast majority of respondents live). Durham represents a valuable vantage point to study Latino immigrant incorporation. The area grew rapidly as part of the national shift in population from Rustbelt to Sunbelt states. The influx of highly educated workers attracted to growing job opportunities in the nearby Research Triangle Park, universities, and other large employers generated intense demand for low-skill service and construction labor. Employers responded by recruiting Latino immigrant laborers from more traditional receiving areas or even directly from Mexico, and a cycle of chain migration began that saw the Latino population surge from a mere 1% of the total population in 1990 to 11.9% by 2007 (Flippen and Parrado 2012).

The precarious position of Latino immigrants in Durham presented unique challenges for approximating a representative sample. Our study relied on a combination of Community Based Participatory Research (CBPR) and targeted random sampling to overcome these difficulties. CBPR is a participatory approach to research that incorporates members of the target community in all phases of the research process. In our case, a group of 14 community members assisted in the planning phase of the study, survey construction and revision, and devising strategies to boost response rates and data quality. In addition, CBPR members were trained in research methods and conducted all surveys. Finally, through ongoing collaborative meetings, they were also influential in the interpretation of survey results. It is difficult to overstate the wealth of culturally grounded understanding that they brought to project findings (for a detailed description see Berry et al. 2013).

At the same time, the relatively recent nature of the Latino community in Durham rendered simple random sampling prohibitively expensive. We therefore employed targeted random sampling techniques. Based on CBPR insights and extensive field work, we identified 49 apartment complexes and blocks that house large numbers of immigrant Latinos. We then collected a census of all apartments in these areas and randomly selected individual units for interview.[1] Using community members as interviewers helped achieve a refusal rate of only 9% and a response rate, which also discounted randomly selected units in which contact was not made after numerous attempts, of over 72% Data collection proceeded in stages. An initial survey, which included 209 women and 472 men between the ages of 18 and 49, was conducted between 2001 and 2002. During 2006 and early 2007, an additional 910 women and 1209 men were interviewed, for total sample sizes of 1,119 and 1,681, respectively.[2] All interviews were conducted in Spanish, usually in the homes of respondents, with interviewers filling out paper surveys that included a mix of closed- and open-ended questions.

Analytic Strategy

The main focus of the analysis is on the transnational support of elders in communities of origin. I examine data collected on remittances, which distinguishes between support provided to different family members. The first step is to simply document the extent of support to parents, in-laws, and other elderly family members (overwhelmingly grandparents and aunts/uncles). I subsequently model the determinants

---

[1] A comparison of our sample with the 2000 Census showed that nearly 80% of Durham's Latinos live in areas similar to our targeted locales, i.e. in blocks that are 25–60% Latino. Moreover, there were no statistically significant differences between data sources on socio-demographic characteristics such as age, employment status, hourly wages, marital status, and year of arrival (Parrado et al. 2005b).

[2] Remittance figures are reported in constant 2007 dollars.

of remittances to elders. For modeling purposes, I divide respondents into three categories: those who do not remit to elders, those who remit lower amounts, and those who remit higher amounts. The dividing point between the latter two categories was the median value among remitters, and was $ 2,400 for women and $ 3,300 for men. As the dependent variable is categorical, I estimated multinomial regressions, separately by gender and marital status.

I also collected a module on wealth and asset accumulation during the second wave of data collection. With a sample size of 339 women and 353 men, this module included a number of questions on common wealth "shocks" among the immigrant Latino population. One question specifically asked respondents whether they had, since migrating, significant expenses apart from regular remittances related to the medical care of family members abroad. I also present data on the incidence of these kinds of one-time, generally unanticipated expenditures that are not typically captured in remittance data. This provides the opportunity to not only capture a type of support not often registered in surveys but also to examine potential gender differences in the types of financial support provided to elders abroad.

Explanatory variables include national origin, age (and a squared term to capture non-linear effects), educational attainment, and work experience in countries of origin as rough measures of human capital. Because nearly all adult men had worked prior to migrating, the latter is included only for women. National origin is captured by a dummy variable indicating Mexican origin, with the reference category being Central American. Educational attainment is measured by a set of dummy variables distinguishing between those with 6 or fewer, 7–9, and 10 or more years of completed schooling, which correspond to primary, secondary, and above secondary education in Mexico. I also include an indicator of whether or not respondents herald from an urban background in Latin America, which could both signal an unmeasured aspect of human capital and correlate with parental need (as rural residents are far more likely than others to lack health insurance and retirement incomes (Angel and Pereira 2013)). Immigration-related characteristics include a variable capturing the number of years of residence in the United States and English ability, which is measured by a dummy variable indicating whether the respondent reported being able to speak English well or very well (as opposed to "more or less" or not at all). Finally, a dummy variable for undocumented status reflects the response from a direct question on legal status.

To assess the impact of family structure on remittance behavior I include measures of marital status, whether the respondent has minor children (under the age of 18) living in the household, and whether the respondent has minor children living abroad. The indicator of marital status differs by gender. Virtually without exception, immigrant women in Durham who are married are also living with their spouses. Thus among women I distinguish between married and unmarried women, the latter of which are predominantly never-married but also include a small number of divorced and widowed women. On the other hand, it is common for married men to migrate without their wives and children. I therefore distinguish between unmarried, accompanied married, and unaccompanied married men.

Descriptive statistics for the sample, by gender, are presented in Table 8.1. The sample is in many ways typical of immigrants in new destinations: respondents are relatively young (averaging roughly 30 years of age), and poorly educated. Averaging just under 8 years of schooling, more than 42 % of immigrant Latinos in Durham did not advance beyond primary school and less than one-third completed 10 or more years of schooling. Respondents are also recently arrived, averaging roughly 6 years in the United States. Reflecting this recent arrival, less than 9 % reported speaking English well or very well and the overwhelming majority, fully

Table 8.1 Descriptive statistics for independent variables

	Women	Men
Mexican	70.2	70
Human capital and immigration characteristics		
Age (mean)	30.4	29.9
	(7.7)	(8.1)
Education (%)		
6 years or less	43.6	41.6
7–9	26.8	33.3
10 or more	29.6	25.1
Worked in country of origin (%)	53.5	–
Years in U.S. (mean)	5.8	7.5
	(3.5)	(5.1)
Good English (%)	7.1	9.0
Undocumented (%)	88.3	88.5
Family and household structure		
Marital status (%)		
Married and accompanied by spouse	80.4	39.2
Married and unaccompanied by spouse	–	22.9
Unmarried	19.6	37.9
Minor child at home (%)	71.9	29.6
Non-resident minor child in country of origin (%)	22.2	34.3
Socioeconomic indicators		
Currently working (%)	59.2	97.1
Weekly wage (mean)	159.83	423.10
(S.D.)	(157.16)	(215.09)
N	1,119	1,681
For accompanied married respondents only		
Spouse's wage (mean)	427.77	137.5
Household earnings (mean)	597.4	592.5
Share of household earnings (mean)	0.21	0.79
N	900	659

88% was undocumented at the time of interview. While the men in the sample averaged slightly better English skills and longer residence in Durham, overall sex differences in human capital and immigration characteristics were remarkably modest. The main factor distinguishing women from men relate to family structure. Just over 80% of women in the sample were married and living with their partner at the time of survey, compared to only 39% for men. This is because a large share of Latino men, 23% were married but migrated without their wives and families. Women were also far more likely than men to be living with children (72 vs. 30%, and somewhat less likely to have left children behind in their countries of origin (22 vs. 34%. And finally, while virtually all men in the sample (97% were working at interview, only 59% of women were employed for pay. Average weekly wages were also substantially lower among women ($ 159 vs. $ 423).

Results

Table 8.2 presents data on remittances sent to Latin America, separately by gender and marital status. To begin, it is instructive to compare undifferentiated remittances with those specifically to elders. Comparing the share of men who report no remit-

Table 8.2 Remittances to family abroad

	Women			Men			
	All	Unmarried	Married	All	Unmarried	Unaccomp. married	Accomp. married
Recipient of remittances							
Parents only	29.0	27.9	29.3	44.1	64.2	3.1	48.7
In-laws only	1.7	–	2.1	1.2	0.2	0.3	2.7
Both sets of parents	1.4	–	1.8	1.0	0.2	0.5	2.0
Other elderly	1.4	0.9	1.6	1.4	2.2	0.0	1.5
Parents and children	4.8	4.5	4.9	4.1	5.7	0.5	4.9
In-laws and children	0.5	0.5	0.6	0.4	0.2	0.3	0.6
Parents and spouse	–	–	–	2.4	–	10.4	–
Wife and/or children	9.6	12.3	8.9	22.7	4.6	79.0	7.3
Other non-elderly	2.9	3.7	2.7	4.3	5.7	0.8	5.0
None	48.6	50.2	48.2	18.4	17.3	5.2	27.3
Mean annual remittance among remitters ($)							
Total	2,459	2,802	2,378	5,382	5,041	7,756	3,948
(S.D.)	(2,493)	(2,658)	(2,448)	(4,774)	(4,084)	(5,358)	(4,409)
To elder	1,950	2,020	1,936	4,527	5,168	4,467	3,778
(S.D.)	(2,078)	(2,048)	(2,087)	(4,489)	(4,275)	(2,132)	(4,694)
N	1,119	219	900	1,681	637	385	659

tances (the inverse of which succinctly gives a sense for undifferentiated remittances) with those who report supporting parents and other elders show important differences by marital status. Consistent with the previous literature, among men, overall remittances are both more common and substantially higher among those with more family abroad and less family in the United States. For instance, only 5.2 % of unaccompanied married men report no remittances, compared to 27.3 % among accompanied married men. Likewise, while unaccompanied married remitters average $ 7,745 in annual remittances, the comparable figure for accompanied married men is only $ 3,948. Single men fall between unaccompanied and accompanied married men both in terms of their likelihood of remitting and average amounts remitted. It is important to note, however, that these patterns are reversed when we consider only remittances to elders. In this case, unaccompanied married men are by far the least likely to provide support, with unmarried men most likely to remit to their parents and accompanied married men occupying the intermediary position. To illustrate, while 64.2 % of unmarried men report sending money to their parents, the figure is 48.7 % for accompanied married men, and a scant 3.1 % among unaccompanied married men. Thus, men whose wives and children continue to reside in their countries of origin send the vast majority of their support to their wives and children, and are very unlikely to help support their parents as well (though it should be noted that just over 10 % of unaccompanied married men support their parents and wives who live together). Differences in the average annual amount remitted to parents across marital status are equally stark among men. Thus, the reconstruction of families in United States is associated with lower remittances overall, but *not* with lower remittances to elders.

Women are clearly less likely to remit both overall and to elders abroad; nearly half of all women (48.6 % report no regular remittances compared to only 18.4 % among men. Likewise, while 44.1 % of men report supporting their parents, only 29 % of women do so. Among those who remit men also average considerably higher annual remittances, both overall and to elders ($ 4,527 per year to elders, on average, relative to only $ 1,950 for women). However, gender differences are more profound than mere differences in levels. In particular, among women remittance behavior is relatively insensitive to marital status, both in terms of total remittances and remittances to elderly relatives in particular. Both married and unmarried women exhibit roughly equal odds of remitting to parents, for instance (29.3 and 27.9 % respectively), and also average similar annual averages in the amount remitted.

The sharp disparities in total-remittance and elder-remittance behavior among men and interaction between gender and marital status both point to the need for greater understanding of the factors that shape men and women's financial support of elderly abroad. Tables 8.3 and 8.4 report results from multinomial logistic regression models predicting remitting to parents or other elders; the coefficients represent the likelihood of remitting lower and higher amounts to elders, with no elder remittances as the reference category. Models are estimated separately for

Table 8.3 Multinomial logit models predicting remitting to parents or other elders, women

	All		Unmarried		Married	
	Remit low	Remit high	Remit low	Remit high	Remit low	Remit high
Human capital and immigration characteristics						
Age	0.25**	0.16	0.28	0.06	0.28**	0.14
	(0.09)	(0.11)	(0.19)	(0.24)	(0.10)	(0.13)
Age sq	0.00**	0.00	0.00	0.00	0.00**	0.00
	(0.00)	(0.00)	(0.00)	(0.00)	(0.00)	(0.00)
Education (ref = 10 years or more)						
6 years or less	−0.53**	−0.52**	−0.50	0.64	−0.56**	−0.76**
	(0.19)	(0.25)	(0.52)	(0.66)	(0.21)	(0.28)
7–9 years	−0.14	−0.64**	−0.17	0.15	−0.14	−0.80**
	(0.20)	(0.29)	(0.54)	(0.70)	(0.22)	(0.32)
Urban origins	0.04	−0.39*	−0.15	−1.10**	0.07	−0.26
	(0.17)	(0.22)	(0.43)	(0.56)	(0.19)	(0.25)
Mexican	−0.12	−0.69**	−0.29	−0.82	−0.06	−0.76**
	(0.20)	(0.24)	(0.48)	(0.54)	(0.22)	(0.27)
Time in U.S.	0.07**	0.12**	0.05	0.11	0.07**	0.12**
	(0.02)	(0.03)	(0.06)	(0.08)	(0.03)	(0.04)
Good English	−0.96**	−0.54	−1.82**	−0.51	−0.69*	−0.58
	(0.37)	(0.44)	(0.84)	(0.80)	(0.42)	(0.54)
Undocu- mented	0.44	0.47	0.30	−0.52	0.53 *	0.98 **
	(0.29)	(0.36)	(0.65)	(0.68)	(0.32)	(0.45)
Family and household structure						
Married	0.24	0.22	–	–	–	–
	(0.25)	(0.29)	–	–	–	–
Co-res. children	0.19	−0.53**	0.25	−1.19**	0.14	−0.38
	(0.21)	(0.24)	(0.47)	(0.58)	(0.24)	(0.28)
Non-res. children	−1.87**	−0.67**	−3.39**	−0.81	−1.60**	−0.62**
	(0.29)	(0.28)	(1.05)	(0.65)	(0.31)	(0.31)
Employment characteristics						
Employed	0.67**	0.59**	0.94	−0.41	–	–
	(0.19)	(0.24)	(0.82)	(1.04)	–	–
Weekly wage	0.00	0.00	0.00	0.00	0.00	0.00
	(0.00)	(0.00)	(0.00)	(0.00)	(0.00)	(0.00)
Share of earnings	–	–	–	–	1.03**	1.48**
	–	–	–	–	(0.39)	(0.52)
Intercept	−5.78**	−4.34**	−6.30**	−1.38	−6.04**	−4.53**
	(1.35)	(1.78)	(3.17)	(3.65)	(1.57)	(2.07)
Chi squared	148.09		48.58		111.66	

$*p<0.10; **p<0.05$

Table 8.4 Multinomial logit models predicting remitting to parents or other elders, men

	All		Unmarried		Accompanied married	
	Remit low	Remit high	Remit low	Remit high	Remit low	Remit high
Human capital and immigration characteristics						
Age	0.06	0.27**	0.14	0.24**	0.08	0.34**
	(0.07)	(0.08)	(0.12)	(0.12)	(0.10)	(0.15)
Age sq	0.00	0.00**	0.00	0.00**	0.00	−0.01**
	(0.00)	(0.00)	(0.00)	(0.00)	(0.00)	(0.00)
Education (ref = 10 years or more)						
6 years or less	0.22	−0.10	0.06	−0.46*	0.38	0.42
	(0.19)	(0.19)	(0.31)	(0.27)	(0.24)	(0.30)
7–9 years	0.04	−0.08	0.24	−0.13	−0.12	0.00
	(0.19)	(0.19)	(0.29)	(0.26)	(0.25)	(0.30)
Urban origins	−0.40**	−0.47**	−0.31	−0.31	−0.52**	−0.66**
	(0.18)	(0.19)	(0.29)	(0.26)	(0.25)	(0.31)
Mexican	−0.48**	−0.15	−0.73**	−0.22	−0.20	0.02
	(0.17)	(0.18)	(0.26)	(0.25)	(0.23)	(0.28)
Time in U.S.	0.03*	−0.02	0.03	−0.05	0.01	−0.03
	(0.02)	(0.02)	(0.03)	(0.03)	(0.02)	(0.03)
Good English	−0.57**	−0.84**	−1.01**	−1.01**	−0.29	−0.70
	(0.25)	(0.28)	(0.42)	(0.39)	(0.31)	(0.44)
Undocumented	0.41*	0.88**	0.52	1.04**	0.35	0.79**
	(0.23)	(0.29)	(0.40)	(0.43)	(0.28)	(0.41)
Family and household structure						
Marital status (ref = unmarried)						
Unacc. married	−0.38*	−1.07**	–	–	–	–
	(0.22)	(0.22)	–	–	–	–
Acc. married	−3.37**	−2.76**	–	–	–	–
	(0.53)	(0.37)	–	–	–	–
Co-res. children	0.39*	−0.40*	–	–	0.27	−0.44*
	(0.21)	(0.23)	–	–	(0.23)	(0.26)
Non-res. children	−1.46**	−1.85**	−1.64**	−2.12**	−1.39**	−1.50**
	(0.22)	(0.23)	(0.36)	(0.34)	(0.29)	(0.36)
Employment characteristics						
Unemployed	−1.31**	−1.32**	–	–	−0.97	−0.36
	(0.50)	(0.59)	–	–	(1.23)	(1.30)
Weekly wage	0.00	0.00**	0.00	0.00**	0.00	0.00**
	(0.00)	(0.00)	(0.00)	(0.00)	(0.00)	(0.00)
Share of earnings	–	–	–	–	0.81	0.12
	–	–	–	–	(0.56)	(0.65)

Table 8.4 (continued)

	All		Unmarried		Accompanied married	
	Remit low	Remit high	Remit low	Remit high	Remit low	Remit high
Intercept	−0.79	−3.90**	−2.09	−4.40**	−2.55	−6.41**
	(1.07)	(1.25)	(1.70)	(1.69)	(1.67)	(2.41)
Chi squared	788.57		147.56		119.79	

*$p<0.10$; **$p<0.05$

women (Table 8.3) and men (Table 8.4),[3] for all men and women and also separately by marital status.

Comparing models for all women and all men (the first columns of Tables 8.3 and 8.4, respectively) shows important similarity between the predictors of elder remittances and those described in the literature on undifferentiated remittances. Indicators of acculturation, such as English ability and time in the United States, are negatively associated with supporting parents abroad. Age and rural origins, which may be associated with greater parental need, are positively associated with elder remittances. Legal status (for men) and education (for women) also predict elder support, as does employment. Higher earnings are associated with higher odds of remitting relatively large amounts to elders among men but has no significant impact on women's elder support.

As in the descriptive table, the relationship between family structure and remittances to elders both differs from the larger literature on undifferentiated remittances and also shows important interactions with gender. For men, family structure is a powerful determinant of transnational elder support. Both accompanied and unaccompanied married men are less likely to remit than their single counterparts, and both the presence of minor children in the U.S. household and abroad is negatively associated with remittances to elders. This latter pattern is directly at odds with the literature on undifferentiated remittances, and highlights the importance of considering elder care separately from other types of remittances. When men and women in the United States have to support children abroad, they are significantly less likely to have resources available to support elderly parents. While children have a similar impact on the elder remittances of men and women, the relationship with marital status shows important variation by gender. As in the descriptive statistics, marital status is a significant predictor of elder remittances for men but not for women.

[3] Since the unmarried typically do not have in-laws, the dependent variable in models for all men and women is restricted to remittances sent to the respondent's own parents or other elderly relatives only, so as not to over-state the impact of marital status on the likelihood of remitting. The models for married respondents include support to either parent. As the number of respondents who report remitting to in-laws is small, substantive results are not sensitive to this specification.

A deeper understanding of gender differences in elder remittances requires examining the predictors of support separately by marital status. When comparing the models restricted to unmarried men and women (the second set of columns on Tables 8.3 and 8.4), we see far few gender differences in the predictors of elder support than were evident among the total population. While fewer human capital and immigration characteristics are significant overall, we again see evidence that greater acculturation is associated with lower remittances to parents; English skills are associated with lower remittances to elders for men and women, while being undocumented is associated with higher remittances among men. Once again, having children abroad reduces unmarried men and women's capacity to contribute to their parents' support, and for women having co-resident children makes it less likely they will be able to provide sizeable support.[4] Once again, wages are important predictors of men's but not women's remittances to elders.

Far more sizeable gender differences in elder remittances are found among accompanied married migrants. For men, the main determinants of elder support for these, most settled, migrants are wages and the competing demands for money. Thus, men with higher incomes remit more to elders, while those with children either in Durham or abroad remit less, on average, than other accompanied married men. Those from urban origins also remit less. The only sign of the importance of acculturation is that undocumented men are particularly likely to remit large amounts to elders. Among women, on the other hand, acculturation is a stronger predictor of remitting; both English ability and legal status discourage remitting while time in the United States is actually positively associated with remitting to elders. The most interesting departure between men and women relates to earnings. For women it is not total household resources that matter; household income, which combines wives' and husbands' weekly wages, is not a significant predictor of remitting to elders. Rather, it is women's *share* of household earnings that is one of the most important determinants of whether and how much money they are able to send to their parents and other elders. This pattern is the opposite of that evidenced among men. Together with the positive effect of time in the United States for women, this suggests the importance of women's negotiating power in relationships (which could increase as women become more familiar and gain greater independence with the U.S. setting) in facilitating women's elder remittances.

The number of unaccompanied married men who remit to elders abroad is too small to estimate reliable models of remittances. In a simple binomial logit model (not reported) of whether or not unaccompanied men remit to elders, few variables attain the level of statistical significance and the predictive power of the model is very low. In addition to small sample size, this could suggest that for these men, it is parental characteristics, rather than their own, that are most determinant of remittance behavior.

---

[4] There were too few unmarried men residing with children to estimate the impact.

Elder Care and Wealth Shocks

While regular remittances are a critical source of support to the elderly in Latin America, and represent a significant expenditure for Latino immigrants in the United States, there are other forms of elder support that are also important. In particular, our CBPR group emphasized the importance of irregular, often unexpected expenditures that also impacted immigrant Latino's financial security but were not regularly included in estimates of remittances. Table 8.5 presents data, by gender and marital status, on unexpected medical expenditures for family members abroad. While these types of expenditures were substantially less common than regular remittances, they were nevertheless sizeable, with roughly one in four respondents reporting sending such funds to parents, and another 5–6% reporting this type of support to other elders. Average amounts are also sizeable; among those who report these types of expenditures, the average amount spent is roughly $ 2,500. Not only was this type of support relatively common, the patterns by gender and marital status are also strikingly different from those evidenced in the remittance data. That is, women are equally likely to provide this type of support as men (26.3 vs. 23.5%, and there is far less variation across marital statuses as well. For instance, while single men are more than 20 times more likely than unaccompanied married men to report remittances to their parents, they are equally likely to report sending money for unexpected medical expenses (46.8 vs. 45.1%. The average amounts remitted also show far less variation by gender than was the case for regular remittances. While our relatively small sample sizes preclude more detailed analyses, these patterns provide further evidence of the importance of women's transnational elder support and highlight the complexity of the relationship between gender, family structure, and remittances to elders.

Table 8.5 Medical expenses

	Women			Men			
	All	Unmarried	Married	All	Unmarried	Unaccomp. married	Accomp. married
Medical expense							
None	58.4	62.1	57.7	52.1	46.8	45.1	60.1
Parents	26.3	22.4	27.1	23.5	29.8	19.7	20.3
In-laws	1.5	–	1.8	1.4	–	1.4	1.9
Other elder	3.9	1.7	4.3	5.1	7.8	1.4	4.4
Child	4.1	8.6	3.2	7.4	1.6	21.1	5.7
Other	5.9	5.2	6.1	10.5	13.7	11.2	7.6
Mean payment for elder	2,508	3,534	2,271	2,487	1,949	2,046	2,951
(S.E.)	(5,145)	(5,179)	(5,142)	(6,736)	(3,339)	(4,054)	(8,739)
N	339	58	281	353	124	71	158

Conclusions

Aging is a pressing issues among U.S. Latinos and throughout Latin America. Migration not only links communities on opposing sides of the border but also simultaneously shapes the aging process in both locales. While dependence on remittances from abroad has been documented among the elderly in Mexico, studies showing the impact of elder support on U.S. immigrant communities are relatively rare. The findings presented above suggest that this impact is substantial. A large share of immigrants regularly send financial assistance to elder relatives abroad, and the average amounts remitted are substantial. More periodic, unexpected expenses are also common and costly. Taken together, this support represents a significant expense for an economically vulnerable immigrant population among whom large numbers approach their own old age with near-zero financial assets for retirement (Flippen 2001; Hao 2007; Smith 1995).

More importantly, the patterns shown above add to our understanding of how families and aging are transformed by migration. Results show important differences between the predictors of total remittances in previous studies, which lump payments to elders with support for wives and children and even investments in housing and productive capital, and the predictors of elder support in particular. While indicators of acculturation such as English ability and time in the United States show the same general patterns across remittance types, the relationship with family structure differs sharply. For undifferentiated remittances, family reunification is clearly associated with falling remittances, while unaccompanied married men remit the most. When it comes to support for elders, in contrast, the relative remittances of accompanied and unaccompanied married men are reversed; it is unaccompanied married men, who often struggle to support themselves and their wives and children abroad, who are least likely to help their elders. These patterns temper the view that acculturation and settlement translate into less support for elders among men. The reconstruction of families in the United States is not incompatible with transnational elder care, which has positive implications for the well-being of the elderly in Latin America but at the same time points to significant strains on working-aged Latino immigrants, who are often "sandwiched" between the needs of their growing children and elderly parents.

This paper also addresses the serious paucity of information available about the remittance behavior of migrant women. Most prior analyses of Latino immigrant remittances either relied on data that essentially excluded women entirely, or failed to give adequate treatment to the interaction between gender and family structure in structuring outcomes. Women remit less than men overall, but their remittance behavior is less sensitive to marital status than is the case for men. Moreover, for women the main financial determinant of elder support is working and the greater say in household financial decisions it seems to imply. Higher household incomes do not help married women remit to their parents; only their own share of family income seems to matter. The opposite is true for accompanied married men. Likewise, weekly wages are less important determinants of elder remittances for single women than they are for single men. And finally, there are also important gender differences in the types of financial support provided to elders. While women are substantially less likely than men to send regular remittances, they are no less likely to have significant outlays for elderly medical care abroad. This type of emergency spending is both common and large, and also less influenced by marital status.

A number of caveats are in order. The survey upon which this analysis was based was not specifically designed to study the connection between migration and aging, and thus lacks important information on the characteristics of parents and siblings that are important determinants of intergenerational transfers. While the results of this study are suggestive, more work is needed to explore the contours of transnational elder care. In addition to more detailed survey data that connects elder remittances to the characteristics of parents and siblings, we also need more qualitative work on the intersection between gender, migration, and transnational elder support in the Americas. A growing body of work has described the phenomenon of "transnational mothering" and how the norms and expectations surrounding the parent-child bond are profoundly altered by migration. Commensurate work is needed on transnational elder care, to explore how the norms surrounding filial obligations and the patterns of support among siblings are shaped by migration.

References

Aguila, E. (2013). Alleviating poverty for older persons: Results of a social welfare program in Mexico. Paper presented at the International Conference on Aging in the Americas.

Amuedo-Dorantes, C., & Mazzolari, F. (2010). Remittances to Latin America from migrants in the United States: Assessing the impact of amnesty programs. *Journal of Development Economics, 91*, 323–335.

Amuedo-Dorantes, C., Bansak, C., & Pozo, S. (2005). On the remitting patterns of immigrants: Evidence from Mexican survey data. Federal reserve bank of Atlanta, *Economic Review, 90*(1), 37–58.

Angel, R., & Pereira, J. (2013). Pension reform, civil society, and old age security in Latin America. Paper presented at the International Conference on Aging in the Americas.

Antman, F. (2009). How does adult child migration affect elderly parent health? Evidence from Mexico. Working paper University of Colorado at Boulder Department of Economics.

Berry, N., McQuiston, C., Parrado, E., & Olmos-Muñiz, J. (2013). CBPR and ethnography: The perfect union. In B. Isreal, E. Eng, A. Schultz, & E. Parker (Eds.), *Methods for community-based participatory research for health* (pp. 305–334). San Francisco: Jossey-Bass.

Blue, S. (2004). State policy, economic crisis, gender, and family ties: Determinants of family remittances to Cuba. *Economic Geography, 80*, 63–82.

Bravo, J., Mun Sim Lai, N., Donehower, G., & Mejia-Guevara, I. (2013). Aging and retirement security: United States, Mexico and Mexican Americans. Paper presented at the International Conference on Aging in the Americas.

de la Briere, B., Sadoulet, E., de Janvry, A., & Lambert, S. (2002). The roles of destination, gender, and household composition in explaining remittances: An analysis for the Dominican Sierra. *Journal of Development Economics, 68*, 309–328.

Desipio, L. 2002. Sending money home … for now: Remittances and immigrant adaptation in the United States. In R. de la Garza & B. L. Lowell (Eds.) *Sending money home: Hispanic remittances and community development* (pp. 157–187). Lanham: Rowman and Littlefield.

Donato, K., Gabaccia, D., Holdaway, J., Manalansan, M., & Pessar, P. (2006). A glass half full? Gender in migration studies. *International Migration Review, 40,* 3–26.

Duany, J. (2010). To send or not to send: migrant remittances in Puerto Rico, the Dominican Republic, and Mexico. *Annals of the American Academy of Political and Social Science, 630,* 205–223.

Duran, J., Kandel, W., Parrado, E., & Massey, D. (1996). International migration and development in Mexican communities. *Demography, 33,* 249–264.

Flippen, C. (2001). Racial and ethnic inequality in homeownership and housing equity. *The Sociological Quarterly, 42*(2), 121–149.

Flippen, C. (2012). Laboring underground: The employment patterns of Latino immigrant men in Durham, NC. *Social Problems, 59,* 21–42.

Flippen, C. (2014). Intersectionality at work: Determinants of labor supply among immigrant Hispanic women in Durham, NC. *Gender and Society, 20,* 1–31.

Flippen, C., & Parrado, P. (2012). The formation and evolution of Latino neighborhoods in new destinations: A case study of Durham, NC. *City and Community, 11,* 1–30.

Flippen, C., & Tienda, M. (2000). Pathways to retirement: Patterns of late age labor force participation and labor market exit by race, Hispanic origin and sex. *Journal of Gerontology, 55B,* S14–S28.

Funkhouser, E. (1995). Remittances from international migration: A comparison of El Salvador and Nicaragua. *The Review of Economics and Statistics, 77,* 137–146.

Hao, L. (2007). *Race, immigration, and wealth stratification in America.* New York: Russell Sage Foundation.

Hondagneu-Sotelo, P. (1994). *Gendered transitions: Mexican experiences of immigration.* Berkeley: University of California Press.

Hondagneu-Sotelo, P. (1997). 'I'm here but I'm there': The meanings of Latina transnational motherhood. *Gender and Society, 11,* 548–571.

Itzigsohn, J., & Giorguli-Saucedo, S. (2002). Immigrant incorporation and sociocultural transnationalism. *International Migration Review, 36,* 766–798.

Kanaiaupuni, S. (2000). *Leaving parents behind: Migration and elderly living arrangements in Mexico.* Madison: Center for Demography and Ecology.

Kim, J., Kim, H., & DeVaney, S. (2012). Intergenerational transfers in the immigrant family: Evidence from the new immigrant survey. *Journal of Personal Finance, 11,* 78–112.

Marcelli, E., & Lowell, L. (2005). Transnational twist: Pecuniary remittances and the socioeconomic integration of authorized and unauthorized Mexican immigrants in Los Angeles County. *International Migration Review, 39,* 69–102.

Markides, K., Rudkin, L., Angel, R., & Espino, D. (1997). Health status of Hispanic elderly. In L. Martin & B. Soldo (Eds.), *Racial and ethnic differentiation in the health of older Americans* (pp. 285–300). Washington, DC: National Academy Press.

Massey, D. (2013). America's immigration Fiasco. *Journal of the American Academy of Arts and Sciences, 142,* 5–15.

Massey, D., & Parrado, E. (1994). Migradollars: The remittances and savings of Mexican migrants to the United States. *Population Research and Policy Review 13,* 3–30.

Menjivar, C., DaVanzo, J., Greenwell, L., & Burciaga Valdez, R. (1998). Remittances behavior among Salvadoran and Filipino immigrants in Los Angeles. *International Migration Review, 32,* 97–126.

Montes de Oca Zavala, V. (2009). Families and intergenerational solidarity in Mexico: Challenges and opportunities. United Nations Programme on Aging report.

Noel-Miller, C., & Tfaily, R. (2009). Financial transfers to husbands' and wives' elderly mothers in Mexico: Do couples exhibit preferential treatment by lineage? *Research on Aging, 31,* 611–637.

Nziramasanga, M., & Yoder, J. (2013). The check is in the mail: Household characteristics and migrant remittance from the US to Mexico. *Applied Economics, 45,* 1055–1073.

Palloni, A., DeVos, S., & Pelaez, M. (1999). Aging in latin american and the caribbean. Center for Demography and Ecology Working Paper No. 99–02.

Parker, S. & Wong, R. (2001). Welfare of male and female elderly in Mexico: A comparison. In E.G. Katz and M.C. Correia (Eds.) *The economics of gender in Mexico* (pp. 249–290). Washington DC: The World Bank.

Parrado, E., & Flippen, C. (2005). Migration and gender among Mexican women. *American Sociological Review, 70*(4), 606–632.

Parrado, E., Flippen C., & McQuiston, C. (2005a). Migration and relationship power among Mexican women. *Demography, 42,* 347–372.

Parrado, E., McQuiston, C., & Flippen, C. (2005b). Participatory survey research: Integrating community collaboration and quantitative methods for the study of gender and HIV risks among Latino migrants. *Sociological Methods and Research, 34*(2), 204–239.

Parreñas, R. (2005). *Children of global migration: Transnational families and gendered woes.* Stanford: Stanford University Press.

Ruiz, I., & Vargas-Silva, C. (2009). To send, or not to send: That is the question. A review of the literature on workers' remittances. *Journal of Business Strategies, 26*(1), 73–98.

Sana, M. (2005). Buying membership in the transnational community: migrant remittances, social status, and assimilation. *Population Research and Policy Review, 24,* 231–261.

Sana, M. (2008). Growth of migrant remittances from the United States to Mexico, 1990–2004. *Social Forces, 86,* 1–31.

Sana, M., & Massey, D. (2005). Household composition, family migration, and community context: Migrant remittances in four countries. *Social Science Quarterly, 86,* 509–528.

Shuey, K., & Hardy, M. (2003). Assistance to aging parents and parents-in-law: Does lineage affect family allocation decisions? *Journal of Marriage and the Family, 65,* 418–431.

Smith, J. (1995). Racial and ethnic differences in wealth in the health and retirement survey. *The Journal of Human Resources, 30,* S158–183.

United Nations. (2001). *World Population Prospects, the 2000 Revision.* New York: United Nations.

Van Hook, J., & Glick, J. (2007). Immigration and living arrangements: Moving beyond economic need versus acculturation. *Demography, 44,* 225–249.

Vertovec, S. (1999). Conceiving and researching transnationalism. *Ethnic and Racial Studies, 22,* 447–462.

Vincent, G., & Velkoff, V. (2010). The next four decades: The older population in the United States 2010 to 2050. U.S. Census Bureau Current Population Report. http://www.aoa.gov/Aging_Statistics/future_growth/DOCS/p25-1138.pdf. Accessed June 2014.

Wong, R. (1999). Transferencias intrafamiliares e intergeneracionales en Mexico. Envejecimiento Demografico de Mexico: Retos y Perspectivas, CONAPO.

Wong, R. & Espinoza, M. (2003). Ingreso y bienes de la poblacion en edad media y avanzada en Mexico (Income and assets of middle-and old-age population in Mexico). *Papeles de Poblacion 9,* 129–166.

Wong, R., & Higgins, M. (2007). Dynamics of intergenerational assistance in middle- and old-age in Mexico. In J. Angel & K. Whitfield (Eds.), *The health of aging Hispanics: The Mexican-origin population.* New York: Springer.

Chapter 9
Dementia Informal Caregiving in Latinos: What Does the Qualitative Literature Tell Us?

Ester Carolina Apesoa-Varano, Yarin Gomez and Ladson Hinton

Background

Latinos are the fastest growing segment of the older adult population in the U.S. and they continue to bear a significant disease burden in the context of persistent health disparities. In the next few decades, the number of older Latinos with Alzheimer's disease and related degenerative brain diseases (ADRD) will increase dramatically. These projections reflect the anticipated growth in the older Latino population in the U.S. and it is believed that the number of older Latinos with ADRD will increase significantly given that the age remains the strongest risk factor for ADRD. For example, among those age 80 and above one third or more may suffer from a major neurocognitive disorder (i.e. dementia) and another third from mild neurocognitive disorder (i.e. mild cognitive impairment). ADRD are characterized by cognitive decline as well as a range of neuropsychiatric or behavioral symptoms that lead to progressive functional decline and growing dependence on others for assistance with day-to-day activities. As a consequence, a growing population of Latinos caring for someone with ADRD is indeed probable in the coming decades. Consequently, there is an urgent need to both advance our understanding and develop more effective approaches to preventing, diagnosing, and treating dementia and mild cognitive impairment (MCI) due to Alzheimer's and other degenerative diseases in order to support Latino families and communities and promote quality of life and wellbeing among care-recipients and caregivers.

Degenerative dementias, conditions characterized by cognitive and functional impairment, are among the most common and debilitating conditions afflicting older adults. Dementia affects 5 % of those ages 65–74 and more than 40 % above the age of 85 (Hebert et al. 2003). Dementia causes increased mortality (Brookmeyer et al. 2002; Larson et al. 2004) and it is the eighth leading cause of death in the

E. C. Apesoa-Varano (✉) · Y. Gomez · L. Hinton
University of California, Davis, CA 95616, USA
e-mail: Ester.Apesoa-Varano@ucdmc.ucdavis.edu

© Springer International Publishing Switzerland 2015
W. A. Vega et al. (eds.), *Challenges of Latino Aging in the Americas,*
DOI 10.1007/978-3-319-12598-5_9

U.S. MCI is a condition characterized by subtle impairments in functioning and increased risk of dementia (DeCarli 2003; Gauthier et al. 2006). Both dementia and MCI are associated with higher levels of behavioral and psychiatric symptoms such as depression, anxiety, agitation, and psychotic symptoms (Lyketsos et al. 2000, 2002). Mild cognitive impairment and dementia are thought to represent transitions to clinically defined disorders along a spectrum of cognitive functioning (DeCarli 2003).

In addition to the noted conditions, cognitive impairment has a broader impact on healthcare and the management of concomitant chronic illnesses. Dementia is associated with increased utilization of health services and costs (Gutterman et al. 1999; Husaini et al. 2003) and dementia behavioral symptoms are particularly difficult for physicians to manage effectively within the context of primary care (Apesoa-Varano et al. 2011; Hinton et al. 2007). There is now a growing awareness of the interplay of cognitive impairment and other chronic medical conditions and how this complex interaction of conditions may worsen health outcomes and lead to increased utilization of health services (Hendrie et al. 2013). Cognitive impairment compromises a person's ability to communicate effectively with healthcare providers, actively engage in health care decision-making, and individually manage their condition in the home setting which impacts how they are also able to cope and manage other co-existing medical conditions.

These diverse set of issues (e.g. management of chronic illnesses, inability to perform daily activities, increased mortality) associated with cognitive impairment are exacerbated in segments of the older adult population who come from disadvantaged ethnic/racial and socio-economic backgrounds. In particular, older Latinos who suffer from well-documented disparities in the burden of risk factors for cognitive impairment (i.e., vascular disease, diabetes, clinical depression) have been associated with lower socioeconomic status and lower levels of formal education (McBean et al. 2003; Rotkiewicz-Piorun et al. 2006; Raji et al. 2007; Karlamangla et al. 2009). However, data on the prevalence ADRD in older Latinos is not unequivocal and some studies have found that Latinos exhibit higher rates of mild cognitive impairment and dementia (Gurland et al. 1999; Royall et al. 2004) while others have not (Haan et al. 2003). It is speculated that this mixed evidence is related to methodological differences between studies such as measurement of cognitive impairment (Manly and Mayeux 2004) as well as heterogeneity among Latinos (e.g., Mexican Americans, Puerto Ricans, Cuban Americans).

Further, some research suggests that community-dwelling Latinos with mild cognitive impairment or dementia suffer higher levels of behavioral symptoms (e.g. depression and agitation) than white non-Hispanics (Hinton et al. 2003; Covinsky et al. 2003). Additionally, Latinos suffer disparities in health care services (i.e., access, quality of care, culturally and linguistically appropriate care) related to cognitive impairment. It has been found that minorities and Latinos receive lower quality of dementia care (Cooper et al. 2010), they are more likely to be diagnosed later in the disease course, indicating less timely access to diagnostic assessment (Espino et al. 2002; Cooper et al. 2010), and they are less likely to be treated with cognitive enhancers (Zuckerman et al. 2008; Kalkonde et al. 2009; Cooper et al. 2010; Mehta et al. 2005).

The Impact on Informal Caregivers

Progressive loss of cognition and functioning causes an increased dependence on caregivers (family members, friends or home assistants). Neuropsychiatric symptoms, such as depression, agitation and anxiety, are common in persons with dementia and are associated with increased Latino family caregiver depression (Hinton et al. 2003). Latino family caregivers provide more care to older adults with functional impairment (Weiss et al. 2005) and report higher levels of depressive symptoms compared with white non-Hispanics (Covinsky et al. 2003). Caregivers are themselves at increased risk of adverse health, mental health and cognitive outcomes (Schulz et al. 1995; Pinquart and Sorensen 2003; Vitaliano et al. 2011). Compared to whites, Latino caregivers, a majority of whom are women, report more depressive symptoms, smaller and less supportive informal support networks, and less satisfied with the support they receive (Valle et al. 2004).

The costs of dementia caregiving are substantial ranging from strained economic resources for the family to adverse health outcomes for the caregiver (Ory et al. 1999). The burden of caregiving becomes intensified for Latino families in the context of increasing geographic fragmentation, economic instability, and loss of traditional systems of intergenerational support (Ortiz et al. 1999). Thus caregiving to persons with cognitive impairment is very resource-intensive; particularly when independence is greatly eroded and increased reliance on family members or other informal caregivers is needed for support with activities of daily living such as eating, bathing, cooking, and transportation. Family members, typically women, often need to reduce work hours and/or juggle multiple caregiving responsibilities to care for persons with dementia.

In addition, Latino caregivers report substantial unmet needs for professional help, particularly with respect to the management of behavioral problems (Hinton et al. 2006a, b). Progress has been made in improving the quality of interventions for Latino caregivers (Belle et al. 2006), yet there are still significant gaps in the development of culturally-tailored interventions appropriate for Latinos (Napoles et al. 2010). Existing intervention models are primarily dyadic in focus and neglect the larger family context in which caregiving occurs. In addition to disparities in formal care, Latinos face greater barriers in knowledge of Alzheimer's disease (Ayalon and Arean 2004) and to hold conceptions of dementia that fall outside the biomedical model (Hinton et al. 2005). Latino families often do not receive much needed information and adequate psychosocial support in their roles as caregivers (Hinton et al. 2006a). While the literature on caregiving and caregiver intervention is vast (Napoles et al. 2010; Connell and Gibson 1997; Dilworth-Anderson et al. 2002; Connell et al. 2001; Roth et al. 2009; Thompson et al. 2007; Burr and Mutchler 1999; Chappell and Reid 2002), a striking gap remains: our lack of understanding and effective policy to support Latino family caregivers, a majority of whom are women and who are more likely to report high levels of burden, depression, and stress compared to other ethnic groups (Napoles et al. 2010; Morano and Sanders 2005; Adams et al. 2002; Navie-Waliser et al. 2001).

In sum, over the coming decades, Latino families will face the challenges of care provision at a time when the nature of intergenerational supports and structures are changing. In the U.S., for example, the "traditional" multigenerational Latino household is becoming less common and family members are becoming more dispersed geographically. Latinas are increasingly entering the workforce and are less available for the type of full-time caregiving that is required when cognitive impairment becomes substantial. With recent economic declines and high levels of unemployment Latino families are also under economic stress that is amplified by the demands of ADRD caregiving. In sum, the increase in older Latinos with dementia occurs at a time of transition and stress for Latino families. To meet these challenges there is an urgent need to develop programs and policies that will more effectively support Latino families who are doing the work of caregiving. Multiple reviews of caregiving in ethnically diverse populations, written over the past 20 years, have underscored the importance of qualitative studies for understanding the sociocultural contexts, experiences, values and priorities of Latino and other minority caregivers (Aranda and Knight 1997; Dilworth-Anderson et al. 2002; Hinton 2002; Janevic and Connell 2001; Knight and Sayegh 2010). The goal of this study was to perform a systematic review of the qualitative literature on caregiving to Latinos with ADRD in order to identify key themes and opportunities to advance the field. The progressive cognitive and functional decline combined with high frequency of distressing neuropsychiatric symptoms set ADRD apart from other chronic medical conditions and underscore its public health importance for aging Latinos and their families. Having a clear understanding of where qualitative literature has contributed and where the gaps are is important in advancing research to address those gaps as well as beginning to translate the knowledge in this literature into useful policy and intervention work (Table 9.1).

Methods

Design

This is a systematic review of qualitative research on the topic of informal/family caregiving to Latinos with cognitive impairment. While we did not use a particular previously published approach, we follow a protocol that is in line with models outlined by Fink (2005) and Walsh and Downe (2005) among others. This literature review focused on empirical studies published since 1980, including both qualitative and mixed-method studies. Criteria for the search included studies conducted in the United States with Latino/Hispanic participants of any national origin (Latino

Table 9.1 Qualitative literature review table

	Author(s)	Objective/aim	Design	Sample	Ethnicity	Main findings
1	Apesoa-Varano et al. (2012)	Address the gap related to developing culturally-tailored intervention by providing a description of the variability and socio-cultural context in which ADRD related behavioral problems emerge and are responded to within the family	Cross-sectional Ethnography	N = 6 (3 discussed in article) Spousal caregivers recruited via relationships with SALSA participants	Mexican-American	Behavioral problems (i.e. aggression, interpersonal violence) were aspects of caregiving experience in 4 of 6 families in this study. There is demonstrated variability in responding to aggressive behavior. The most successful management of aggressive behavior was exhibited by the family who utilized a biomedical view of dementia, conceptualized the demented behavior as not volitional, and utilized family support
2	Borrayo et al. (2007)	Explore Latino caregivers' cultural explanatory models of caring for an older adult with ADRD	Cross-sectional Focus Group	N = 33 Caregivers resided in three states: Minnesota, Colorado, and Florida	33.3 % Colombian 18.2 % Mexican 9.1 % Puerto Rican 9.1 % Peruvian 6.1 % U.S. origin 3.0 % Dominican 3.0 % Panamanian 19.0 % did not report	Results suggest that Latino caregivers' role strain is similar to that of European Americans but it is mitigated by the cultural explanatory models. Latino caregivers spend considerable time caring for their older relatives and fail to recognize effective ways to seek help from other family members or outside sources. Latino caregivers lacked the necessary medical understanding of ADRD and the skills needed to manage the disease

Table 9.1 (continued)

	Author(s)	Objective/aim	Design	Sample	Ethnicity	Main findings
3	Calderon and Tennstedt (1998)	Detect differences in the way care-givers in three ethnic groups describe their experiences with and reactions to caregiving	Cross-sectional Semi-structured Interviews	Caregivers recruited from Springfield Elder Project (population-based sample of older adults and their caregivers)	African American Puerto Rican White	White women and African American and Puerto Rican men expressed feelings of frustration and anger during difficult times in their caregiving situations. Women, particularly African Americans and Puerto Ricans, used somatic complaints as outlets for those feelings. In addition, African American caregivers described their caregiving as an extremely time-consuming activity. Puerto Rican female caregivers described their caregiving situation as one which fostered social isolation. Resignation, denial, respect and faith in religion were ways these caregivers dealt with the burden of their caregiving responsibilities. These findings suggest that African American and Puerto Rican caregivers are experiencing burden, but expressing it in different ways than White caregivers and available measures of caregiver burden do not adequately measure the impact of caregiving on minority caregivers

Table 9.1 (continued)

	Author(s)	Objective/aim	Design	Sample	Ethnicity	Main findings
4	Flores et al. (2009)	To explore the nuances of an ethics of care that constitute one caregiver's experience and to explore the tensions generated by divergent cultural mandates	Cross-sectional Ethnography	N = 1 Recruited from Larger Hinton study (2003)	Mexican American family caregiver to mother	Shows a second generation working-class Latina daughter's struggle negotiating the cultural mandates of familism while providing compassionate care for her mother. Her understanding of family values is rooted in her culture of origin, and informs her interactions as a caregiver and the health system. Her cultural influence constrains her desire to be assertive and more individualistic, often feeling trapped between her family background and the Americanized individual experience
5	Gelman (2010)	To present the experiences of 10 Latino AD caregivers to explicate caregivers' experiences dealing with mental and physical health effects and the special challenges Latinos face	Cross-sectional Semi-structured interviews/ counseling sessions	N = 10 New York University Caregiver Intervention with Latinos (NYUCI)	5 Spanish only speaking 4 Spanish and English speaking 1 English only speaking All of Latino origin	Emerging Themes: Lack of knowledge about AD and appropriate diagnostic and treatment services within the Latino community Structural obstacles to care (i.e. language, financial and legal status barriers, and the experiences of multiple biopsychosocial problems (physical illness co-morbidity) The cultural value of familismo contrasts with the reality of stressed extended families inability or unwillingness to provide emotional, physical and financial supports to primary caregivers

Table 9.1 (continued)

	Author(s)	Objective/aim	Design	Sample	Ethnicity	Main findings
6	Gelman (2012)	Explore the discrepancy between the long-held view that Latinos' and reliance on family leads to greater involvement of extended family in caring of sick members and reduced perception of burden; and research reports of low levels of social support and high levels of distress among Latino caregivers.	Cross-sectional Standardized measures and Semi-structured interviews	N = 41 Family caregivers recruited through two separate studies	Latino 34 Dominican 17 Puerto Rican 12 South American 2 Cuban 34 US born	For some caregivers familism facilitates caregiving in the traditional, expected manner. Other caregivers disavow its current relevance. Yet others feel a contrast between familism, which they may value in a general, abstract way and more personal, immediate negative feelings they are experiencing from caregiving
7	Henderson and Gutierrez-Mayka (1992)	Identify major ethno-cultural themes in Hispanic health belief symptoms	Longitudinal Semi-structured interviews	N = 37 Caregivers residing in Tampa, FL identified from senior-serving organizations	Mostly Cuban, Spanish, and Puerto Rican Latinos (no exact breakdown of sample)	Caregiving is tied with cultural definitions of sex roles: caregiving duties of a frail elder fall first on a daughter, then on a daughter-in-law. Younger cohorts of caregivers are changing these sex-stereotyped trends

Table 9.1 (continued)

	Author(s)	Objective/aim	Design	Sample	Ethnicity	Main findings
8	Herrera et al. (2008)	Explore cultural attitudes toward caregiving and long-term care and their influence on patterns of long-term care use among Mexican American family caregivers of relatives aged 50 and older	Cross-sectional Semi-structured Interviews	N=66 Family caregivers in San Diego, CA	Mexican American	Anderson behavioral model of health service utilization was used to examine familism, gender roles, acculturation, religiosity, and knowledge and perceptions of long-term care as factors in usage. Caregivers with greater long-term care use displayed lower levels of familism, were knowledgeable about services, had a care recipient with health insurance, shared caregiving responsibilities, and were less acculturated. Medicaid coverage for low-income care recipients was associated with higher long-term care use. Familism may deter service use, but caregivers with resources and knowledge or Medicaid coverage are inclined to use long-term care services
9	Hinton et al. (2009)	To describe the nature and frequency of Latino family caregiver attributions for dementia-related neuropsychiatric symptoms	Cross-sectional Open-ended questions and Interview using the Neuro-psychiatric Inventory (NPI)	N=30 Family caregivers recruited from SALSA	Latino elderly and their caregivers residing in the greater Sacramento, CA area	121 explanations for NPI domains were obtained. Seven categories were developed: Alzheimer's disease, interpersonal problems, other medical conditions, personality, mental aging, and genetics. AD was the most frequent attribution category but accounted for less than 30% of the total attributions. These results indicate that Latino caregivers are often more likely to attribute NPI symptoms to causes other than AD or a related dementia

Table 9.1 (continued)

	Author(s)	Objective/aim	Design	Sample	Ethnicity	Main findings
10	Hinton et al. (2006a)	To examine dementia Neuropsychiatric symptom (1) severity, (2) help seeking patterns, and (3) associated family unmet needs for professional help	Cross-sectional Open-ended questions and Interview using Neuropsychiatric Inventory	$N = 38$ Family caregivers recruited from SALSA	38 Latino caregivers	The majority of Neuropsychiatric Symptoms reported were moderate to severe. The majority of caregivers in this study reported that they need more professional help especially in the realm of counseling and education
11	Hinton et al. (2006b)	Highlight behavior changes of people with dementia as observed and responded to by families and primary care providers with a special focus on the 'borderlands' of care (i.e. boundaries of different care settings)	Cross-sectional Semi-structured interviews	$N = 2$ Larger Study	2 dyads involving a family physician and care recipient with dementia; one care recipient is non-Hispanic white and the other recipient is Mexican-American	The two interviews demonstrate borderlands of primary care constrained by economic and bureaucratic structures. Behavior changes in this analysis have gone unrecognized, are misunderstood, or are not discussed in primary care offices. The influence of ethnic/cultural and gender differences and the power inherent in the physician role, caregiver, and health system can be acknowledged, recognized, and renegotiated to create collaboration sites

Table 9.1 (continued)

	Author(s)	Objective/aim	Design	Sample	Ethnicity	Main findings
12	Hinton et al. (1999)	A narrative approach is used to show how family caregivers draw on their cultural and personal resources to create stories about the nature and meaning of illness and to ask how ethnic identity may influence the kinds of stories family caregivers tell.	Cross-sectional Semi-structured interviews	N=40 Convenience sample of narratives from four papers analyzing caregivers of elders with dementia	African-American Chinese-American Irish-American Latino	Subset of African-American, Irish-American, and Chinese-American caregivers: Alzheimer's as a disease that erodes the core identity of a loved one and deteriorates their minds Subset of Chinese caregivers: emphasized how families managed confusion and disabilities, changes ultimately construed as an expected part of growing old Subset of Puerto Rican and Dominican families: while using the biomedical label of Alzheimer's disease or dementia, placed the elder's illness in stories about tragic losses, loneliness, and family responsibility Caregivers drew upon both biomedical explanations and other cultural meanings of behavioral and cognitive changes in old age. Their stories challenge us to move beyond the sharp contrast between ethnic minority and non-ethnic minority views of dementia-related changes, to local clinics and hospitals as sites where biomedical knowledge is interpreted, communicated, discussed, and adapted to the perspectives and lived realities of families
13	Hinton et al. (2005)	Describe caregiver conceptions of dementia using a previously developed typology to examine the correlates of conceptions of dementia in a multi-ethnic sample	Cross-sectional Semi-structured interviews	N=92 Boston, MA-area caregivers recruited from senior serving organizations	19 African American 30 Asian American 16 Latino 27 Anglo	More than half of the caregivers held explanatory models of dementia that combined folk and bio-medical elements. Ethnicity, lower education, and sex were associated with explanatory model type

Table 9.1 (continued)

	Author(s)	Objective/aim	Design	Sample	Ethnicity	Main findings
14	Jolicoeur and Madden (2002)	Explore the dynamics of informal care of the elderly in Mexican-American families	Cross-sectional Structured questionnaire and Semi-structured interviews	N=39 Southern CA area caregivers recruited through social service agencies for the elderly	Mexican-American Caregivers divided into two groups: more assimilated (English-speaking) and less assimilated (Spanish-speaking)	This study highlights the impact of acculturation in the experience of Hispanic caregivers. Spanish-speaking (less assimilated) caregivers provided elders with more assistance with activities of daily living. Spanish-speaking group spent more time caregiving than the English-speaking group. English-speaking group identified physical work and guilt as the main challenges whereas Spanish-speaking group identified the emotional difficulty of mediating between demands of two generations (their children and the elderly)
15	Karlawish et al. (2011)	Discover whether Latino Puerto Rican and non-Latino communities differ in the words they use to talk about Alzheimer's disease (AD)	Cross-sectional Semi-structured interviews	N=120 Caregivers recruited from longitudinal cohort of the ADCC at the University of Pennsylvania. Non-caregivers recruited from clinics where Penn ADCC recruits and distributes outreach materials	Latino Puerto Ricans Non-Latino Whites	Both Latino Puerto Ricans and non-Latino Whites recognize AD as a disease of memory loss and other cognitive problems. Although both groups used the term "sadness" to describe AD, non-Latino Whites did not feature emotional, behavioral, or psychological problems as among the causes of AD. Although all the groups' descriptions of a person who lives with and cares for a person with AD shared the word "loving," Latino Puerto Ricans focused on a good spouse who exercises intelligence, patience, and attention on behalf of the person with AD and did not use the term "caregiver." In contrast, non-Latino Whites typically used the term "caregiver." Both groups' lists shared words that describe research as presenting harms to an AD patient and requiring a commitment of time. Latino Puerto Ricans' lists suggested an understanding of research benefits akin to clinical care

Table 9.1 (continued)

	Author(s)	Objective/aim	Design	Sample	Ethnicity	Main findings
16	Levkoff et al. (1999)	Understand the help seeking of minority family caregivers and the role of religious and ethnic factors	Cross-sectional Semi-structured interviews	N=40 Family Caregivers Recruited from Exploratory Center on Culture and Ethnicity at Harvard Medical School Department of Social Medicine	10 African American 10 Chinese American 10 Irish American 10 Puerto Rican	Chinese American and Puerto Rican narratives contained themes of language barriers to care, and a lack of culturally competent services. Both Irish-American and African-American narratives showed themes of alienation from religious groups on the one hand, and using prayer to cope on the other. Narratives from all groups contained themes of religious and/or ethnic imperatives for providing care. Religious/ethnic factors may both aid and impede the help seeking of caregivers
17	Mastel-Smith and Stanley-Hermanns (2012)	Examined the educational needs of caregivers of older adults with chronic illnesses living in the community, or how they prefer to learn about caregiving and care-giving skills by focusing on: (a) caregivers' tasks, (b) the way caregivers learned about caregiving, (c) the information caregivers needed or wished they had, and (d) their preferred learning methods	Cross-sectional Focus groups and one personal interview	N=29	76% White 17% African American 3% Latino 3% Asian	Themes arose from verbatim data transcriptions and coded themes. Four categories of educational needs were identified: (a) respite, (b) caregiving essentials, (c) self-care, and (d) the emotional aspects of caregiving. Advantages and disadvantages of learning methods are discussed, along with reasons for and outcomes of attending caregiver workshops. An informed caregiver model is proposed. Health care providers must assess educational needs and strive to provide appropriate information as dictated by the care recipient's condition and caregiver's expressed desires. Innovative methods of delivering information that are congruent with different caregiving circumstances and learning preferences must be developed and tested

Table 9.1 (continued)

	Author(s)	Objective/aim	Design	Sample	Ethnicity	Main findings
18	Neary and Mahoney (2005)	Explore the phenomenon of dementia caregiving in an ethnically diverse group of Latino caregivers, with the goal of identifying cultural influences on the caregiving experience	Cross-sectional Semi-structured interviews	$N=11$ Recruited from research and Training Institute of the Hebrew Rehabilitation Center for the Aged, Massachusetts	All Latino 2 Colombian 3 Puerto Rican 3 Argentinian 1 Guatemalan 1 Cuban 1 Dominican	More similarities than differences between groups regarding understanding symptoms, beliefs about caregiving role, and the factors associated with ongoing care. A lack of knowledge about dementia, rather than culturally influenced beliefs was the major deterrent to the recognition of initial symptoms. Family centered care is a culturally-embedded value—placement was considered when care became impractical
19	Ortiz et al. (1999)	Prove that spatially and culturally constituted definitions of personhood, the moral life, and justice shape perceptions of normative aging, the agency of the demented persons and their place in the community, the appropriate care of the aged and demented, as well as partially determine the concrete resources which will be available to elders and their families	Cross-sectional Semi-structured interviews		Latino Irish-American	We review how ties to homelands and neighborhood institutions act as mediators and shapers of anticipatory grief, caregiver burdens, and caregiver resources, serving as a buffer against exhaustion and despair for some families (primarily the Irish-American sample), and as an additional site of loss or stress for others (primarily the Latino sample)

Table 9.1 (continued)

	Author(s)	Objective/aim	Design	Sample	Ethnicity	Main findings
20	Radina (2007)	To look at the process of caregiver preparation and designation among Mexican American families	Cross-sectional Semi-structured interviews	20 Members of MA sibling dyads (10 dyads)	Mexican American	Assuming the role of the caregiver: role designation did not happen at either the individual or family level exclusively. Usually the primary role began at the individual level and was then designated by family Individual-level decision-making: Natural extension and personal choice are the main factors of the decision to pursue caregiver role Family-level processes: Majority of cases held 'family meetings' to give indications about who would have responsibility for parent care Major themes for caregiving: all children should be involved, caregiving was the responsibility of older siblings, caregivers and care recipients should be of the same gender Reasons for sibling non-involvement: geographic distance, perceived responsibility related to work and family, personal illness, other caregiving Participants' perceptions of reasons for care providing: proximity/co-residence to the recipient, reciprocity, child vs. in-law in care recipient, nature of relationship with recipient

Table 9.1 (continued)

	Author(s)	Objective/aim	Design	Sample	Ethnicity	Main findings
21	Radina et al. (2009)	Investigates relationships among use of implicit versus explicit decision-making strategies, generation since immigration, and orientation toward family	Cross-sectional Semi-structured interviews	Unavailable	Unavailable	Recent US immigrants were more likely than those who immigrated longer ago to describe family decision making about care as implicit. Those classified as using mainly implicit decision making were found to be not significantly different from those classified as mainly explicit with regard to their overall support of a collective orientation toward family. There were exceptions to this, however, regarding two specific aspects of orientation toward family: engagement in shared activities with family members and avoiding family conflict by making choices that are consistent with the family values. In both cases those categorized as implicit expressed stronger endorsement of these specific values. These findings suggest that the relationship between family decision-making strategies, generation since immigration, and orientation toward family is more complex than suggested previously in the literature. Implications for future research and practice are offered
22	Simpson (2010)	Explore the lived experience of two Hispanic caregivers and the phenomenon of mastery	Cross-sectional Semi-structured interviews	$N = 2$ Family caregivers who self-identified as Hispanic	No further information	Caregivers derive meaning of their experience through a process of reconciliation of the self between the roles of parental caretaker, respectful daughter, and caregiver. This reconciliation of self is a form of mastery, and the process of reconciliation is grounded in cultural values of intergenerational reciprocity and familism

Table 9.1 (continued)

	Author(s)	Objective/aim	Design	Sample	Ethnicity	Main findings
23	Vickrey et al. (2007)	Elicit perceptions of the caregiving experience from informal caregivers of persons with dementia across different ethnicities	Cross sectional Focus group	N=47	African American Chinese American Euro American Hispanic American	Caregiving roles, concern about person with dementia, and unmet information and resource needs were expressed amongst all groups similarly. However, perspectives differed across ethnic groups on stigma surrounding dementia, benefits of caregiving, spirituality/religion to ease caregiving burden, and language barriers and discrimination. Findings suggest that interventions to reduce disparities in dementia care quality are needed to address ethnic variations in caregiving experiences
24	Winslow and Flaskerud (2009)	Illustrate the unique concerns that a small subsample of ethnically and racially diverse family caregivers expressed as they discussed their experiences with placement decision-making	Longitudinal Semi-structured interviews	N=12 Subsample of ethnically and racially diverse family caregivers from study of 46 caregivers	3 African American 8 Hispanic/Latino 1 Middle Eastern	We must listen with unbiased attention to expressed will of families. What a family will want/need is unknown. People of color do not always have adequate family support nor do they want to continue caregiving or feel able to do so Culturally sensitive and appropriate resources and facilities need to be known. Nurses need to provide emotional support and willingness to assist in developing an acceptable and safe plan of care Advocacy in the larger long-term care arena to support more culturally appropriate services/facilities is an important professional mandate

only or as part of a multi-ethnic study), and if the main focus was on the perspectives and experiences of informal, kin/non-kin caregivers of elders with Alzheimer's disease, dementia, or other mild cognitive impairment (MCI). Sample: Articles published from year 1980 to 2013 matching keywords "Latino," "Hispanic," and "Alzheimer's," "cognitive impairment," "dementia," "mild cognitive impairment," and "qualitative," "mixed-method."

Instruments

Scholarly search engines PubMEd, Google Scholar, Cumulative Index to Nursing and Allied Health Literature (CINAHL), PsychINFO, and Sociological Abstracts were used to conduct the literary search for qualifying abstracts.

Review Criteria

Inclusion criteria comprised: (1) empirical, data-driven studies (i.e. reviews were excluded), (2) study relies on qualitative methods (i.e. focus groups, in-depth interviews, single open-ended questions embedded in a survey, ethnography) alone or in combination with quantitative methods (i.e. mixed method study), (3) focus on some aspect of the experience of informal/family caregivers, (4) focus on care for elders with Alzheimer's or a related type of degenerative dementia, and/or mild cognitive impairment, (5) Latino/Hispanic sample alone or as part of comparative multi-ethnic study (comparative studies must include cross-ethnic analyses and/or report results separately by ethnic group), (6) conducted in the United States, (7) journal articles in English, and (8) not interventions, report of intervention development, or intervention results. Literature that was not a journal article or a book chapter, studies of caregiving not specific to ADRD, international and non-English studies were excluded from this review.

Data Collection Procedures and Analysis

Study staff conducted database searches following strict protocol with pre-established keywords. A list of articles meeting keyword criteria was generated and indexed with all relevant information (i.e. authors, source of publication, abstract). The review of the literature and analysis to identify main themes involved three phases. During phase I, database searches were conducted in order to identify relevant literature. Following the keyword search, references were compiled with abstracts, which were individually reviewed to further identify those that fully met criteria for inclusion in review. Phase II involved verification of inclusion criteria and compilation of references into an annotated table of evidence which included design, sample, and findings of each of the studies that met criteria for inclusion. Lastly, phase III involved a close review of all articles in order to identify themes and subthemes. Once this initial round of coding of themes was completed, we constructed a matrix per main thematic categories in order to tally frequencies and group references accordingly.

Results

Description of Studies Meeting Criteria

Phase I of the search yielded over 500 articles and we found that 24 met full criteria for review. Of those 24 studies, 21 used qualitative methods alone while the remaining four utilized a combination of quantitative and qualitative (mixed) methods (Gelman 2012; Hinton et al. 2006a, 2009; Jolicoeur and Madden 2002). Two studies were identified as longitudinal (Henderson and Gutierrez-Mayka 1992; Winslow and Flaskerud 2009), with the rest being cross-sectional. In terms of the data collection methods, two studies were based on ethnographic observations (Apesoa-Varano et al. 2012; Flores et al. 2009), five used a combination of open-ended questions embedded in structured questionnaires or a structured questionnaire and semi-structured interviews (Gelman 2012; Hinton et al. 2006a, 2009; Jolicoeur and Madden 2002), 14 utilized in-depth semi-structured interviews (Calderon and Tennstedt 1998; Gelman 2010; Henderson and Gutierrez-Mayka 1992; Herrera et al. 2008; Hinton et al. 1999, 2005, 2006b; Karlawish et al. 2011; Levkoff et al. 1999; Neary and Mahoney 2005; Ortiz et al. 1999; Radina 2007; Radina et al. 2009; Simpson 2010), and two were based on focus groups (Borrayo et al. 2007; Vickrey et al. 2007). In terms of the samples described in the studies that met inclusion criteria, 9 out of 18 were composed of multi-ethnic/racial participants, including Latinos/ Hispanics of various national origins, African Americans, Asian Americans, and non-Hispanic whites. There was a range of data analysis approaches such as narrative analysis and categorical analysis, with a majority of them utilizing thematic analysis to identify main topics in the qualitative data. In what follows we describe the main three themes we identified in our review of the qualitative literature on Latino informal/family caregiving for Alzheimer's disease and related-dementias.

Theme I: The Caregiving Experience

Fourteen of the 24 studies examined some aspect of the caregiver experience (Borrayo et al. 2007; Calderon and Tennstedt 1998; Flores et al. 2009; Gelman 2010, 2012; Hinton et al. 1999, 2006b; Jolicoeur and Madden 2002; Levkoff et al. 1999; Neary and Mahoney 2005; Ortiz et al. 1999; Simpson 2010; Vickrey et al. 2007; Winslow and Flaskerud 2009). Most of the studies focusing on this theme often reported that caregivers disclosed having ambivalent feelings about their role and what was expected of them. These studies reported that for Hispanic caregivers, a majority of whom were women, the stress of caregiving was not only related to the actual demands of this invisible work but also to these contradictory feelings, which in turn were often internalized as negative self-perceptions, guilt, failure to uphold normative expectations, and shame (often characterized in the literature as "dissonance" or "role strain"). These studies shed light, for example, on the tensions between filial piety and caregivers' insights into their own needs such as attending to their own physical and mental health or dedicating time to their paid work or recreational activities.

Studies that conducted ethnic comparisons found Hispanics to report more pronounced discontent and ambivalent feelings when describing their caregiving experience. The literature also reports that Hispanic caregivers more commonly described "being on a crisis mode" and having to deal with and deescalate ongoing family conflicts and divisions that often were exacerbated by the emergent health issues related to the care recipient or the progression of ADRD. Studies often found that Hispanics relied on religion (e.g. going to church, praying at home, reading the bible) to help them cope with their difficult experiences as caregivers.

Among studies that generally focused on the subjectivity of caregiving, the formal health care system also emerged as a dimension that often colored the experience of those caring for someone with ADRD. Based on reported findings, caregivers described unpleasant encounters in the clinical setting or throughout their attempts to manage or coordinate services on behalf of those they cared for. Hispanic caregivers often felt neglected, distrustful, or unprepared to deal with these tasks, often citing language barriers and lack to timely and adequate interpreting services during clinical encounters. In addition, studies also found Hispanic caregivers to feel distanced from the practitioners they came in contact with; caregivers felt that practitioners did not know or understood their culture and world views. These strained encounters with the formal health care system were found to often cause further distress and add to the burden that caregivers experienced and had to reconcile. In sum, qualitative studies in this area have described the subjective experience of Hispanic caregivers, privileging their lay accounts and highlighting how the difficulties and challenges caregivers routinely face shape their perceptions of caregiving burden.

Theme II: Caregivers' Knowledge of Dementia

Nine of the 24 studies also examined how much caregivers know about ADRD (Gelman 2010; Herrera et al. 2008; Hinton et al. 2005, 2006a, b, 2009; Karlawish et al. 2011; Mastel-Smith and Stanley-Hermanns 2012; Neary and Mahoney 2005). Most of these studies focused on caregivers' biomedical knowledge while others sought to describe the caregivers "explanatory models" or lay explanations about the causes, symptoms, and progression of disease. Typically, these studies found that caregivers lacked biomedical knowledge of ADRD. However, studies reported that this lack of understanding coexisted with a range of folk (or what has been characterized as "cultural" or "ethnic" explanations) knowledge characterizing dementia as normal aging, old timers' disease, "susto," stress, other illnesses, or medications. Some of the reported findings suggest that Hispanics (and arguably other groups) held a variety of combinations of both biomedical and folk knowledge about ADRD, rather than simply lacking the former while exclusively relying on the latter.

A smaller subset of these studies also shed light on the knowledge caregivers had of available services provided in mainstream health care settings such as primary care clinics. Studies found that Hispanic caregivers were typically unaware of social services support and are thus less likely to seek help when facing caregiving challenges. The reviewed literature also reported that this lack of awareness about the nature and availability of caregiver support resources was also reflected on caregivers' expressed views about nursing home placement for those whom they care for. These studies suggested that Hispanic caregivers' reported resistance to placement was neither absolute nor unchangeable, even in the context of strong filial expectations and norm. Based on the reviewed studies, Hispanics caregivers often reported that they preferred providing caregiving at home, yet simultaneously recognized that nursing home placement, for example, was appropriate when home care was inadequate or unmanageable. The studies reviewed thus suggested that there is a diversity of views among Hispanic caregivers on the topic of caregiving in the home versus institutional placement based on the needs and circumstances of the care recipient.

Theme III: The Caregiving Division of Labor

Four out of the 24 studies reviewed, examined in detail the way in which caregiving as a set of tasks and responsibilities to be accomplished was structurally organized (Apesoa-Varano et al. 2012; Henderson and Gutierrez-Mayka 1992; Radina 2007; Radina et al. 2009). These few studies outlined the beginning of a complex landscape involving the informal organization of care in the home for those with ADRD. Some of these studies provided a cursory look at the negotiations involved in this process, while others focused on who was identified to be responsible for caregiving in the non-expert setting. Studies that reported on this topic found that family members were preferred caregivers, with women across the generational continuum (e.g. wives first, then daughters followed daughters-in law, etc.) typically identified as the main individuals to bear sole responsibility. In addition, studies also found that physical proximity was an important factor shaping decisions of whom and why someone was recognized as a primary caregiver for someone with ADRD. The studies that were reviewed also reported on that Hispanic caregivers described to be part of the extended networks, but that these networks often had relatively loose ties and were unreliable. Studies that examined the nature of these networks found that caregivers attributed this unreliability to family conflict (e.g. the myth of "familismo"), geographic dispersion of family members, and limited socio-economic resources. In short, the studies that covered this topic found that cultural expectations such as those of "familismo" were not always related to positive views and perceptions of support on the part of Hispanic caregivers.

Discussion and Conclusion

As we described in this review of the qualitative literature, sociocultural context and processes are fundamental in our understanding of informal caregiving for dementia among Hispanics. The studies we reviewed have provided rich descriptions of the experience of ADRD Hispanic caregivers' and their views of the costs and burden of caregiving throughout the ADRD illness trajectory. These studies have also begun to paint the contours of the social mechanisms involved in how ADRD Hispanic caregivers define, make sense of, and enact their role on a routine basis in the context of everyday life. Based on this literature, it is clear that beyond the physical demands of caregiving, Hispanic caregivers frequently report feeling caught in the tension between cultural expectations of familial piety and their ability to fulfill caregiving demands. Additionally, the studies included in this review have found a gap in formal biomedical knowledge of ADRD, while bringing to our attention the rich folk knowledge and various explanatory models of the illness held by Hispanic caregivers. Along with the reported lack of biomedical knowledge, these studies have identified a lack of awareness of available caregiver support services and poor levels of service utilization. Equally important, this qualitative literature has started to unpack the fluidity in conceptions of long-term care institutions and placement among ADRD Hispanic caregivers in the context of normative expectations informed by cultural values such as "familismo."

Lastly, the studies reviewed here have documented the existence of extended social networks centered on the Hispanic primary caregiver and through which caregiving labor is seemingly distributed. While these networks appear important, the studies reviewed here have started to unpack the nature of relationships within those networks. Based on our review, there is some initial data on the complex set of dynamics shaping how these networks actually work as evidenced by Hispanic caregivers reported lack of coordination, support, and significant levels of interpersonal conflict among network participants. In sum, the qualitative literature on Hispanic caregiving for ADRD has so far pointed to both cultural and structural factors such as gendered division of labor in the household, cultural values (e.g. familismo), conceptions of illness, available resources, and geographic location of extended family (among others) to explain caregiving processes in this group. In this sense, the reviewed body of literature has provided solid leads in the role of social class, gender, and ethnicity as a complex intersection of macro and micro level factors shaping the caregiving trajectory in Hispanics.

Our literature review findings are partially congruent with the extant body of literature on dementia caregiving in other racial/ethnic groups. Specifically, there are similarities and important cross-cultural differences to be considered. For example, the experience of caregiving across other racial/ethnic Black and Asian groups is shaped by the predominance of filial piety as main motivator for caregiving for older adults. However, there are racial/ethnic differences in how that value translates in perceptions of burden and rewards as well as who is supposed to fulfill that expectation. Prior literature suggests that among Blacks filial piety includes both kin and non-kin relationships and caregiving is typically assumed by a network of individuals linked through church affiliations that are likely to include older women and men (e.g. Porter et al. 2000). This kin/non-kin involvement as means to maintain community cohesion and support has been explained in light of the historical burden imposed by slavery and more contemporary socioeconomic disadvantage on Black families. In contrast, while a highly ethnically diverse based on regional and national origins, Asian groups express how levels of adherence to traditional conceptions of filial piety involving kin children's duty to care for ill parents (e.g. Zhan 2004). In this group, the fulfillment of this duty has been related to caregiver pride, respect, and honor, and emotional reward. In this context, caregiving is more commonly confined to the home and the family based on kinship ties. Alternatively, the larger literature on Hispanic caregiving for dementia continues to document higher levels of emotional distress and burden compared to these other groups and it is well established that women (wives and daughters) are typically expected to provide caregiving in the home. Further, there is now an emergent understanding that a significant factor in reported caregiving burden in this group is due to the "myth of familismo," whereby expectations of family unity, support, and cooperation go unmet in light of dementia caregiving demands and changing generational orientations (e.g. Simpson 2010). In short, the interplay of sociocultural and historical factors is an important source of variation in the caregiving experience and trajectory. Social and health policies might prove more effective if they take into account the specific factors (and the degree to which they are influential) affecting one group more than others. Lack of comprehensive models that more systematically integrate the symbolic, structural and contextual factors involved in ADRD Hispanic caregiving will remain limited at best.

Our findings hence provide a springboard for identifying new areas of inquiry in regards to ADRD Hispanic caregiving. Based on this review of the qualitative literature, it appears that there is a gap in how we evaluate Hispanic caregivers' perceptions, for example, against their actual behavior. Research remains to be done in terms of direct observational research (e.g. ethnography) to fully understand and inform intervention and policy that will be effective in changing Hispanic caregiver outcomes. In a similar light, there is a gap in our efforts to pursue more longitudinal research (both quantitative and qualitative) in order to identify changes over time, pivotal events, and causal pathways shaping the ADRD Hispanic caregiver trajectory. Likewise, other fruitful areas to develop include better understanding how the caregiving division of labor crystallizes and how it shapes a caregiver's trajectory in Hispanic caregiver networks and households (e.g. how are multiple caregivers integrated and how the family as a unit mobilizes to care for an elder with ADRD). Finally, more research examining how local contexts and environments (e.g. urban versus rural) shape Hispanic caregiving experiences, role, and trajectories would be helpful in developing policy and interventions. Investigating these areas further will promote the dissemination and implementation of work to address disparities and challenges that Latino caregivers face in their communities.

Our review also provides a foundation for how and why qualitative inquiry contributes to advancing our understanding of ADRD Hispanic caregiving. Prior reviews of Latino caregiving have recommended that qualitative studies can be used to inform the development of culturally-tailored policy and interventions, to modify or tailor existing ones for use in older Latinos, or to inform recruitment approaches (Pinquart and Sorensen 2005; Napoles et al. 2010). In community-based participatory research and in international health studies, qualitative methods may be essential to identify the perspectives, priorities and values of local communities and stakeholders. This initial work can be essential to develop sustainable social and health policy and interventions that meet local needs. Therefore, qualitative studies may inform policy and interventions in a variety of ways that include data collection prior to the intervention to elicit stakeholder perspectives and priorities, to data collection during a clinical or effectiveness trial to better describe the process of the intervention or to assess outcomes, to data collection at the conclusion of an intervention to assess participant experience of the intervention and identify factors that may have influenced outcomes (e.g. sources of "unexplained variance") and to inform future theoretical and empirical work. In addition, qualitative research may be used as part of a program of research that unfolds over time moving from observational studies to interventions (e.g. Apesoa-Varano and Hinton 2013). It is clear, however, that more in-depth and nuanced theorizing on how cultural and structural factors play out over time would help develop more effective programs and policies to enhance community care and support for Hispanic caregivers.

There are limitations of this study. Challenges to reviewing literature on Latino caregivers to a person with ADRD are that qualitative studies are difficult to identify, studies published in books may have been missed, and there were difficulties in finding intervention studies with a qualitative component. This systematic review was limited to the qualitative literature on Alzheimer's disease and related dementias and mild cognitive impairment and excluded qualitative studies on Latino family caregiving that did focus specifically on issues related to Alzheimer's disease and related dementias. We choose to focus on Alzheimer's disease because of the unique challenges associated with this condition. However, some of the qualitative studies we excluded may address issues that are relevant to the experience of caregivers more generally, including family caregivers to persons with Alzheimer's disease and related dementias. In addition, we excluded studies in Spanish language that may be relevant to this review.

In sum, future directions for research must include well-designed studies (quantitative, qualitative, mixed-method) to further uncover mechanisms, contextual factors, and contradictions and consistency between reported subjective experience and actual behavior. These studies might focus on sources of strength and resilience in Hispanic kinship and communities as well as the experiences of burden and rewards, social networks, and the caregiving division of labor. In a similar light, more longitudinal research to better identify caregiving trajectories and temporal trends is needed to theorize and model potential points of interventions. More cross-cultural work might also help to tackle the complexity of caregiving as a phenomenon, especially to go beyond static notions of ethnicity/race to incorporating a more sophisticated rubric for charting how sociocultural factors intersect at the point of caregiving.

Acknowledgments We thank Dr. Charles S. Varano for his invaluable feedback on several versions of this manuscript. We also thank Vanessa Santillan, Emily He, and Rachel Turner for their assistance with the literature search.

Support This study was supported by grant R01-MH080067 from the National Institute of Mental Health. Ester Carolina Apesoa-Varano and Ladson Hinton received support from the UC Davis Latino Aging Research Resource Center (Resource Center for Minority Aging Research) under NIH/NIA Grant P30-AG043097.

References

Adams, B., Aranda, M. P., Kemp, B., & Takagi, K. (2002). Ethnic and gender differences in distress among Anglo American, African American, Japanese American, Mexican American spousal caregivers of person with dementia. *Journal of Clinical Geropsychology, 8*(4), 340–354.

Apesoa-Varano, E. C., Barker, J. C., & Hinton, L. (2011). Curing and caring: The work of primary care physicians with dementia patients. *Qualitative Health Research, 21*(11), 1469–1483.

Apesoa-Varano, E. C., Barker, J. C., & Hinton, L. (2012). Mexican-American families and dementia: An exploration of "work" in response to dementia-related aggressive behavior. In J. L. Angel, F. Torres-Gil, & K. S. Markides (Eds.), *Aging, health, and longevity in the Mexican-origin population* (pp. 277–292). New York: Springer.

Apesoa-Varano, E.C., & Hinton, L. (2013). The promise of mixed-methods for advancing Latino health research. *Journal of Cross-Cultural Gerontology, 28*(3), 267–282.

Aranda, M. P., & Knight, B. G. (1997). The influence of ethnicity and culture on caregiver stress and coping process: A sociocultural review and analysis. *Gerontologist, 37*(3), 342–354.

Ayalon, L., & Arean, P. A. (2004). Knowledge of Alzheimer's disease in four ethnic groups of older adults. *International Journal of Geriatric Psychiatry, 19*(1), 51–57.

Belle, S. H., Burgio, L., Burns, R., Coon, D., Czaja, S. J., Gallagher-Thompson, D., et al. (2006). Enhancing the quality of life of dementia caregivers from different ethnic or racial groups: A randomized, controlled trial. *Annals of Internal Medicine, 145,* 727–738.

Borrayo, E. A., Goldwaser, G., Vacha-Hasse, T., & Hepburn, K. W. (2007). An inquiry into Latino caregivers' experience caring for older adults with Alzheimer's disease and related dementias. *Journal of Applied Gerontology, 26,* 486–507. doi:10.1177/0733464807305551.

Brookmeyer, R., Corrada, M. M., Curriero, F. C., & Kawas, C. (2002). Survival following a diagnosis of Alzheimer disease. *Archives of Neurology, 59*(11), 1764–1767.

Burr, J., & Mutchler, J. (1999). Race and ethnic variation in norms of filial responsibility among older persons. *Journal of Marriage and the Family, 61,* 674–688.

Calderon, V., & Tennstedt, S. L. (1998). Ethnic differences in the expression of caregiver burden. *Journal of Gerontological Social Work, 30*(1–2), 159–178.

Chappell, N. L., & Reid, R. C. (2002). Burden and well-being among caregivers: Examining the distinction. *The Gerontologist, 42*(6), 772–780.

Connell, C. M., & Gibson, G. D. (1997). Racial, ethnic, and cultural differences in dementia caregiving: Review and analysis. *Gerontologist, 37*(3), 355–364.

Connell, C. M., Janevic, M. R., & Gallant, M. P. (2001). The costs of caring: Impact of dementia on family caregivers. *Journal of Geriatric Psychiatry and Neurology, 14*(4), 179–187.

Cooper, C., Tandy, A. R., Balamurali, T. B. S., & Livingston, G. (2010). A systematic review and meta-analysis of ethnic differences in use of dementia treatment, care, and research. *The American Journal of Geriatric Psychiatry, 18*(3), 193–203.

Covinsky, K. E., Newcomer, R., Fox, P., Wood, J., Sands, L., Dane, K., & Yaffe, K. (2003). Patient and caregiver characteristics associated with depression in caregivers of patients with dementia. *Journal of General Internal Medicine, 18*(12), 1006–1014.

DeCarli, C. (2003). Mild cognitive impairment: Prevalence, prognosis, aetiology, and treatment. *Lancet Neurology, 2*, 15–21.

Dilworth-Anderson, O., Willliams, I. C., & Gibson, B. E. (2002). Issues of race, ethnicity, and culture in caregiving research: A 20-year review (1980–2000). *Gerontologist, 42*(2), 237–272.

Espino, D. V., DelAguila, D., Mouton, C. P., Alford, C., Parker, R. W., Miles, T. P., et al. (2002). Characteristics of Mexican-American elders with dementia presenting to a community-based memory evaluation program. *Ethnicity & Disease, 12*(4), 517–521.

Fink, A. (2005). *Conducting research literature reviews: From the internet to paper*. Thousand Oaks: Sage.

Flores, Y. G., Hinton, L., Franz, C. E., Barker, J. C., & Velasquez, A. (2009). Beyond familism: Ethics of care of Latina caregivers of elderly parents with dementia. *Health Care for Women International, 30*(12), 1055–1072.

Gauthier, S., Reisberg, B., Zaudig, M., Peterson, R. C., Ritchie, K., Broich, K., et al. (2006). Mild cognitive impairment. *Lancet, 367*(9518), 1262–1270.

Gelman, C. R. (2010). "La lucha": The experiences of Latino family caregivers of patients with Alzheimer's disease. *Clinical Gerontologist, 33*(3), 181–193.

Gelman, C. R. (2012). Familismo and its impact on the family caregiving of Latinos with Alzheimer's disease: A complex narrative. *Research on Aging, 00*(0), 1–32.

Gurland, B. J., Wilder, D. E., Lantigua, R., Stern, Y., Chen, J., Killeffer, E. H., & Mayeux, R. (1999). Rates of dementia in three ethnoracial groups. *International Journal of Geriatric Psychiatry, 14*(6), 481–493.

Gutterman, E. M., Markowitz, J. S., Lewis, B., & Fillit, H. (1999). Cost of Alzheimer's disease and related dementia in managed-medicare. *Journal of the American Geriatrics Society, 47*(9), 1065–1071.

Haan, M. N., Mungas, D. M., Gonzalez, H. M., Ortiz, T. A., Archarya, A., & Jagust, W. J. (2003). Prevalence of dementia in older Latinos: The influence of type 2 diabetes mellitus, stroke and genetic factors. *Journal of American Geriatrics Society, 51*, 169–177.

Hebert, L. E., Scherr, P. A., Bienias, J. L., Bennett, D. A., & Evans, D. A. (2003). Alzheimer disease in the US population: Prevalence estimates using the 2000 census. *Archives of Neurology, 60*(8), 1119–1122.

Henderson, J. N., & Gutierrez-Mayka, M. (1992). Ethnocultural themes in caregiving to Alzheimer's disease patients in Hispanic families. *Clinical Gerontologist, 11*, 59–74.

Hendrie, H. C., Lindgren, D., Hay, D. P., Lane, K. A., Gao, S., Purnell, C., et al. (2013). Comorbidity profile and healthcare utilization in elderly patients with serious mental illnesses. The American *Journal of Geriatric Psychiatry, 21*(12), 1267–1276.

Herrera, A. P., Lee, J., Palos, G., & Torres-Vigil, I. (2008). Cultural influences in the patterns of long-term care use among Mexican American family caregivers. *Journal of Applied Gerontology, 27*(2), 141–165.

Hinton, L. (2002). Improving care for ethnic minority elderly and their family caregivers across the spectrum of dementia severity. *Alzheimer Disease and Associated Disorders, 16*(Suppl 2), S50–S55.

Hinton, W. L., Fox, K., & Levkoff, S. (1999). Exploring the relationships among aging ethnicity, and dementing illness. *Culture, Medicine and Psychiatry, 23,* 403–413.

Hinton, L., Haan, M., Geller, S., & Mungas, D. (2003). Neuropsychiatric symptoms in Latino elderly with dementia and mild cognitive impairment without dementia and factors that modify their impact on caregivers. *Gerontologist, 43,* 669–677.

Hinton, L., Franz, C., Yeo, G., & Levkoff, S. (2005). Conceptions of dementia in a multi-ethnic sample of family caregivers. *Journal of the American Geriatric Society, 53,* 1405–1410. doi:10.1111/j.1532-5415.2005.53409.x.

Hinton, L., Chambers, D., Velasquez, A., Gonzalez, H., & Haan, M. (2006a). Dementia neuropsychiatric symptom severity, help-seeking patterns, and family caregiver unmet needs in the Sacramento area Latino study on agining (SALSA). *Clinical Gerontologist, 29*(4), 1–15.

Hinton, L., Flores, Y., Franz, C. E., Hernandez, I. M., & Mitteness, L. (2006b). The borderlands of primary care: Family and primary care physician perspectives on "troublesome" behaviors of people with dementia. In A. Leibing & L. Cohen (Eds.), *Thinking about dementia—Culture, loss, and the anthropology of senility* (pp. 43–63). New Brunswick: Rutgers University Press.

Hinton, L., Franz, C., Reddy, G., Flores, Y., Kravtiz, R. L., & Barker, J. C. (2007). Practice constraints, behavioral problems, and dementia care: Primary care physicians' perspectives. *Journal of General Internal Medicine, 22*(11), 1487–1492.

Hinton, L., Chambers, D., & Velásquez, A. (2009). Making sense of behavioral disturbances in persons with dementia: Latino family caregiver attributions of neuropsychiatric inventory domains. *Alzheimer Disease and Associated Disorders, 23*(4), 401–405.

Husaini, B. A., Sherkat, D. E., Moonis, M., Levine, R., Holzer, C., & Cain, V. A. (2003). Racial differences in the diagnosis of dementia and its effects on the use and costs of health care services. *Psychiatric Services, 54,* 92–96.

Janevic, M. R., & Connell, C. M. (2001). Racial, ethnic, and cultural differences in the dementia caregiving experience: Recent findings. *Gerontologist, 41*(3), 334–347.

Jolicoeur, P. M., & Madden, T. (2002). The good daughters: Acculturation and caregiving among Mexican-American women. *Journal of Aging Studies, 16*(2), 107–120.

Kalkonde, Y. V., Pinto-Patarroyo, G. P., Goldman, T., Strutt, A. M., York, M. K., Kunik, M. E., et al. (2009). Ethnic disparities in treatment of dementia in veterans. *Dementia and Geriatric Cognitive Disorders, 28,* 145–152.

Karlamangla, A. S., Miller-Martinez, D., Aneshensel, C. S., Seeman, T. E., Wight, R. G., & Chodosh, J. (2009). Trajectories of cognitive function in late life in the US: Demographic and socioeconomic predictors. *American Journal of Epidemiology, 170*(3), 331–342.

Karlawish, J., Barg, F. K., Augsburger, D., Beaver, J., Ferguson, A., & Nunez, J. (2011). What Latino Puerto Ricans and non-Latinos say when they talk about Alzheimer's disease. *Alzheimer's & Dementia, 7*(2), 161–170.

Knight, B. G., & Sayegh, P. (2010). Cultural values and caregiving: The updated sociocultural stress and coping model. *Journals of Gerontology: Series B, 65B*(1), 5–13.

Larson, E. B., Shadlen, M. F., Wang, L., McCormick, W. C., Bowen, J. D., Teri, L., et al. (2004). Survival after initial diagnosis of Alzheimer disease. *Annals of Internal Medicine, 140*(7), 501–509.

Levkoff, S., Levy, B., & Weitzman, P. F. (1999). The role of religion and ethnicity in help seeking of family caregivers of elders with Alzheimer's disease and related disorders. *Journal of Cross-Cultural Gerontology, 14,* 333–356.

Lyketsos, C. G., Steinberg, M., Tschanz, J. T., Norton, M. C., Steffens, D. C., & Breitner, J. C. (2000). Mental and behavioral disturbances in dementia: Findings from the cache county study on memory and aging. *American Journal of Psychiatry, 157*(5), 708–714.

Lyketsos, C. G., Lopez, O., Jones, B., Fitzpatrick, A. L., Breitner, J., & DeKosky, S. (2002). Prevalence of neuropsychiatric symptoms in dementia and mild cognitive impairment: Results from the cardiovascular health study. *The Journal of the American Medical Association, 288*(12), 1475–1483.

Manly, J. J., & Mayeux, R. (2004). Ethnic differences in dementia and Alzheimer's disease. In N. B. Anderson, R. A. Bulatao, & B. Cohen, (Eds.), *Critical perspectives on racial and ethnic differences in health in late life* (pp. 95–142). Washington, DC: National Academies Press.

Mastel-Smith, B., & Stanley-Hermanns, M. (2012). "It's like we're grasping at anything": Caregivers' education needs and preferred learning methods. *Qualitative Health Research, 22*(7), 1007–1015.

McBean, A. M., Huang, Z., Virnig, B. A., Lurie, N., & Musgrave, D. (2003). Racial variation in the control of diabetes among elderly medicare managed care beneficiaries. *Diabetes Care, 26,* 3250–3256.

Mehta, K. M., Yin, M., Resendez, C., & Yaffe, K. (2005). Ethnic differences in acetylcholinesterase inhibitor use for Alzheimer disease. *Neurology, 65*(1), 159–162.

Morano, C. L., & Sanders, S. (2005). Exploring differences in depression, role captivity, and self-acceptance in Hispanic and non-Hispanic adult children caregivers. *Journal of Ethnic & Cultural Diversity in Social Work, 14*(3): 27–46.

Napoles, A. M., Chadiha, L., Eversley, R., & Moreno-John, G. (2010). Developing culturally sensitive dementia caregiver interventions: Are we there yet? *American Journal of Alzheimer's Disease and Other Dementias, 25,* 389–406.

Navie-Waliser, M., Feldma, P. H., Gould, D. A., Levine, C., Kuerbis, A. N., & Donelan, K. (2001). The experiences and challenges of informal caregivers: Common themes and differences among Whites, Blacks, and Hispanics. *Gerontologist, 41*(6), 733–741.

Neary, S. R., & Mahoney, D. F. (2005). Dementia caregiving: The experiences of Hispanic/Latino caregivers. *Journal of Transcultural Nursing, 16*(2), 163–170.

Ortiz A., Simmons, J., & Hinton W. L. (1999). Locations of remorse and homelands of resilience: Notes on grief and sense of loss of place of Latino and Irish-American caregivers of demented elders. *Culture, Medicine and Psychiatry, 23*(4), 477–500.

Ory, M. G., Hoffman, R. R. 3rd, Yee, J. L., Tennstedt, S., & Schulz, R. (1999). Prevalence and impact of caregiving: A detailed comparison between dementia and nondementia caregivers. *Gerontologist, 39,* 177–185.

Pinquart, M., & Sorensen, S. (2003). Differences between caregivers and noncaregivers in psychological health and physical health: A meta-analysis. *Psychology and Aging, 18*(2), 250–267.

Pinquart, M., Sorensen, S. (2005). Ethnic differences in stressors, resources and psychological outcomes of family caregiving: A meta-analysis. *The Gerontologist, 45,* 90–106.

Porter, L. J., Ganong, L. H., & Armer, J. M. (2000). The church family and kin: An older rural Black woman's support network and preferences for care providers. *Qualitative Health Research, 10,* 452–470.

Radina, M. (2007). Mexican American siblings caring for aging parents: Processes of caregiver selection/designation. *Journal of Comparative Family Studies, 38*(1), 143–168.

Radina, M. E., Gibbons, H. M., & Lim, J. (2009). Explicit versus implicit family decision-making strategies among Mexican American caregiving adult children. *Marriage & Family Review, 45*(4), 392–411.

Raji, M. A., Reyez-Ortiz, C. A., Kuo, Y. F., Markides, K. S., & Ottenbacher, K. J. (2007). Depressive symptoms and cognitive change in older Mexican Americans. *Journal of Geriatric Psychiatry and Neurology, 20,* 145–152.

Roth, D. L., Perkins, M., Wadley, V. G., Temple, E. M., & Haley, W. E. (2009). Family caregiving and emotional strain: Associations with quality of life in a large national sample of middle-aged and older adults. *Quality of Life Research, 18*(6), 679–688.

Rotkiewicz-Piorun, A. M., Al Snih, S., Raji, M., Kuo, Y. F., & Markides, K. S. (2006). Cognitive decline in older Mexican Americans with diabetes. *Journal of the National Medical Association, 98*(11), 1840–1847.

Royall, D. R., Espino, D. V., Polk, M. J., Palmer, R. F., & Markides, K. S. (2004). Prevalence and patterns of executive impairment in community dwelling Mexican Americans: Results from the Hispanic EPESE Study. *International Journal of Geriatric Psychiatry, 19*(10), 926–934.

Schulz, R., O'Brien, A. T., Bookwala, J., & Fleissner, K. (1995). Psychiatric and physical morbidity effects of dementia caregiving: Prevalence, correlates, and causes. *Gerontologist, 35*(6), 771–791.

Simpson, C. (2010). Case studies of Hispanic caregivers of persons with dementia: Reconciliation of self. *Journal of Transcultural Nursing, 21*(2), 167–174. doi:10.1177/1043659609357630

Thompson, C. A., Spilsbury, K., Hall, J., Birks, Y., & Adamson, J. (2007). Systematic review of information and support interventions for caregivers of people with dementia. *BMC Geriatrics, 7*, 18.

Valle, R., Yamada, A. M., & Barrio, C. (2004). Ethnic differences in social network help-seeking strategies among Latino and Euro-American dementia caregivers. *Aging & Mental Health, 8*(6), 535–543.

Vickrey, B. G., Strickland, T. L., Fitten, L. J., Adams, G. R., Ortiz, A., & Hays, R. D. (2007). Ethnic variations in dementia caregiving experiences. *Journal of Human Behavior in the Social Environment, 15*(2), 425–434.

Vitaliano, P. P., Murphy, M., Young, H. M., Echeverria, D., & Borson, S. (2011). Does caring for a spouse with dementia promote cognitive decline? A hypothesis and proposed mechanisms. *Journal of American Geriatrics Society 59*(5), 900–908.

Walsh, D., & Downe, S. (2005). Meta-synthesis method for qualitative research: A literature review. *Journal of Advanced Nursing, 50*(2), 204–211.

Weiss, C. O., Gonzalez, H. M., Kabeto, M. U., & Langa, K. M. (2005). Differences in amount of informal care received by non-Hispanic Whites and Latinos in a nationally representative sample of older Americans. *Journal of the American Geriatrics Society, 53*(1), 146–151.

Winslow, B. W., & Flaskerud, J. H. (2009). Deciding to place a relative in long-term care: "We really don't do that". *Issues in Mental Health Nursing, 30*(3), 197–198.

Zhan, L. (2004). Caring for family members with Alzheimer's disease: Perspectives from Chinese American caregivers. *Journal of Gerontological Nursing, 30*, 19.

Zuckerman, I. H., Ryder, P. T., Simoni-Wastila, L., Shaffer, T., Sato, M., Zhao, L., et al. (2008). Racial and ethnic disparities in the treatment of dementia among medicare beneficiaries. *Journal of Gerontology, 63*(5), S328–S333.

Chapter 10
Prevalence and Determinants of Falls Among Older Mexicans: Findings from the Mexican National Health and Nutrition Survey

María P. Aranda, Mariana López-Ortega and Luis Miguel Gutiérrez Robledo

Background

Sustaining a fall is a common occurrence in older persons worldwide. According to the World Health Organization (2007), about 28–35 % of persons 65 years of age or older fall each year with the rate increasing to 32–42 % for the over 70 age group. Falls are the most common cause of unintentional injury and the incidence of falls varies among countries: 6–31 % in China, 20 % in Japan, and 21.6–34 % in Latin American and the Caribbean region (WHO 2007; Reyes-Ortiz et al. 2005).

Falls and falls-related injuries are a global phenomenon with global consequences. Although much is known about falls in late-life in developed countries, less is known about falls in low- or middle-income countries. Examining the rates of falls, the consequences of falls, and prevention and primary intervention strategies are public health concerns that need to be attended to given the growth of the older adult population worldwide. In 2012 the number of older persons increased to almost 810 million, and will increase to 2 billion by 2050 with the greatest growth occurring in low to middle-income countries (UNFPA and HelpAge International 2012).

Looking beyond global population aging, there may be regional variations in fall risk factors due to cultural and environmental determinants that vary across countries (WHO 2007). Such variations can be the result of region-specific perceptions of falls, expectations regarding developmental expectations of older adulthood

---

M. P. Aranda (✉)
School of Social Work, University of Southern California, MRF #214, Los Angeles,
CA 90089-0411, USA
e-mail: aranda@usc.edu

M. López-Ortega · L. M. G. Robledo
National Geriatrics Institute, National Institutes of Health, Mexico City, Mexico

M. P. Aranda · M. López-Ortega
USC Edward R. Roybal Institute on Aging, University of Southern California,
Los Angeles, CA, USA

© Springer International Publishing Switzerland 2015   171
W. A. Vega et al. (eds.), *Challenges of Latino Aging in the Americas,*
DOI 10.1007/978-3-319-12598-5_10

and aging processes, environmental influences (housing and community conditions, health and social resources; WHO 2007) and variation in behavioral risk factors such as smoking, substance use, and physical activity (UNFPA 2012).

The health, psychosocial, and economic consequences of falls have been well documented. First, falls are the leading cause of both fatal and non-fatal injuries in older people (Sterling et al. 2001). Although not all falls are injurious, falls represent the main cause of accidents in adults 60+ years of age with roughly 10–15 % resulting in injuries, disability, and death (Adams et al. 2011; Sterling et al. 2001; Tinetti and Williams 1998; WHO 2007). Falls are considered the main cause for 40 % of all injury deaths (Rubenstein 2006), and although women have higher rates of falls, males are more likely to die from a fall (Stevens et al. 2006; WHO 2007). Second, falls can increase psychosocial stress in the form of fear of falling, social isolation, and increased functional dependency. For example, fear of falling in older persons is described as the persistent, often exaggerated, concern about falling resulting in activity restriction, changes in posture and gait, future falls and loss of independence (Gagnon and Flint 2003; Gillespie et al. 2012; Institute of Medicine 2012). Third, falls are a relevant economic burden to individuals and society. The economic burden of falls in late-life is actually quite high given that older adults are five times more likely to be hospitalized for a fall than from other causes (Alexander et al. 1992). Systematic reviews comparing costs found that fall-related costs ranged between 0.85 and 1.5 % of all health care expenditures within the US, Australia, the European Union, and the UK (Heinrich et al. 2010). Non-fatal and fatal falls cost $ 23.3 billion in the US annually, and $ 1.6 billion in the UK (Davis et al. 2010). The magnitude of costs is apparent worldwide as the average health system cost per one fall injury episode for people 65+ ranges from $ 1,049 in the US to US$ 3,611 in Finland and Australia (WHO 2007).

Factors Associated with Falls

As with many health-related events, the risk of falling is a result of multiple risk factors which directly or in combination can lead to falling (Todd and Skelton 2004). These factors can be intrinsic or extrinsic to the person and are categorized across four dimensions: biological, behavioral, environmental and socioeconomic (see Fig. 10.1; WHO 2007).

Biological risk factors include a diverse array of demographic factors (age, gender), chronic medical illnesses, cognitive and affective capacities, and gait problems. Female gender, age, race, chronic diseases, chronic pain, depressive symptoms, cognitive impairment, and abnormalities with balance and gait have been identified as determinants of falls (Anstey et al. 2006; Cesari et al. 2002; Himes and Reynolds 2012; Eggermont et al. 2012; Lin et al. 2011; Tinetti et al. 1988; Tromp et al. 1998; NAP 1990). Sensory impairment, quality of vision and visual impairment have been shown to be important factors determining fall risk (Lin et al. 2011; Lord and Dayhew 2001; NAP 1990).

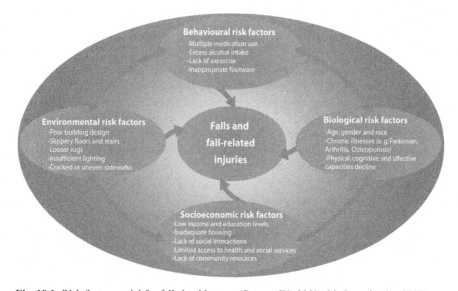

Fig. 10.1 Risk factor model for falls in older age. (Source: World Health Organization 2007)

Additional biological determinants include gait, balance and weakness issues which have been measured through objective performance (e.g., timed walk speed, chair stands, etc.) and anthropometric measures (body weight, height, calf circumference, etc.). Previous work has found that older adults with lower scores on physical performance measures are at higher risk for falls (Chan et al. 2007; Tromp et al. 1998), and that physical functioning (a determinant of falls) can be determined by muscle mass, leg strength, and fat mass in older adults (Bouchard et al. 2011), and calf circumference (Chien and Guo 2014; Stewart et al. 2002).

Behavioral factors include individual behaviors that may lead to falls such as the use of medications (especially with multiple medications, and psychoactive drugs), excess alcohol intake, physical inactivity, and inappropriate footwear (Bloch et al. 2011; Hartikainen et al. 2007; Gregg et al. 2000; Himes and Reynolds 2012; Eggermont et al. 2012; Moreno-Martinez et al. 2005; Nevitt 1990; O'Loughlin 1993; Skelton 2001; Tinetti et al. 1988; Tromp et al. 1998). On the other hand, environmental factors such as poor building design, slippery or uneven walking surfaces, insufficient lighting including reduced sunlight exposure, and inappropriate walking devices are also implicated in falling (Dean and Ross 1993; Lord et al. 2001; Nilson et al. 2014; Studenski et al. 1994). Lastly, falls may be attributed to socioeconomic factors such as low income and education, inadequate housing, lack of social interactions, and limited access or availability of health and community resources (Todd and Skelton 2004). Most causes—about 70%—are due to multifactorial interactions while 30% are attributed to an external event that would cause most people to fall, or a single identifiable cause such as syncope (Campbell and Robertson 2006).

Older Adults in Mexico: Health Care Needs

The aging population in Mexico is growing rapidly, and several health-related indicators in this population highlight the need to address falls as a public health concern. Census data indicate that 26 % of Mexican households have at least one adult 60 years or older, and the older adult population is expected to grow from 9 % in 2010 to 30 % in 2050 (Hernández López et al. 2013). Similar to the general Mexican population, older adults live primarily in urban areas, are married or have a partner, and have no formal education or very low education levels (Hernández López et al. 2013). Major health problems affecting this population group are diabetes mellitus and ischemic heart disease (representing 16.5 and 16.2 % of total deaths in this age group, respectively), and visual impairment affecting almost one-half of older adults (INEGI 2013). About 26 % or 2.1 million adults 60 years and older in Mexico report having limitations in at least one activity of daily living. The most likely reported limitations were walking or moving, talking or communicating, personal care, and visual impairments, with the age group of 80 years and older presenting the most limitations (López Ortega 2012).

As the older Mexican population increases, falls and related consequences such as hospitalization, fractures and fatal injuries pose severe challenges to individuals, families, as well as health care systems. Research is needed to accurately inform prevention, diagnostic, and management strategies tailored to the Mexican context (de Santillana Hernández et al. 2002; Reyes-Ortiz et al. 2005; Tinetti et al. 2006). To date, previous work has focused primarily on higher income countries. Understanding the public health burden of falls and their determinants in Mexico fills this gap in knowledge while exploring determinants of falls in the context of rapid demographic shifts.

Based on a national population study of health and nutrition, this chapter addresses the 12-month prevalence and determinants of falls among older persons 60 years and older in Mexico. To date, the prevalence of falls and factors associated with falls has not been well established in Mexico since the few studies that exist focus mainly on small regional samples that have addressed the association of falls with other health conditions or factors. These include studies that examined: (1) factors associated with falls among cognitively intact hospital patients in Mexico City (de Santillana Hernández et al. 2002; Reyes-Ortiz et al. 2005); (2) the association of balance with falls as an outcome in Merida, Yucatan (Estrella-Castillo 2011a); (3) the incidence of falls in a university rehabilitation unit among patients with balance problems (Estrella-Castillo 2011b); (4) the association of falls with functional dependency as an outcome among beneficiaries of a means-tested, national cash assistance program (Manrique-Espinosa et al. 2011) and (5) rates of accidental injuries (including falls) among older residents from marginalized neighborhoods in four Mexican cities (Ruelas Gonzalez and Salgado de Snyder 2008). In sum, data from these studies are largely consistent with studies in other countries: (1) Between 33.5 and 40.0 % of older adults fall each year; (2) Falls are the main cause of accidental injuries in older adults; and (3) Common risk factors include female gender, multiple chronic diseases, depressive symptoms, functional limitations, problems with vision and/or hearing, and use of multiple medications.

The inclusion for the first time of a nationally representative sample of adults 60 years and older in Mexico creates a unique opportunity to investigate the prevalence of falls and associated biological, behavioral, environmental and socioeconomic risk factors. Although the global burden of years lived with disability (YLDs) due to falls is higher in developing countries (66 %) than developed countries (34 %), falls prevention strategies are implemented primarily in high income countries. Population-based data is urgently needed to inform the development of relevant and effective programs into the Mexican public health infrastructure.

Methods

Data

We use data from a nationally representative health and nutrition study in Mexico, the National Health and Nutrition Survey–ENSANUT 2012. This is the first time that persons 60 years of age and older were included in the National System of Health Surveys. Details of the sampling structure design and methodology can be found at http://www.ensanut.insp.mx. Surveys are cross-sectional and based on multi-staged, stratified cluster sampling. Information in each survey covers a wide range of health and nutrition topics in addition to socioeconomic and demographic information at the individual and household level. For the year 2012, the National Health and Nutrition survey gathered information from 50,528 households. The total sample of older adults 8,874 which represents 10,695,704 adults 60 years and older, or approximately 9.0 % of the total Mexican population (Gutiérrez et al. 2012).

Measures

The outcome variable represents the number of falls in the 12-months previous to the interview. Specifically, the questionnaire includes the question: "How many times have you fallen in the last 12 months?", with response options of none or number of total falls within the reference period. This item has been used extensively in previous population-based studies (Blake et al. 1988; Campbell et al. 1981; Hestekin et al. 2013; Lord et al. 2001; Reyes-Ortiz et al. 2005).

Following WHOs risk factor model for falls (2007), we include covariates in the model that represent main risk factors across four dimensions:

Biological We include sex, age, diagnosed chronic illnesses, having suffered a stroke, difficulty performing basic and instrumental daily activities (physical decline), hearing and vision impairment (0 = excellent/very good/good; 1 = fair/ poor). Chronic disease comorbidity is included as a dummy variable indicating if the older adult has no chronic disease or from one or more chronic illnesses: diabetes mellitus, hypertension and cardiovascular disease. Functional decline in basic (ADL) or instrumental activities of daily living (IADL) is included as two separate dummy variables indicating no difficulties or difficulty with at least one basic or instrumental activity of daily living, respectively.

ENSANUT interview protocols include face-to-face interviews that include standardized anthropometric measures administered by trained personnel according to standard international protocols and procedures (Gutiérrez et al. 2012). In order to examine if these factors are associated with falls in our sample of older adults, we include body mass index, walking speed (timed 4-m walk), and calf circumference as covariates in the analysis. Studies in different countries have showed how anthropometric measures are related to geriatric syndromes, including falls. In their study using five waves of the Health and Retirement Study (HRS), Himes and Reynolds (2012) note a relationship between falls and obesity, with higher obesity related to higher risk of falls. While daily life gait characteristics have been found to be associated with fall history, gait analysis is regarded as a key measure in predicting the occurrence of falls (Rispens et al. 2014; Mignardot et al. 2014; Maki 1997). Body measurements such as arm and calf circumference have been associated with under-nutrition, physical mobility disturbances, sarcopenia and frailty, which in turn can affect mobility, strength, and ultimately the risk of falling (Cesari et al. 2002; Landi et al. 2014; Chien and Guo 2014; Velázquez Alva et al. 2013; Saka et al. 2010).

We include two additional variables to measure psychological functioning: Depressive symptoms is based on a modified version of the CES-D (Radloff 1977), namely a six-item scale based on depressed mood, loss of interest, concentration difficulties, fatigue, sleep difficulties, and sadness. A dummy variable was generated to classify those with and without any depressive symptoms. Cognitive functioning is measured by one item on self-reported memory difficulties. Specifically, the question asks: "Have you had difficulties with your memory which have been a problem for you?" with the following response options: no memory difficulties; mild to moderate or infrequent; and severe, frequent or persistent.

Behavioral We constructed a dummy variable indicating if the respondent takes medications for their diagnosed chronic diseases (0 = none; 1 = takes one medication, 2 = takes two to three medications).

Socioeconomic Measures of educational attainment and health insurance are included. We added marital/partner status as a social interaction measure given that being in a committed relationship is highly associated with co-residence.

Environmental We include urban/rural residence to account for possible variations in residential and community risk factors (poor housing conditions, inadequate walking paths or surfaces): rural = population <2500; urban = population ≥2500.

Statistical Analysis

Models Similar to most clinical and health outcomes, the phenomenon of falls is discrete, non-negative, and most likely comprised by non-normal distributions. Therefore, in order to estimate falls we use count data models such as the Poisson regression model (PRM) defined by the Poisson distribution for the number of counts or occurrences of an event. However, given that the main assumptions of the PRM—the equidispersion property of equality of mean and variance, and that all events occur independently over time— are both difficult to hold, the fit of the PRM has to be tested for over dispersion (Cameron and Trivedi 2008). If the tests fail, an alternate distribution to treat over dispersion is the negative binomial regression model (NBRM), which includes unobserved individual heterogeneity as an error term to the Poisson model (Cameron and Trivedi 2005; Long and Freese 2006). Several tests have been developed in order to establish which model best fits the data. On the one hand, nested models such as the PRM and NBRM can be tested using likelihood ratio tests, while the Akaike information criterion and Bayesian information criterion tests can be used to compare both nested and non-nested models (Vuong 1989).

Even when these models have proved better estimation and reduced biases in the estimation of health outcomes including falls, few studies have actually used these models to estimate falls (Bauer et al. 2012, Ullah et al. 2010, Byers et al. 2003). An in-depth study on count data models for falls recommends that futures studies routinely use the NBRM in preference to the Poisson or zero-inflated models (Ullah et al. 2010).

A descriptive analysis of falls in the sample was generated as a first step in order to characterize fallers and non-fallers according to the multifactorial risk model described above. Then, tests for differences in falling status were performed using the chi-square test for categorical variables, and the Kruskal-Wallis Test for continuous scores. After conducting an exploratory analysis of our outcome variable, our data indicated that the variance is higher than the mean, and that our data is highly skewed (10.37), with values ranging from 0 to 65 falls, a mean of 0.880, and standard deviation of 2.35.

For the regression analyses, the first step included exploring the frequencies and distribution of falls in order to check for over dispersion and probable excess zeroes in the data. Given that a substantial proportion of the respondents reported no falls and that the distribution of this variable is highly skewed, count data models were generated to explore determinants of falls in this sample of the Mexican older adults. Both PRM and NBRM and the aforementioned tests were performed in order to explore the best fit for the data. Both analytical and graphical approaches were used to compare and select models. The likelihood test, Akaike Information Criterion (AIC) and Bayesian Information Criterion (BIC) tests were used to compare the PRM and NBRM. All analyses were performed using STATA 12 (StataCorp 2011) statistical software.

Results

Descriptive Statistics

The total sample of ENSANUT 2012 adults 60 years and older is 8894 individuals of which 54.6 % were women. Due to missing data on falls for 64 respondents, the final working sample for the study is 8830 adults 60 years and older. Of the total working sample, 3011 older adults (34.1%) reported having one or more falls in the past 12 months.

Table 10.1 presents the sample characteristics. Given that older adults in Mexico are largely in the younger cohort (60–69), we observe the highest concentration of falls in this age group. With the exception of size of locality, most socioeconomic status and demographic factors including sex, age, educational attainment and marital/partner status, show statistically significant differences between fallers and non-fallers. In addition, hearing and visual impairment, higher levels of chronic disease comorbidity, reporting difficulties in performing basic and instrumental daily activities, presenting depressive symptoms and using multiple medicines for chronic diseases were also all statistically significant between the groups ($p < 0.001$). Regarding anthropometric and performance measures, walking speed and calf circumference differ significantly between the groups, while BMI is not significantly different.

Results of the PR and NBR models, the tests of over dispersion of the PRM and of the overall fit of the models, gave strong evidence of over dispersion. The tests of the fit of the PRM vs. NBR showed a much better fit of the NBRM (AIC, BIC and the likelihood ratio test favor the NBRM over the PRM). This is also observed in the plot of the fit of the models. Thus, the NBRM is preferred over the PRM and only the results of the NBRM estimates for falls are presented.

Our data indicate that a combination of biological or clinical factors were associated with falls in the previous 12-month period (Table 10.2). As expected, factors that address individual functioning emerged as significant determinants of falls and in the expected direction. Self-reported memory impairment either of low ($p = 0.000$) to severe intensity or frequency ($p = 0.000$) was associated with falling. Difficulties in performing ADLs ($p = 0.000$) and depressive symptoms were also significant ($p = 0.000$). Reporting a history of stroke ($p = 0.019$), one or more chronic diseases ($p = 0.015$), vision ($p = 0.000$) or hearing impairment ($p = 0.009$) were significantly associated with falling. Our medication factor was not significant. Although the relationship between falls and each of the anthropometric measures was in the expected direction, there were no significant associations in the models with walk speed and calf circumference (Table 10.2).

Turning to demographic factors we found that older age—either in the 70–79 group ($p = 0.019$) or 80+ group ($p = 0.000$), and being insured was associated with falling. On the other hand, being male was negatively associated with falling. Our marital/partner status variable was not significant.

Table 10.1 Sample characteristics and test of differences between groups in the 2012 national health and nutrition survey (ENSANUT, Mexico)

Variable	No falls	One or more falls	p
	$n=5,819$	$n=3,011$	
Demographic			
Gender			
Female	48.7	38.9	<0.001
Age group (years)			
60–69	53.8	44.5	<0.001
70–79	32.0	35.0	
80+	14.2	20.5	
Locality			
Urban	64.5	62.4	N.S.
Marital status			
Single/divorced/separated	14.3	14.0	<0.001
Married or in a union	57.2	50.7	
Widowed	28.5	35.3	
Educational attainment			
No formal schooling	26.4	31.8	<0.001
Completed elementary	55.3	54.9	
Completed secondary and higher	18.3	13.3	
Health			
Fair/poor	39.6	50.0	<0.001
Vision			
Fair/poor	44.8	55.3	<0.001
ADL difficulties			
Difficulty in at least one activity	19.6	36.4	<0.001
IADL difficulties			
Difficulty in at least one activity	16.6	30.4	<0.001
Comorbidity			
One or more chronic diseases	47.6	57.9	<0.001
Stroke (in last year)			
Yes	2.8	4.3	<0.001
Medication use			
No medications	55.5	46.8	<0.001
One or more medications	44.4	53.2	
Depressive symptoms			
Yes	41.5	59.8	<0.001
Memory problems			
No memory problems	64.2	50.0	<0.001
Low/moderate intensity or non-frequent	32.3	41.4	
Severe intensity, frequent or persistent	3.5	8.6	

ADLs walking across the room, bathing/showering, getting into and out of bed and dressing, *IADLs* preparing a hot meal, grocery shopping, taking medicines, managing own money

None of our socioeconomic or environmental factors was significant although all showed associations in the expected direction.

Discussion

With an increasingly aging society, the prevalence and determinants of falls in Mexico are a public health concern. Our study is the first to document the 12-month prevalence and determinants of falls based on a national probability sample of Mexicans 60 years of age and older in Mexico. Twelve-month prevalence estimates indicate that about 1 out of 3, or 34 % of the sample reported falling. Women reported higher prevalence than men, or 38.1 vs. 29.2 %, respectively. The overall prevalence closely mirrors that of older respondents in Mexico City (33.5 %), Santiago, Chile (34 %), and the Coastal and Andes Mountains regions of Ecuador (34.7 %; Orces 2013), yet is higher than in Latin American cities such as Sao Paulo, Brazil (29 %), Buenos Aires, Argentina (28.5 %), Montevideo, Uruguay (27 %), and Havana, Cuba (24 %) (see Reyes-Ortiz et al. 2005). The overall prevalence of falls was slightly higher than older adults of Mexican descent living in the US (30.8 %; Reyes-Ortiz et al. 2005). Similar rates among non-Hispanic white populations have been found (see WHO 2007; Blake et al. 1988; Campbell et al. 1981; Tinetti et al. 1988; Tromp et al. 2001). All of these studies used a similar survey item to ascertain falls during a 12-month period.

Our results indicate that biological and health-related correlates have the strongest association with occurrence of falls (Table 10.2). Specifically, advanced age, being female, higher ADL limitations, hearing impairment, increased chronic disease comorbidity, and stroke are factors that have been shown to predict falls across diverse regions in the world (Campbell et al. 1981; de Santillana Hernández et al. 2002; Damián et al. 2013; Hestekin et al. 2013; Lord and Dayhew 2001; Moreno-Martinez et al. 2005; de Santillana Hernández et al. 2002; Orces 2013; Reyes-Ortiz et al. 2005; Todd and Skelton 2004; WHO 2007).

Psychological determinants also played a significant role in determining 12-month fall rates. In particular, both self-reported memory problems and depressive symptoms were determinants of falls. These findings echo previous work which found affective and cognitive determinants of falls in older people (Eggermont et al. 2012; Kvelde et al. 2013; Reyes-Ortiz et al. 2005; Whooley et al. 1999). Although the causal link between depression, cognition and falls is yet to be determined, hallmark symptoms of both depression and cognitive impairment include poor balance and slowing of gait, slowed information processing and reaction time, low energy and activity levels, which in turn, can increase fall risk (Kvelde et al. 2013).

None of the physical performance or anthropometric measures was significantly correlated with falls, although the direction of the coefficients was in the expected direction. It could be that our measures such as the timed short walk and calf circumference were not the best measures for physical performance or nutritional health and stability, respectively. Perhaps more detailed assessments—in conjunction with

Table 10.2 Determinants of falls, 2012 Mexican health and nutrition survey (ENSANUT, Mexico)

Falls	Coef.	(95% Conf. Interval)		$p>z$
Male	−0.252	−0.380	0.123	0.000
70–79 years old~	0.152	0.025	0.278	0.019
80 years and older	0.386	0.195	0.576	0.000
Urban	0.008	−0.114	0.130	0.898
Insured	0.258	0.084	0.431	0.004
Married or in a union~	−0.111	−0.278	0.055	0.190
Widowed	−0.121	−0.298	0.057	0.183
Completed primary~	0.110	−0.024	0.245	0.107
Secondary or higher	0.009	−0.188	0.207	0.926
Hearing impairment	0.159	0.040	0.278	0.009
Vision impairment	0.223	0.106	0.340	0.000
ADL difficulties	0.504	0.355	0.653	0.000
IADL difficulties	0.090	−0.069	0.249	0.267
One or more chronic diseases	0.254	0.049	0.459	0.015
Stroke	0.405	0.066	0.743	0.019
One medicine~	−0.086	−0.294	0.122	0.418
Two to three medicines	−0.015	−0.254	0.224	0.901
Low/moderate intensity or non-frequent memory problems	0.261	0.139	0.383	0.000
Severe intensity, frequent or persistent memory problems	0.783	0.543	1.023	0.000
Depressive symptoms	0.424	0.309	0.540	0.000
Walk speed	0.006	−0.005	0.017	0.268
Calf circumference	−0.002	−0.015	0.010	0.736
BMI	0.004	−0.008	0.017	0.480
_cons	−1.387	−1.863	−0.912	0.000
/lnalpha	0.932	0.846	1.018	
Alpha	2.540	2.331	2.768	

~Reference categories: 60–69 years old, single, no educational attainment, takes no medicines, no self-reported memory problems

other measures such as static balance tests, and waist circumference—can provide better ascertainment of fall risk in this sample (Crimmins et al. 2008; Rubenstein 2006). Conversely, we did not ascertain potential gender-specific differences in our variables of interest which may have informed our performance-based analyses.

Medication use was not a significant determinant of falls. Perhaps the measure we used was less-than-ideal since the number of medications may not have been an adequate determinant of falls. It would be important to include not only the number of medications but classes of medications since not all types of medications are associated with falls. Some medications such as psychoactive drugs are known to increase the risk of falls while others are not (see Bloch 2010). We were unable to ascertain this information as no categories or types of medications were recorded in the national survey.

Net of all other variables in our model, being insured was actually associated with falling, a result we did not expect. It is unclear why having health insurance increases the rate of falls. One possible explanation is that persons with insurance in Mexico are more likely to receive health services and thus more likely to have a medical encounter identified as a consequence of "a fall," or to receive follow-up services for falls, or subsequent falls. Otherwise, not receiving formal services may underestimate the rate of falling. Another explanation could be that having insurance may increase pharmacological treatments for medical and psychiatric conditions, thus increasing the risk of medication-related falls.

A discussion of the study's strengths and limitations are in order. First, this is the first study to document national estimates of falls in a sample of older Mexicans in Mexico. Second, the study used analytic methods that are more appropriate to study count data such as falls. Third, we have identified specific factors that can inform the identification of at-risk groups and strategies that can focus potentially modifiable factors salient to the Mexican health, social service, and long-term care systems. For instance, adopting routine depression and cognitive screen assessments can serve to identify potential risk for falls especially for older persons living with comorbid medical conditions and functional limitations. Fourth, psychological issues such as self-rated depressive symptoms and cognitive complaints can be assessed and managed by guideline concordant strategies (Lichtenberg 2010).

Limitations include a cross-sectional design, self-report measures, and the lack of attention to the consequence of falls (mortality, change of living arrangements). Other limitations point to the definitional issues that plague falls research, namely the problems with how respondents perceive the phenomenon of falls (tripping during exercise; underreporting due to embarrassment). It is unclear how falls are perceived culturally by older Mexicans given the lack of in-depth information on phenomenological issues related to individual and societal fall perceptions. Measures such as medication use should include classes of medications that are known to be associated with falls such as hypnotics, opiates, and hypertensives. The timed short walk and calf circumference measures for physical performance may have not been the best measures to ascertain body strength and flexibility, or may perform differently in determining falls when comparing males and females (Smee et al. 2012). Lastly, we did not include specific measures that could add to the current knowledge gaps—alcohol use and misuse, and environmental indicators such as physical design of the person's residence and outlying community, and additional anthropometric and nutritional measures. Future work should address these methodological limitations and include the types of circumstances that led to the fall(s), fall-related outcomes such as injuries and deaths, economic costs due to emergency room and hospital utilization, family caregiver-related burden, and associated costs due to decreased quality of life.

Our results call for targeted interventions for older persons who possess the characteristics that were significantly associated with falls—e.g., persons with mobility and functional difficulties, affective conditions, and sensory deficits. Early detection of potential at-risk groups may decrease falls incidence, disability, and even death. Complex clinical syndromes such as falls in older adulthood necessitate interdisciplinary approaches to assessment, prevention and management. Interventions for preventing falls in older community-dwelling older adults can offer programs that include (1) falls education strategies for patients, families, and providers; (2) group and home-based exercise programs; (3) home safety assessment and home modifications; and (4) multifactorial assessment and interventions (Gillespie et al. 2012; Rubenstein 2006).

We offer several policy recommendations that have direct implications for the health and welfare of older Mexicans. First, research is needed to promote a national agenda on falls and falls-related injury prevention. Now that the magnitude of the problem has been at least partially addressed, subsequent empirical efforts are needed to examine more fully how falls affect specific subgroups of individuals (e.g., young-old vs. older age groups); regional variations, the social and economic burden associated with falls, future projections, and evidenced based prevention strategies.

Second, the clinical designation of falls syndrome should be included in Mexico's National Chronic Disease Program which is an integral part of the Ministry of Health's standards and clinical guidelines. This would permit the inclusion of best practices for falls prevention and treatment of falls-related injuries for older adults with varying levels functioning (Gutiérrez Robledo and Caro López 2012). Third, given the multifactorial nature of fall risk, policy makers and legislators must appropriate funding for service infrastructure directed to multiple government sectors typically affected by fall-related injuries: emergency medical services and first responders; primary care and community health; hospital services; housing and public works; social welfare and community-based services; transportation; etc. (Gutiérrez Robledo 2012).

Fourth, a national public education and awareness campaign should focus on the preventable or modifiable fall risk factors to reduce the incidence of falls and re-occurring falls in older people. Of particular importance, is the need to increase knowledge and self-care strategies that older adults and their caregivers can engage in to promote health and wellness.

Fifth, interprofessional education for health and social service providers should include curricula on falls and falls-related injury prevention, treatment, and rehabilitation. For example, interprofessional training can include assessments by physicians, nurses, occupational and physical therapists, social workers, and pharmacists to identify individual and environmental risk factors such as in-home hazards and unmanaged chronic diseases.

When one of three persons over the age of 60 suffer a fall in any given year, promoting public health, clinical and community-based policies will help delay or avoid unnecessary injury, disability, and burden on individuals, their caregivers and society.

References

Adams, P. F., Martinez, M. E., Vickerie, J. L., & Kirzinger, W. K. (2011). Summary health statistics for the U.S. population: National health interview survey. *Vital Health Statistics, 10,* 1–118.

Alexander, B. H., Rivara, F. P., & Wolf, M. E. (1992). The cost and frequency of hospitalization for fall-related injuries in older adults. *American Journal of Public Health, 82,* 1020–1023.

Anstey, K. J., Von Sanden, C., & Luszcz, M. A. (2006). An 8-year prospective study of the relationship between cognitive performance and falling in very old adults. *Journal of the American Geriatrics Society, 54,* 1169–1176. doi: 10.1111/j.1532-5415.2006.00813.x.

Bauer, T. K., Lindenbaum, K., Stroka, M. A., Engel, S., Linder, R., & Verheyen, F. (2012). Fall risk increasing drugs and injuries of the frail elderly—Evidence from administrative data. *Pharmacoepidemiology and Drug Safety, 21,* 1321–1327.

Blake, A. J., Morgan, K., Bendall, M. J., Dallosso, H., Ebrahim, S. B. J., Arie, T. H. D., Fentem, P. H., & Bassey, E. J. (1988). Falls by elderly people at home: Prevalence and associated factors. *Age and Ageing, 17,* 365–372.

Bloch, F., Thibuaud, M., Dugue, B., Breque, C., Riguad, A-S., & Kemoun, G. (2011). Psychotropic drugs and falls in the elderly people: Updated literature review and meta-analysis. *Journal of Aging and Health, 23*(2), 329–346. doi:10.1177/0898264310381277.

Bouchard, D. R., Heroux, M., & Janssen, I. (2011). Association between muscle mass, leg strength, and fat mass with physical function in older adults: Influence of age and sex. *Journal of Aging and Health, 23,* 313–328. doi:10.1177/0898264310388562.

Byers, A. L., Allore, H., Gill, T. M., & Peduzzi, P. N. (2003). Application of negative binomial modelling for discrete outcomes: A case study in aging research. *Journal of Clinical Epidemiology, 56,* 559–564.

Cameron, A., & Trivedi, P. (2005). *Microeconometrics: Methods and applications.* Cambridge: Cambridge University Press.

Cameron, A., & Trivedi, P. (2008). *Regression analysis of count data.* Cambridge: Cambridge University Press.

Campbell, A. J., & Robertson M. C. (2006). Implementation of multifactorial interventions for fall and fracture prevention. *Age and Ageing, 35*(Suppl. 2), ii60–ii64.

Campbell, A. J, Reinken, J., Allan, B., et al. (1981). Falls in old age: A study of frequency and related clinical factors. *Age and Ageing, 10,* 264–270.

Cesari, M., Landi, F., Torre, S., Onder, G., Lattanzio, F., & Bernabei, R. (2002). Prevalence and risk factors for falls in an older community-dwelling population. *The Journals of Gerontology. Series A Biological Sciences and Medical Sciences, 57*(11), M722–M726.

Chan, B. K., Marshall, L. M., Winters, K. M., Faulkner, K. A., Schwartz, A. V., & Orwoll, E. S. (2007). Incident fall risk and physical activity and physical performance among older men: The Osteoporotic Fractures in Men Study. *American Journal of Epidemiology, 165,* 696–703. doi: 10.1093/aje/kwk050.

Chien, M.-H., & Guo, H.-R. (2014). Nutritional status and falls in community-dwelling older people: A longitudinal study of a population-based random sample. *PLoS ONE, 9*(3), e91044. doi:10.1371/journal.pone.0091044.

Crimmins, E., Guyer, H., Langa, K., Ofstedal, M. B., Wallace, R., & Weir, D. (2008). Documentation of physical measures, anthropometrics and blood pressure in the health and retirement study. Survey Research Center, University of Michigan, Ann Arbor. http://hrsonline.isr.umich.edu/sitedocs/userg/dr-011.pdf. Accessed 22 March 2013.

Damián J, Pastor-Barriuso R, Valderrama-Gama E, & de Pedro-Cuesta J. (2013). Factors associated with falls among older adults living in institutions. *BMC Geriatrics, 13,* 6.

Davis, J. C., Robertson, M. C., Ashe, M. C., Liu-Ambrose, T., Khan, K. M., & Marra, C. A. (2010). International comparison of cost of falls in older adults living in the community: A systematic review. *Osteoporosis International, 21*(8), 1295–1306.

Dean, E., & Ross, J. (1993), Relationships among cane fitting, function, and falls. *Physical Therapy, 73,* 494–504.

de Santillana Hernández, S. P., Alvarado Moctezuma, L. E., Medina Beltrán, G. R., Gómez Ortega, G., & Cortés González, R. M. (2002). Caídas en el adulto mayor. Factores intrínsecos y extrínsecos. *Revista Medica IMSS, 40,* 489–493.

Eden, J., Maslow, K., Le, M., & Blazer, D. (Eds.), and the Committee on the Mental Health Workforce for Geriatric Populations, Board on Health Care Services (2012).The mental health and substance use workforce for older adults. In whose hands? *Institute of Medicine of the National Academies.* Washington, DC. PMID: 24851291

Eggermont, L. H. P., Penninx, B. W. J. H., Jones, R. N., & Leveille, S. G. (2012). Depressive symptoms, chronic pain, and falls in older community-dwelling adults: The MOBILIZE Boston Study. *Journal of the American Geriatrics Society, 60,* 230–237.

Estrella-Castillo, D. F., Euán-Paz, A., Pinto-Loría, M. L., Sánchez-Escobedo, P. A., & Rubio-Zapata, H. A. (2011a). Alteraciones del equilibrio como predictoras de caídas en una muestra de adultos mayores de Mérida Yucatán, México. *Rehabilitación (Madr), 45,* 320–326.

Estrella-Castillo, D. F., Rubio Zapata, H. A., Sánchez Escobedo, P., Aguilar Alonzo, P., & Araujo Espino, R. (2011b). Incidencia de caídas en una muestra de adultos mayores de la Unidad Universitaria de Rehabilitación de Mérida Yucatán. *Revista Mexicana de Medicina Física y Rehabilitación, 23,* 8–12.

Gagnon, N., & Flint, A. J. (2003). Fear of falling in the elderly. *Geriatrics & Aging, 6,* 15–17.

Gillespie, L. D., Robertson, M. C., Gillespie, W. J., Sherrington, C., Gates, S., Clemson, L. M., Lamb, S. E. (2012). Interventions for preventing falls in older people living in the community. Cochrane Database of Systematic Reviews, Issue 9. Art. No.: CD007146. doi:10.1002/14651858. CD007146.pub3.

Gregg, E. W., Pereira, M. A., & Caspersen, C. J. (2000) Physical activity, falls and fractures among older adults: A review of the epidemiologic evidence. *Journal of the American Geriatrics Society, 48,* 883–893.

Gutiérrez, J. P., Rivera-Dommarco, J., Shamah-Levy, T., Villalpando-Hernández, S., Franco, A., Cuevas-Nasu, L., Romero-Martínez, M., & Hernández-Ávila, M. (2012). Encuesta Nacional de Salud y Nutrición 2012. Resultados nacionales. 2a. ed.Cuernavaca, México: Instituto Nacional de Salud Pública (MX), 2013. http://ensanut.insp.mx/informes/ENSANUT2012ResultadosNacionales2Ed.pdf

Gutiérrez Robledo, L. M. (2012). La Academia Nacional de Medicina, el envejecimiento y la salud de los mexicanos. In L. M. G. Robledo & D. K. Stalnikowitz (Eds.), *Envejecimiento y Salud: Una propuesta para un plan de acción* (pp. 17–25). México: Academia Nacional de Medicina de México, Academia Mexicana de Cirugía, A.C., Instituto de Geriatría, Universidad Nacional Autónoma de México.

Gutiérrez Robledo, LM., & Caro López, E. (2012). Recomendaciones para la acción. Propuesta para un plan de acción en envejecimiento y salud. In L. M. G. Robledo & D. K. Stalnikowitz (Eds.), *Envejecimiento y salud: Una propuesta para un plan de acción* (pp. 321–347). México: Academia Nacional de Medicina de México, Academia Mexicana de Cirugía, A.C., Instituto de Geriatría, Universidad Nacional Autónoma de México.

Hartikainen, S., Lonnroos, E., & Louhivuori, K. (2007). Medication as a risk factor for falls: Critical systematic review. *Journals of Gerontology Series A: Biological Sciences and Medical Sciences, 62,* 1172–1181.

Heinrich, S., Rapp, K., Rissmann, U., Becker, C., & Konig, H. H. (2010). Cost of falls in old age: A systematic review. *Osteoporosis International, 21*(6), 891–902.

Hernández López, M. F., López Vega, R., & Velarde Villalobos, S. I. (2013). La situación demográfica en México. Panorama desde las proyecciones de población. In: La situación demográfica de México. Mexico: CONAPO.

Hestekin, H., O'Driscoll, T., Stewart Williams, J., Kowal, P., Peltzer, K., & Chatterji, S. (2013). Measuring prevalence and risk factors for fall-related injury in older adults in low- and middle-income countries: Results from the WHO Study on Global AGEing and Adult Health (SAGE). AGE Working Paper No. 6. http://www.who.int/healthinfo/sage/SAGEWorkingPaper6_Wave-1Falls.pdf. Accessed 12 May 2013.

Himes, C. L., & Reynolds, S. L. (2012). Effect of obesity on falls, injury, and disability. *Journal of the American Geriatrics Society, 60,* 124–129.

Instituto Nacional de Estadística y Geografía (INEGI). (2013). Estadísticas a propósito del Día Internacional de las Personas de Edad. http://www.inegi.org.mx. Accessed 10 February 2013.

Kvelde, T., McVeigh, C., Toson, B., Greenaway, M., Lord, S. R., Delbaere, K, & Close, J. C. (2013). Depressive symptomatology as a risk factor for falls in older people: Systematic review and meta-analysis. *Journal of the American Geriatrics Society, 61*(5), 694–706. doi:10.1111/jgs.12209.

Landi, F., Onder, G., Russo, A., Liperoti, R., Tosato, M., Martone, A. M., Capoluongo, E., & Bernabei, R. (2014). Calf circumference, frailty and physical performance among older adults living in the community. *Clinical Nutrition, 33*(3), 539–544.

Lichtenberg, P. A. (Ed.). (2010). *Handbook of assessment in clinical gerontology* (2nd ed.). New York: Wiley.

Lin, C.-H., Liao, K.-C., Pin, S.-J., Chen, Y.-C., & Liu, M.-S. (2011). Associated factors for falls among the community-dwelling older people assessed by annual geriatric health examinations. *PLoS ONE, 6,* e18976, 1–5.

Long, J., & Freese, J. (2006). *Regression models for categorical dependent variables using Stata.* College Station: Stata Press.

López Ortega, M. (2012). Limitación funcional y discapacidad: Conceptos, medición y diagnóstico. Una introducción a la situación en México. In L. M. G. Robledo & D. K. Stalnikowitz (Eds.), *Envejecimiento y Salud: Una propuesta para un plan de acción* (pp. 215–227). México: Academia Nacional de Medicina de México, Academia Mexicana de Cirugía, A.C., Instituto de Geriatría, Universidad Nacional Autónoma de México.

Lord, S. R., & Dayhew, J. (2001). Visual risk factors for falls in older people. *Journal of the American Geriatrics Society, 49,* 508–515.

Lord, S. R., Sherrington, C., & Menz, H. B. (2001). *Falls in older people: Risk factors and strategies for prevention.* Cambridge: Cambridge University Press.

Maki, B. E. (1997). Gait changes in older adults: Predictors of falls or indicators of fear. *Journal of the American Geriatrics Society, 45*(3), 313–320

Manrique-Espinosa, B., Salinas-Rodriguez, A., Moreno-Tamayo, K., & Tellez-Rojo, M. M. (2011). Prevalencia de dependencia funcional y su asociación con caídas en una muestra de adultos mayores pobres en Mexico. *Salud Publica de Mexico, 53,* 26–33.

Mignardot, J. B., Deschamps, T., Barrey, E., Auvinet B., Berrut G., Cornu C, Constans T, & de Decker L. 2014. Gait disturbances as specific predictive markers of the first fall onset in elderly people: A two-year prospective observational study. *Frontiers in Aging Neuroscience, 6*(22), 1–13, doi:10.3389/fnagi.2014.00022.

Moreno-Martinez, N. R., Ruiz-Hidalgo, D., Burdoy-Joaquim, E., & Vazquez-Mata, G. (2005). Incidencia y factores explicativos de las caídas en ancianos que viven en la comunidad. *Revista Española de Geriatría y Gerontología, 40*(Suppl. 2), 11–17.

NAP (1990). Institute of medicine (U.S.). Division of health promotion and disease prevention. The second fifty years: promoting health and preventing disability. *Institute of Medicine.* http://www.nap.edu/catalog/1578.html.

Nevitt, M. C. (1990). Falls in older persons: Risk factors and prevention. In R. L. Berg & J. S. Casells (Eds.), *The second fifty years: Promoting health and preventing disability* (pp. 263–290). Washington, DC: National Academy Press.

Nilson, F., Moniruzzaman, S., & Andersson, R. (2014). A comparison of hip fracture incidence rates among elderly in Sweden by latitude and sunlight exposure. *Scandinavian Journal of Public Health, 42*(2), 201–206. doi:10.1177/1403494813510794.

O'Loughlin, J. L., Robitaille, Y., Boivin, J. F., Suissa, S. (1993). Incidence of and risk factors for falls and injurious falls among the community-dwelling elderly. *American Journal of Epidemiology, 137*(3), 342–354.

Orces, C. H. (2013). Prevalence and determinants of falls among older adults in ecuador: An analysis of the SABE I survey. *Current Gerontology and Geriatrics Research, 2013,* 495468. doi:10.1155/2013/495468.

Radloff, L. S. (1977).The CES-D scale: A self-report depression scale for research in the general population. *Applied Psychological Measurement, 1*(3), 385–401.

Reyes-Ortiz, C. A., Al Snih, S., & Markides, K. S. (2005). Falls among elderly persons in Latin America and the Caribbean and among elderly Mexican-Americans. *Revista Panamericana de Salud Publica, 17,* 362–369.

Rispens, S. M., van Schooten, K. S., Pijnappels, M., Daffertshofer, A., Beek, P. J., & van Dieën, J. H. (2014). Identification of fall risk predictors in daily life measurements: Gait characteristics' reliability and association with self-reported fall history. *Neurorehabilitation and Neural Repair.* [Epub ahead of print].

Rubenstein, L. Z. (2006). Falls in older people: Epidemiology, risk factors and strategies for prevention. *Age Ageing, 35*(Suppl. 2), ii37–ii41.

Ruelas Gonzalez, M. G., & Salgado de Snyder, V. N. (2008). Lesiones accidentales en adultos mayores: Un reto para los sistemas de salud. *Salud Publica de Mexico, 50,* 463–471.

Saka, B., Kaya, O., Ozturk, B., Erten, N., & Akif Karan, M. (2010). Malnutrition in the elderly and its relationship with other geriatric syndromes. *Clinical Nutrition, 29,* 745–748.

Skelton, D. A. (2001). Effects of physical activity on postural stability. *Age and Ageing, 30*(Suppl. 4), 33–39.

Smee, D. J., Anson, J. M., Waddington, G. S., & Berry, H. L. (2012). Association between physical functionality and falls risk in community-living older adults. *Current Gerontology and Geriatrics Research, 2012,* Article ID 864516, 6 pages. doi:10.1155/2012/864516.

StataCorp. Stata statistical software (2011). *Release 12. 2011.* College Station: StataCorp LP.

Sterling, D. A., O'Connor, J. A., & Bonadies, J. (2001). Geriatric falls: Injury severity is high and disproportionate to mechanism. *Journal of Trauma–Injury, Infection and Critical Care, 50,* 116–119.

Stevens, J. A., et al. (2006). Fatalities and injuries from falls among older adults, United States, 1993–2003 and 2001–2005. *Morbidity and Mortality Weekly Report, 55,* 1221–1224.

Stewart, A. D., Stewart, A., & Reid, D. M. (2002). Correcting calf girth discriminates the incidence of falling but not bone mass by broadband ultrasound attenuation in elderly female subjects. *Bone, 31*(1), 195–198.

Studenski, S., Duncan, P. W., Chandler, J., Samsa, G., Prescott, B., Hogue, C., et al. (1994). Predicting falls: The role of mobility and nonphysical factors. *Journal of the American Geriatrics Society, 42,* 297–302.

Tinetti, M. E., & Williams, C. S. (1998). The effect of falls and fall injuries on functioning in community-dwelling older persons. *The Journal of Gerontology: Medical Sciences, 53 A,* M112–M119.

Tinetti, M. E., Speechley, M., & Ginter, S. (1988). Risk factors for falls among elderly persons living in the community. *The New England Journal of Medicine, 319,* 1701–1707.

Tinetti, M. E., Gordon, C., Sogolow, E., Lapin, P., & Bradley, E. H. (2006). Fall-risk evaluation and management: Challenges in adopting geriatric care practices. *Gerontologist, 46,* 717–725.

Todd, C., & Skelton, D. (2004) What are the main risk factors for falls among older people and what are the most effective interventions to prevent these falls? Copenhagen, WHO Regional Office for Europe (Health Evidence Network report). http://www.euro.who.int/document/E82552.pdf. Accessed 22 March 2013.

Tromp, A. M., Smit, J. H., Deeg, D. J. H., Bouter, L. M., & Lips, P. (1998). Predictors for falls and fractures in the Longitudinal Aging Study Amsterdam. *Journal of Bone and Mineral Research, 13,* 1932–1939.

Tromp, A. M., Pluijm, S. M. F., Smit, J. H., et al. (2001). Fall-risk screening test: a prospective study on predictors for falls in community-dwelling elderly. *Journal of Clinical Epidemiology, 54*(8), 837–844.

Ullah, S., Finch, C. F., & Day, L. (2010). Statistical modelling for falls count data. *Accident Analysis and Prevention, 42,* 384–392.

United Nations Population Fund (UNFPA) and HelpAge International. (2012). Ageing in the twenty-first century: A celebration and a challenge. New York, and HelpAge International,

London. http://www.unfpa.org/webdav/site/global/shared/documents/publications/2012/Ageing-Report_full.pdf. Accessed 30 Apri 2013.

Velázquez Alva, M. C., Irigoyen Camacho, M. E., Delgadillo Velázquez, J., & Lazarevich, I. (2013). The relationship between sarcopenia, undernutrition, physical mobility and basic activities of daily living in a group of elderly women of Mexico City. *Nutricion Hospitalaria, 28,* 514–521.

Vuong, Q. (1989). Likelihood ratio tests for model selection and non-nested hypotheses. *Econometrica, 57,* 307–333.

Whooley, M. A., Kip, K. E., Cauley, H. A., Ensrud, K. E., Nevitt, M. C., & Browner, W. S. (1999). Depression, falls, and risk fracture in older Women. *Archives of Internal Medicine, 159,* 484–490.

World Health Organization (WHO). (2007). WHO global report on falls prevention in old age. Geneva, Switzerland. Ageing and life course, family and community health report. http://www.who.int/ageing/publications/Falls_prevention7March.pdf. Accessed 22 march 2013.

Part III
Binational, Transnational Migration Perspectives: Mexico, Latin America, and the USA

Chapter 11
Binational Migration Perspectives: Mexico, Latin America, and the USA

William A. Vega and Stipica Mudrazija

This section of the volume connects the worlds of migrating and non-migrating people of Mexican heritage in the U.S. and Mexico, and examines selected issues accompanying consequences of resettlement and rebuilding lives and lifestyles. The chapters reflect major domains of transnational research in aging: impacts of migration and social adjustment, processes of support in families and social networks, and the impact of changing demography and residential status for foreign-born migrants in the U.S. The authors examine important lifestyle factors affecting geographic mobility, environmental conditions associated with health and functioning, and how older persons' needs are met through reciprocal support processes in families. Of particular interest is the U.S.-Mexico international border, specifically how it represents a *physical and conceptual* parameter for isolating mobility patterns and specific associations of older people's social adjustment and well-being. While certainly imposing practical problems on many older individuals, the international border does not seem to constitute an insurmountable barrier in decisions of older people to cross it in response to their changing needs for moral support, material sufficiency, or lifestyle change. To the extent that the border actively impedes family visitation and provision of support, receipt of needed services, and eligibility for specific health and social benefits such as Medicare, older persons on both sides face inconvenience and hardships.

The process of transiting geographic space and vacating former lives, including disconnecting from family and cultural forms of social relationships, and assuming the risks that accompany these transformational experiences, is an enduring point of interest. We are in an era of accelerated global migration due to the upheavals of war, changing demographics, urbanization, sustained economic inequities, and climate change (International Migration Institute 2011; United Nations 2009). Mexico has a long border with the U.S. that is unique for many historical reasons, and

W. A. Vega (✉) · S. Mudrazija
University of Southern California, Los Angeles, CA, USA
e-mail: williaav@usc.edu

© Springer International Publishing Switzerland 2015
W. A. Vega et al. (eds.), *Challenges of Latino Aging in the Americas,*
DOI 10.1007/978-3-319-12598-5_11

perhaps the most enduring aspect are cyclical migration patterns driven by "push–pull" migration effects of financial insufficiency in Mexico and U.S. labor markets.

A useful research approach for studying migration patterns and resettlement is explained by Portes and Rumbaut (2006), and simply restated here with a few key assumptions. We should improve our understanding of the migration processes by examining the "contexts" of premigration experiences including the environmental conditions that motivated migration (e.g. civil strife, poverty, family circumstances) and personal preparedness (material and social) to facilitate the physical transition. Knowing the social profile of migrants provides insights for investigating the resettlement process. Another key assumption is that examining "contexts of reception," such as structural factors and social resistance to migrants encountered in local areas of resettlement, offer perspectives for examining outcomes such as social assimilation and labor incorporation. "Segmented assimilation" is a related concept that captures social, economic, and political factors associated with migrant children's social adjustment and mobility outcomes in "contexts of reception" (Zhou 1997). The chapters in this section contain information relevant to these topics but focused on older Hispanics living in Mexico and the U.S., people who are frequently linked to transnational family networks.

While there remain many unanswered questions about migrant and non-migrant older persons in Mexico and the U.S., the studies presented in this section benefit from a higher quantity and quality of cross-sectional and longitudinal survey data to study migratory patterns and regional variation in resettlement, differences in societal integration, and the connection of these patterns with health and healthcare. More recently gathered data sets offer the opportunity for comparative studies of migrants who settled more or less permanently in the U.S., migrants who returned to Mexico, as well as those residing in Mexico who never migrated, using relatively uniform information from national or regional population surveys. More broadly, this trend toward a unified focus reflects a recognition that there is a vast social and cultural system undergirding transnational migration that should be acknowledged and addressed with empirical research to more fully understand its association with health and mortality.

Only a few decades ago research on older Hispanics was considered a low yield investment due to the age structure of Latin America and of Hispanics in the U.S. These populations were younger and characterized by high fertility rates; indeed this profile is now quickly changing with declining fertility rates corresponding to women's higher educational attainment (Shorrocks et al. 2013). With an increased interest on life course aging and the rapid increase in the proportions and numbers of aging people across the Americas, as suggested in the chapter by Sáenz, both scientific interest and public policy implications have engendered wider support for research. Sáenz presents a demographic and socioeconomic profile of older individuals in the U.S. and Mexico as well as projections of future trends. His analysis shows that the growth of both the U.S. Mexican-origin older population and the older population in Mexico is much faster than the growth of the U.S. non-Hispanic white and African American older populations. This change in the racial and ethnic composition of older populations of the U.S. is important given that racial and ethnic minorities have fewer economic resources and suffer from more disabilities than non-Hispanic whites. The U.S. Mexican-origin population without U.S. citizenship or permanent residency is particularly vulnerable as they are more likely to lack healthcare coverage or to qualify for other social-support programs. In Mexico, a major issue facing the aging population with marginal assets is a weak organizational structure to deliver needed services to older persons.

The chapter by Mejia, Ham-Chande, and Coubes considers the migration flow of older persons from Mexico to the U.S. and reports a significant increase of aging people migrating north in recent years. Many study respondents expressed the intention to spend an extended period of time residing in the U.S. This is counterintuitive to the broader reduction of border crossings widely reported over the same time frame. The reasons are not confirmable at this point but the possibilities include decreased security in Mexico, the inadequacy of personal supports in Mexico, or inability of adult migrant children, especially if undocumented in the U.S., to return to Mexico to visit and care for older parents. The latter would represent an unintended consequence of the increased immigration-related security-enforcement efforts. On the other hand, it is notable that many older Mexicans are increasingly legally entering the U.S., suggesting greater flexibility for family engagement and a higher quality of life rather than solely economic necessity as the central driver for migration.

Today the topics of inequality and opportunity structures, housing, stigma, community safety and strengthening, education, and access to social and health services are foremost on the public agenda of the Americas, and have a direct bearing on the expanding dependency of aging and accommodation of a growing demand for public support. The political implication is clear: cost and state exposure to future liability. The asset base for the bottom wealth quartile of the aging population is marginal in the U.S. and Mexico (Shorrocks et al. 2013). In the U.S. the asset base for Hispanics over 65 is much lower than for non-Hispanic Whites (Gassoumis 2012; Mudrazija and Angel 2014), suggesting high economic vulnerability of Hispanics. Combined with inadequate public pension support levels in the U.S. and Mexico and a lifetime of inadequate access to preventive or other health services, life over 65 is too often accompanied by high levels of disease burden and undertreatment by current care systems in both nations (González-González et al. 2014; González et al. 2009).

In the chapter by Ward, we learn about housing issues facing low-income aging Hispanics living in *colonias* in Texas. While offering them an opportunity to own a home and have flexible housing arrangements, poor housing conditions in *colonias* may be an important contributing factor to poor health outcomes in this population. Ward documents high vulnerability of older residents in two *colonias* in Central Texas to chronic illnesses including diabetes as well as cardiovascular problems and respiratory ailments. Inadequate housing conditions, especially the lack or suboptimal functioning of heating and cooling systems, can substantially exacerbate these conditions. Policy intervention to improve the housing conditions of poor older individuals living in *colonias* may have substantial positive impact on their health outcomes.

The chapter by Guerrero, Khachikian, Kong, and Vega describes growing need for effective substance abuse prevention and treatment programs targeting older Hispanics. This population has experienced substantial growth in alcohol and drug use, and by 2020 the rate of substance use disorders for Hispanics age 50 and over will double from its 2006 level. The implications of these trends for their long-term health and well-being are profound. Major challenges in the treatment of this population include linguistic and cultural barriers that make Hispanics an especially hard-to-reach population. Based on the information from Southern California, Guerrero and colleagues find that older Hispanics in substance abuse treatments have both higher substance use severity and higher use of opiates than the general Hispanic population. Older Hispanic's sobriety is positively affected by retention in treatment, and negatively with counselors' Spanish language proficiency. While the latter finding appears counterintuitive, it might be reflecting the underlying lack of treatment resources in smaller community clinics rather than insufficient provision of care.

Migration is an important factor to consider in examining the care issue among older Hispanics. History has demonstrated that migrants are flexible and adaptive, and they may move between the U.S. and Mexico as their personal preferences and physical requirements change, especially when relatives alter their care arrangements. The options available are moderated by availability of family members to offer specific forms of assistance. While a very high proportion of low income older Hispanics in the U.S. and Mexico live in multigenerational families, these families often have very limited space in their residential dwellings and are struggling to manage household expenses, as described by Ward. This situation, combined with the "weathering" effects of caregiving, can create strains and conflicts for older persons and for their family members (Pinquart and Sorensen 2005). Women in the Americas are the historic caregivers yet they are entering the workforce at unprecedented rates, and their availability for caregiving is decreasing as a consequence. These situations can result in shifts of time and resource contributions including transferring caregiving responsibilities to other family members and family units, and possibly to grandchildren, and may require elders to change residence, including the possibility of transnational relocation.

These family care burdens will increase with longer life expectancies, better medical management and reduced mortality from chronic diseases, and higher rates of dementia in later life. This scenario, which is already in motion, sets the stage for increasingly adverse conditions for older persons with activity limitations and requiring home assistance. These critical issues for older Hispanic populations have remained understudied, in part because reciprocal (time-ordered) processes of aging—health trajectories and caregiving are difficult to estimate and accurately interpret. We need more and better information on how families make decisions about support and health care for older family members. This knowledge is useful for aligning healthcare and social services to optimize quality of life of older persons and reduce the substantial costs currently borne by low-resource family support systems.

With this in mind, the chapter by Montes-de-Oca, Ramírez, Santillanes, García, and Sáenz represents an important contribution in understanding the obstacles in access to health care faced by frail older Mexican-origin immigrants in the U.S. and the strategies deployed by their families to support them. Data suggest that among Mexico-born population about 25% lacks any health insurance coverage, with the proportion particularly high (around 40%) for non-U.S. citizens, recent immigrants, and individuals ages 60–64. In this context, living in extended households is often used as a strategy to provide for older family members in need who cannot rely on formal health care systems. In fact, Montes-de-Oca and colleagues find 44% of all older Mexican-origin immigrants in the U.S. living in extended family households, with the proportion even higher for individuals without health insurance and non-U.S. citizens.

The chapter by Quashie provides information on another important aspect of family support: provision of financial and non-financial support from adult non-coresident children to older parents. Using data from Mexico City, Quashie explores patterns of financial and non-financial transfers from adult children to their older parents with particular attention to the geographic location of children and their siblings and the implications this has for parental care provision. The results suggest that non-coresident children are more likely to provide support, in particular financial support, to their older parents if they have a sibling living with parents who presumably can provide them with support with activities of daily living. This finding is consistent with children coordinating their support efforts according to their comparative advantage. Single children are particularly likely to provide all types of support to parents.

This section contains information about a wide range of issues affecting older Hispanics and their daily life requirements. We have entered an era of astonishing similarities in demand for new aging services confronting nations across the Americas. Despite substantial differences in historical development, political systems across the Americas are facing major challenges in restructuring and expansion of inadequate systems of education, medical care and social care, and design or redesign of dependency support programs for both children and older persons. The economic scope of this revolution in systems of personal support and health is massive and it will take multiple decades to implement the needed changes. But one fact is indisputable, there will be many more millions of older people in the years ahead that are socially and economically marginal and need basic housing, nutrition, and health resources for their families and for themselves. For researchers this situation provides an unprecedented opportunity to produce badly needed information that is useful for public and health policy formulation—a point further developed in the final section of this volume. It will be essential to train new and more investigators in global research in key areas identified throughout the volume.

References

Gassoumis, Z. D. (2012). *The recession's impact on racial and ethnic minority elders: Wealth loss differences by age, race and ethnicity*. Los Angeles: USC Edward R. Roybal Institute on Aging.

González, H. M., Ceballos, M., West, B. T., Bowen, M. E., Tarraf, W., & Vega, W. A. (2009). The health of older Mexican Americans in the long run. *American Journal of Public Health, 99*(10), 1879–1885.

González-González, C., Samper-Ternent, R., Wong, R., & Palloni, A. (2014). Mortality inequality among older adults in Mexico: The combined role of infectious and chronic diseases. *Revista Panamericana de Salud Pública, 35*(2), 89–95.

International Migration Institute. (2011). Global 'Megatrends' for Future International Migration (IMI Policy Brief 11-9). http://www.imi.ox.ac.uk/pdfs/projects/gmf-pdfs/pb-11-9-global-megatrends-for-future-international-migration/view. Accessed June 2, 2014.

Mudrazija, S., & Angel, J. L. (2014). Economic security of older Americans: The implications of growing minority elderly population. In: Hudson, R. B. (ed.), *The new politics of old age policy* (3rd ed.). Baltimore: The Johns Hopkins University Press.

Pinquart, M., Sorensen, S. (2005). Ethnic differences in stressors, resources, and psychological outcomes of family caregiving: A meta-analysis. *The Gerontologist, 45*(1), 90–106.

Portes, A., & Rumbaut, R. G. (2006). *Immigrant America: A portrait* (3rd ed.). London: University of California Press.

Shorrocks, A., Davies, J. B., & Lluberas, R. C. S. (2013). *Credit suisse global wealth databook report* (pp. 79–102). Zurich: Credit Suisse Group AG.

United Nations, Department of Economic and Social Affairs, Population Division. (2009). International migration report 2006: A global assessment. http://www.un.org/esa/population/publications/2006_MigrationRep/report.htm.

Zhou, M. (1997). Segmented assimilation: Issues, controversies, and recent research on the new second generation. *International Migration Review, 31*(4), 975–1008.

Chapter 12
The Demography of the Elderly in the Americas: The Case of the United States and Mexico

Rogelio Sáenz

The Demography of the Elderly in the Americas: The Case of the United States and Mexico

Many countries around the world are experiencing a dramatic aging of their populations due to major demographic shifts (Fishman 2010). As countries traverse through the demographic transition, they are characterized by stable growth (or decline) and older populations. Thus, while the elderly comprise 17% of the population of developed countries, they only make up 6% of the population of developing nations (Population Reference Bureau 2013). The elderly already account for one-fourth of Japan's population and slightly more than one-fifth of those of German and Italy. Japan illustrates well the demographic forces that lead to massive aging of the population. Japan's total fertility rate dropped significantly from 4.5 in 1947 to 1.4 in 2012, while the life expectancies of men and women soared from 50 in 1947 to 79 in 2012 among men and from 54 to 86 among women (Aoki 2012; Population Reference Bureau 2013). Other countries around the world will experience accelerated rates of aging in the coming decades due to historical as well as contemporary demographic trends associated with fertility, mortality, and migration (Khoo 2012; McNicoll 2012).

The United States and Mexico represent two countries that currently are at different stages in the aging transition, though both will experience a significant graying of their populations over the coming decades. For example, the U.S. elderly population will expand significantly during the 2011–2029 period as baby-boomers reach age 65. Mexico, too, will experience a significant aging of its population due to major declines in fertility alongside outmigration of its workforce over the last half century. While much research examines the aging in these two countries separately, it is important to shift our analytical lens to capture the binational context in which aging occurs in the United States and Mexico. International migration

R. Sáenz (✉)
University of Texas at San Antonio, San Antonio, TX, USA
e-mail: rogelio.saenz@utsa.edu

© Springer International Publishing Switzerland 2015
W. A. Vega et al. (eds.), *Challenges of Latino Aging in the Americas,*
DOI 10.1007/978-3-319-12598-5_12

involving the movement of people from Mexico to the United States, in particular, has played an important role in the structure and the velocity in which aging is actualized in both countries. One of the unique features of Mexican immigration to the United States is that it has been relatively constant for an extended period of time extending back to the early twentieth century but particularly over the last three decades of the century (Massey and Pren 2012), although the volume of movement from Mexico to the United States is declining over the last several years (Passel et al. 2012, 2013).

As such, among Mexican-origin elderly in the United States, some were born in this country, while others have lived most of their lives in the United States, and still others have resided in this nation for a shorter period of time. It is also important to understand that there is great diversity in the elderly population and their socioeconomic standing, especially in the United States. Thus, it is not appropriate to view the elderly as a homogeneous population. Indeed, their cumulative advantages/disadvantages, associated with race and ethnicity, structure the health and socioeconomic well-being of people during their elderly period. It is important, then, to identify particular segments of the elderly that are particularly disadvantaged to call attention to their plight.

Moreover, for the most part until recently, Mexican immigration involved relatively easy transnational movement with Mexican immigrants keeping a foot in each country. However, in the last couple of decades the transnational movement of Mexican immigrants has been stemmed by the criminalization of immigrants, the post-9/11 militarization of the border, and the enhancement of the detention and deportation of immigrants (Douglas and Sáenz 2013). These obstacles have complicated and muddled the binational context in which aging takes places in the face of the splitting of families between Mexico and the United States.

Hence, given these barriers in caring for elderly and the diverse nature of the elderly in the United States and Mexico, there is a need for a comprehensive examination of the elderly population. To this end, the analysis undertaken here provides a profile of the demographic trends and socioeconomic standing of four groups of elderly living in the United States and Mexico: U.S. white elderly, U.S. black elderly, U.S. Mexican-origin elderly, and Mexico's elderly. The analysis will also identify factors that stratify Mexican elderly in the United States as well as in Mexico, resulting in certain segments of Mexican elderly in both countries facing particularly challenging social and economic conditions during their senescence period of life. The demographic and socioeconomic profile of diverse groups of elderly in a binational context will help us identify important public policy issues that need to be addressed in order to provide for the needs of an elderly population that will increasingly be Mexican, that will increasingly have major socioeconomic limitations, and that will increasingly undergo aging across two countries due to a variety of demographic changes and immigration realities. These public policy needs will be addressed at the conclusion of the paper.

Before turning to the analysis, however, we provide a framework for understanding how the aging transition arises.

The Demographic Transition

The demographic transition model conceptualized originally by Warren S. Thompson (1929) nearly a century ago and developed further by Notestein (1945) continues to be a useful perspective for understanding the timing of demographic shifts brought about by alterations in fertility and mortality rates (Stokes and Preston 2012). The model describes how countries traverse the different stages comprising the demographic transition from stage 1 characterized by high rates of fertility and morality, population stability, and a young age structure to stage 3 depicted by low rates of fertility and mortality, population stability and eventual population decline, and an older age structure (Diggs 2008). As originally conceived, the timing at which countries started the demographic transition depended on their level of socioeconomic development. As such, countries in the most socioeconomically developed region of the world, northern and western Europe, were the first to experience significant population growth due to declining mortality rates alongside high fertility rates in the late eighteenth and nineteenth centuries (Coale 1972). In contrast, most African countries have undergone this demographic transition only in the last several decades.

Despite its usefulness as a model for understanding demographic changes across countries, the demographic transition theory has been criticized on various fronts (Diggs 2008; Friedlander et al. 1999; Kammeyer 1970; Teitelbaum 1975). For example, it has been criticized for the failure to account for the exact role of fertility and mortality in producing population change, being ethnocentric with the model based on the experience of European countries used to understand demographic changes in vastly different countries of the world, the failure to incorporate the role of international migration in the model, the role that technology has played in the rapidity at which demographic shifts occur today, and so forth.

Nonetheless, one of the most interesting newer formulations of the demographic transition theory concerns the subsequent transitions that take place as countries pass through the different stages of the demographic transition. Thus, the traditional demographic transition associated with declining mortality followed by declining fertility spawns ensuing demographic transitions (Weeks 2012). These include five such transitions. First, the urban transition occurs as population pressures and limited land and economic opportunities in rural areas propel people to migrate to urban areas. Second, the marriage transition materializes as educational opportunities and shifting attitudes result in people delaying or foregoing marriage, entering into alternative forms of marriage such as cohabitation, and increasingly ending marriage through divorce (Hiu and Zhang 2013; Lin and Brown 2012). Third, the household transition arises as lower fertility, increasing life expectancy, and marriage shifts lead to smaller households, a rise in single-person households, and an increase in female-headed households (Hiu and Zhang 2013). Fourth, the aging transition emerges as the combination of low fertility and high levels of life expectancy bring about a greater share of elderly among the population. Fifth, the race/ethnic transition surfaces from two sources: externally, as people migrate from

Table 12.1 Demographic characteristics of Mexico, the United States, and three U.S. racial/ethnic groups, 2010–2012[a]

Demographic characteristics	Mexico	U.S.	U.S.		
			White	Black	Mexican
Crude birth rate	19[a]	13[b]	11[c]	15[c]	18[c]
Crude death rate	4[a]	8[b]	10[d]	7[d]	4[d]
Rate of natural increase	1.5[a]	0.5[b]	0.1	0.8	1.4
Total fertility rate	2.2[a]	1.9[b]	1.8[c]	2.0[c]	2.3[c]
Life expectancy					
Total	77[a]	79[b]	79[d]	75[d]	81[d]
Male	75[a]	77[b]	76[d]	71[d]	79[d]
Female	79[a]	81[b]	81[d]	78[d]	84[d]
Pct. population					
Less than 15	30[a]	19[b]	16[e]	22[e]	31[e]
65 and older	6[a]	14[b]	16[e]	9[e]	5[e]

[a]The data for Mexico and the United States are for 2012; the data for the U.S. white, black, and Mexican populations are for 2010
[b]Population Reference Bureau (2013)
[c]Martin et al. (2012)
[d]Murphy et al. (2013). Note also that the data for Mexicans are based on all Latinos
[e]2010 American Community Survey Public-Use File (Ruggles et al. 2010)

developing countries to developed countries; and internally, as the majority white population due to its superior socioeconomic resources goes through the transition before minority populations with the result being an aging white population and a more youthful and increasing minority population (Dowd and Bengston 1978; Khoo 2012). Coleman (2006) refers to the race/ethnic transition as the third demographic transition.

The United States and Mexico differ on their stage in the demographic transition. This is also the case among the largest three racial and ethnic groups in the United States. Table 12.1 provides a demographic profile to assess the stage in the demographic transition and the current relative size of the elderly population in Mexico, the United States, and the three racial/ethnic groups in the United States. The United States and particularly the U.S. white population are more advanced in the demographic transition in various respects such as their slow rate of natural increase, low fertility rate, and a high percentage of elderly. Mexico and the U.S. Mexican population are in less advanced stages of the demographic transition as evidenced by a faster rate of natural increase, relatively higher fertility, and a larger share of younger inhabitants. The U.S. black population falls between these groups with respect to its position in the demographic transition.

In certain respects, however, there are a couple of areas in which differences have narrowed significantly across countries and groups. For example, the total fertility rate between Mexico and the United States has narrowed greatly, with the U.S. TFR being only 0.3 lower. There has also been a major reduction in the fertility gap between U.S. Mexicans and U.S. whites (Martin et al. 2012). In addition, life expectancy between Mexico and the United States has narrowed markedly as well, with the U.S. life expectancy being only 2 years higher than that of Mexico. Yet, what is particularly noteworthy is that the life expectancy across all groups is highest for U.S. Latinos (a group that consists of all persons identifying as Latinos or Hispanics, a category in which approximately 63 % are of Mexican origin) (Arias 2012). Indeed, U.S. Latino newborn males are expected, on average, to outlive U.S. white newborn males by 2 years while U.S. Latina newborn females are expected to outlive U.S. white newborn females by 3 years. This pattern is associated with the Latino paradox (also known as the epidemiological paradox, Mexican paradox, and Mexican immigrant paradox) (Markides and Coreil 1986; Sáenz and Morales 2012).

The Latino paradox relates to the low mortality of Latinos despite their low socioeconomic status. The Latino population as a group is characterized by low levels of education, low levels of income, high levels of poverty, high prevalence of the lack of health insurance coverage, and a relatively high percentage of workers employed in physically demanding and dangerous jobs. Yet, Latinos have lower mortality rates and higher life expectancies than do whites, a population that has comparatively higher socioeconomic levels. Various hypotheses have been proposed to explain the paradox (Palloni and Arias 2004; Sáenz and Morales 2012). For example, it has been suggested that Mexican immigrants are selected from the healthiest segment of Mexico's population, with the result being that they also have favorable health and mortality outcomes in the United States. In addition, it has been argued that Mexican cultural factors (e.g., familism, social networks, diet, lower prevalence of smoking and drinking, etc.) lead to more favorable health and mortality outcomes among Mexican immigrants. Indeed, the longer Mexican immigrants are in the United States, the more likely that their health is to deteriorate and their mortality risk is to ascend. Finally, it has been suggested that statistical and measurement shortcomings tend to underestimate the level of mortality—e.g., inconsistencies in the way that people are classified racially/ethnically in census records and death records (Rosenberg et al. 1999), as well as the possibility that Mexican immigrants who become seriously ill may return to Mexico, where if they die their death is recorded in Mexico rather than the United States, a phenomenon commonly referred to as the "salmon bias." There is a growing volume of research suggesting that the Mexican paradox is real rather than due to a statistical or methodological artifact (Elo et al. 2004; Hummer et al. 2007).

The ensuing demographic transitions stemming from reductions in fertility and increasing longevity have altered the characteristics of the elderly population. Indeed, as individuals live longer lives, they are more likely to do so with health and physical disabilities along with a high prevalence of chronic diseases (Guralnik et al. 1996). Moreover, the family resources that elderly hold are increasingly limited due to marital disruption (e.g., divorce, widowhood, etc.) (Franks et al. 2004; Goldman et al. 1995; Grundy 2001; Hays 2002; Manzoli et al. 2007; Peters and Liefbroer 1997), alternate partnership arrangements (cohabitation) (Brown et al. 2006; Koskinen et al. 2007), along with the increasing geographic mobility of offspring which tend to sever social ties and networks that become important at the older stages of life. Immigration between the United States and Mexico alongside the militarization of the border have significantly torn familial networks that become increasingly important in the context of aging in a binational setting.

Furthermore, as the elderly population is increasingly comprised of women and minority group members, a greater share of aged individuals have socioeconomic limitations and health concerns that have accumulated over the life course due to gender and, particularly, racial and ethnic disparities (Geronimus et al. 2007). The double jeopardy hypothesis argues that people who are racial or ethnic minorities or immigrants are at risk of less favorable aging due to inequities associated with their racial/ethnic minority status or immigrant status (Bird and Rieker 1999; Dowd and Bengston 1978; House et al. 2005; Khoo 2012; Seeman and Crimmins 2001; Williams 1999; Williams and Jackson 2005). Thus, elderly who simultaneously hold multiple minority statuses are especially vulnerable to less favorable aging experiences. For example, in the United States, elderly Mexican immigrants who are not naturalized citizens are likely to have much more limited social and economic resources to deal with the challenges of the older years compared to their counterparts with fewer minority statuses. Massey and Pren (2012) have illustrated the major economic, social, and political difficulties that undocumented Mexicans have faced, especially since 9/11. In fact, Massey and Pren (2012) suggest that under such conditions we are seeing the making of undocumented Mexicans into an underclass. Massey and Pren (2012) note:

> In the absence of U.S. policies, the only possible outcome for the United States is the creation of a large underclass that is permanently divorced from American society and disenfranchised from its resources, with little hope of upward mobility. Evidence suggests the process of underclass creation is already underway. Levels of Hispanic residential isolation are increasing…; Mexican wages have stagnated and fallen behind those of non-Hispanic whites and even blacks…; Hispanic household wealth fell by 66% from 2005 to 2009 …; and Latino poverty rates have risen steadily to equal those of African Americans…. On virtually every measure of socioeconomic well-being, Hispanics in general and Mexicans in particular, have fallen from their historical in the middle of the American socioeconomic distribution … to a new position at or near the very bottom…. With huge fractions of Latinos lying outside the protections of the law and even larger shares related to people who lack legal protections, and with most rights stripped away from all non-citizen foreigners, the Hispanic population has never been more vulnerable and its position in America more precarious (p. 15).

Historical Context of the U.S. and Mexico's Elderly

In order to understand such inequities, it is important to examine the temporal and social context in which the elderly in the United States and Mexico have existed. The large majority of elderly in the analysis presented below was born between 1926 and 1945 (88% of U.S. elderly and 89% of Mexico's elderly). Elderly born in the United States include the depression cohort and the pre-Baby Boom cohort born between the late 1920s and early 1940s, a relatively small cohort given the low fertility rates and the postponement of marriage at that time. It has been shown that this group experienced favorable economic conditions when they entered the work force due to their small numbers, the relative absence of immigrant labor as competitors, and the favorable U.S. economy following the end of WWII (Carlson 2008; Easterlin 1980). However, the favorable economic conditions were not evenly distributed, as it was whites—rather than Latinos and African Americans—who experienced upward social and economic mobility (Gassoumis et al. 2010; Telles and Ortiz 2008).

Many U.S.-born Mexican American and Mexico-born elderly were also impacted by two significant events—the Mexican Revolution and the Great Depression. The Mexican Revolution which took place between 1910 and the early 1920s took a severe toll on the Mexican population with countless numbers of Mexicans fleeing to the United States during and after the revolution. As an indicator of this ascension in Mexican immigration to the United States, the number of Mexicans obtaining legal permanent residence rose steadily from 31,188 in 1900–1909 (before the revolution) to 185,334 in 1910–1919 (during the better part of the revolution) to 498,945 in 1920–1929 (during the end of the revolution and the post-revolution period) (U.S. Department of Homeland Security 2013). Even Mexican Americans born in the United States were significantly impacted by the Mexican Revolution as many of their families came to this country during this period and retained ties to Mexico. In addition, during the depression, the infamous Repatriation Program deported to Mexico approximately 500,000—one-third of the Mexican population enumerated in the 1930 U.S. Census—Mexican immigrants living in the United States alongside persons of Mexican origin born in this country (Balderrama 1995; Guerin-Gonzales 1994; Hoffman 1974; Sáenz and Murga 2011).

The elderly included in the analysis below reached adulthood primarily between 1946 and 1965. As noted above, this was a favorable economic period in the United States as the country's position in the world stage rose significantly following the end of WWII, although as noted above whites were much more likely to benefit socioeconomically during this time compared to Latinos and blacks (Gassoumis et al. 2010; Telles and Ortiz 2008). In addition, the shortage of labor during WWII resulted in the establishment of the Bracero Program, which brought contract laborers to the United States for specified periods of time. Due to its great popularity among U.S. employers, the program was extended to 1964, way past the conclusion of WWII. Approximately 4.7 million Mexican men came as braceros during this period with many others coming as undocumented workers (Ueda 1994). Mexico was undergoing major population growth due to declining mortality alongside high levels of fertility, creating the need for immigration as a safety valve. The total fertility rate of Mexico stood at 7.3 in 1960 (Frank and Heuveline 2005).

Two other important policy events took place in the United States in the mid-1960s which addressed racial inequality and immigration in the country during a period when many of today's elderly were in the labor force. The first of these, the Civil Rights Act of 1964, outlawed discrimination on the basis of race, ethnicity, national origins, religion, and gender. The second policy, the Immigration Act of 1965 (Hart-Celler Act of 1965) banned the immigration quota system, set immigration ceilings for all countries, and established a preference system for immigration, including a provision for family reunification. This immigration policy significantly changed the face of immigration in the United States, as the source of immigration turned from Europeans being the major group of immigrants to the United States to Latin Americans and Asians, with Mexicans becoming the largest source of immigration after the passage of the policy (Saenz et al. 2004).

Racial and Ethnic Inequality

These significant events—the Mexican Revolution, the Great Depression, the Repatriation Program, WWII, the Bracero Program, the Civil Rights Act of 1964, and the Immigration Act of 1965—had varying levels of impact on today's elderly in Mexico and the United States. Certainly, the experiences of the elderly diverge significantly on the basis of their race and ethnicity as well as national origin. For example, while the mass media and social scientists laud the favorable post-WWII economy that Americans enjoyed which saw massive increases in higher education and homeownership through the G.I. Bill along with the large-scale movement to suburban areas, this for the most part reflected the experiences of whites. Indeed, people of color, as a whole, did not enjoy this prosperity (Gassoumis et al. 2010; Telles and Ortiz 2008). Many Mexican American and African American soldiers who fought valiantly in WWII returned after the war to their hometowns which remained unchanged, places where they were segregated and excluded from the opportunity structure. For example, in 1949 a local mortuary in Three Rivers, Texas, denied service to the body of Felix Longoria, a Mexican American soldier killed in the Philippines, because "whites would not like it" (Carroll 2003). In addition, Ira Katznelson (2005) in his book titled *When Affirmative Action Was White: An Untold History of Racial Inequality in Twentieth-Century America* demonstrates that whites were the disproportionate beneficiaries of the G.I. Bill which opened vast opportunities for whites but not for minorities, especially African Americans. As has been suggested, racism, as evidenced by racial inequality and discrimination, has an impact on the lives of persons of color through blocked access to valuable societal resources (e.g., education, housing, healthcare, job opportunities, etc.) and health problems associated with direct experience with racism and discrimination (Williams 1999; Williams and Jackson 2005).

Moreover, on a global scale, there are major distinctions between the United States and Mexico with respect to socioeconomic levels of the population. As such, throughout the life course, Mexicans living in Mexico and Mexican immigrants in the United States who grew up in Mexico faced less favorable access to healthcare and more limited socioeconomic conditions compared to their U.S. counterparts, especially the more advantaged U.S. white population. In addition, within Mexico as well as other parts of Latin America, there are significant disparities not only along social class but also race. In particular, the indigenous population of Mexico has had comparatively more limited social and economic resources as well as access to healthcare compared to the majority mestizo population (Telles and Bailey 2013; Van Cott 2007). Thus, as was the case in the United States involving African Americans and Mexican Americans, indigenous elderly in Mexico grew up in more vulnerable conditions compared to the majority group.

While we call attention to indigenous elderly, a group that has been largely neglected in the literature, we do not imply that the minority experience of the indigenous in Mexico is comparable to that of U.S. minorities, such as Latinos and African Americans. Indeed, the indigenous population of Mexico has deep historical roots in the country, as the original inhabitants of Mexico when the Spaniards arrived and colonized the country. The indigenous represent the segment of people of Mexico that did not disappear as a result of *mestizaje*—the offspring of Spanish men and indigenous women. Moreover, while Latinos and, particularly, African Americans are physically segregated from whites in the United States (Britton and Goldsmith 2013; Lichter et al. 2012; Park and Iceland 2011; Perez 2012), the level of segregation between the indigenous and mestizo population is quite distinct in Mexico, involving the continued concentration of the indigenous in certain parts of the country, although as indigenous people have migrated to urban centers, they are situated afar from mestizos (Acharya and Codina 2012). Moreover, on a comparative basis, the social, economic, and political gaps between Mexico's indigenous and mestizo population are greater than those between U.S. minority groups and the white population (see Gall 2004; Heaton et al. 2007; Martínez Nova 2003; París Pombo 2002; Vargas Becerra and Flores Dávila 2002). Moreover, the roots of racism against the indigenous in Mexico tend to be viewed as a problem of ignorance, intolerance, or modernization (Gall 2004; París Pombo 2002; Vargas Becerra and Flores Dávila 2002), as opposed to more structural explanations seen in the United States (Bonilla-Silva 2013).

The life experienced in the elderly years is related to the types of advantages and disadvantages that people possessed throughout the earlier stages of the life course (Crystal and Shea 1990; Dannefer 2003; Ross and Wu 1996). One of the primary indicators of one's standing in the socioeconomic ladder is educational attainment (Ross and Wu 1996), which tends to be associated with parental educational attainment, e.g., the reproduction of educational inequality (Saenz et al. 2007; Saenz and Siordia 2012). It has been observed that more educated individuals tend to have more social and economic resources at their disposal (e.g., access to healthcare, nutritious diet, exercise, and lower levels of smoking) which result in more favorable health outcomes compared to people with lower levels of schooling. Over the life course, health disparities along levels of education tend to expand as people who are more advantaged socioeconomically fare progressively better in their health compared to their less socioeconomically endowed peers (Ross and Wu 1996), illustrative of the adages indicating that "success breeds success" (Huber 1998), "the rich get richer," and "the poor get poorer" (Dannefer 2003; Entwisle et al. 2001).

The analysis conducted below examines demographic, socioeconomic, and health characteristics of three elderly groups in the United States (whites, blacks, and Mexicans) and the elderly in Mexico. Furthermore, we examine the diversity that exists within the population of each country by examining the demographic and socioeconomic attributes of Mexicans in the United States on the basis of nativity and U.S. citizenship and in Mexico based on indigenous status.

Methods

Data from the 2010 American Community Survey (ACS) Public-Use File and the 2010 Mexico Population and Housing Census Public-Use File are used to conduct the analysis. The data were obtained from the IPUMS-USA (see Ruggles et al. 2010) and IPUMS-International (see Minnesota Population Center 2013) websites, respectively. The sample used in the analysis includes persons 65 years of age and older for three groups in the United States (non-Hispanic whites, non-Hispanic blacks, and Hispanics of Mexican origin) and all persons 65 and older in Mexico. The unweighted U.S. sample contains 455,592 persons (403,183 non-Hispanic whites, 36,613 non-Hispanic blacks, and 15,796 Hispanics of Mexican origin) and the unweighted Mexico sample contains 858,595 individuals. The sample data are weighted to obtain population estimates for the analysis conducted below.

The initial part of the analysis focuses on the comparison of the four primary groups of interest (U.S. whites, blacks, and Mexicans, and Mexicans in Mexico) across the broad dimensions: demographic patterns, socioeconomic status, and health. Subsequently, the analysis examines internal variations among the U.S. Mexican population along the lines of nativity and citizenship status (U.S.-born Mexicans, Mexican immigrants who are U.S. naturalized citizens, and Mexican immigrants who are not naturalized citizens) as well as Mexicans in Mexico along the lines of race (indigenous population versus non-indigenous population). Note that the group of U.S. Mexicans who are not naturalized citizens is a broad group consisting of people who are immigrants holding permanent legal residence and undocumented immigrant, with the former eligible for governmental health care programs but not the latter. Finally, the last part of the analysis examines population projections of the elderly population in the United States (U.S. Census Bureau 2012) and Mexico (U.S. Census Bureau 2003). All of the data analysis undertaken below is descriptive in nature.

Results

Demographic Patterns

We begin with an overview of the demographic patterns associated with the four groups of primary interest. There are significant variations across the groups with respect to the relative size of the elderly population. As noted earlier, the U.S. white population is the oldest with approximately one-sixth of the population being 65 and older in 2010 (Fig. 12.1). On the other hand, the U.S. Mexican population and

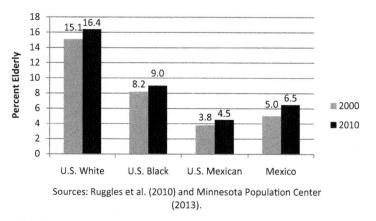

Sources: Ruggles et al. (2010) and Minnesota Population Center
(2013).

Fig. 12.1 Elderly as a percentage of the total population for selected groups, 2000 and 2010

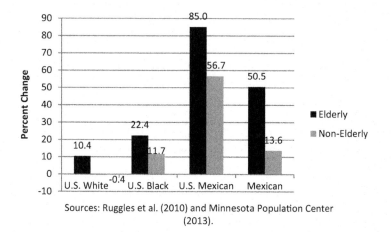

Sources: Ruggles et al. (2010) and Minnesota Population Center
(2013).

Fig. 12.2 Percentage change in elderly population and non-elderly population for selected groups,
2000–2010

the Mexico population are the youngest with the elderly constituting one of every
22 persons in the U.S. Mexican population and one of every 15 in Mexico's population in 2010. However, the percentage of the elderly population is rising the most
rapidly in Mexico's population with the percentage increase being 30 % (increase
of 5.0 to 6.5 %) followed by the U.S. Mexican population with the percentage rising
by 18 % (from 3.8 to 4.5 %).

Similarly, we can examine the growth of the elderly across the four groups from
another angle. Figure 12.2 provides the percentage change in the numbers of elderly
(age 65 and older) population and the non-elderly (ages 0–64) population between
2000 and 2010. Across the four groups, the elderly population grew significantly
faster than the non-elderly population. The growth in the elderly population was
the most prominent among the U.S. Mexican elderly population which posted an
increment of 85 % followed by an increase of nearly 51 % among Mexico's elderly

population. In contrast, the growth of the elderly was much slower among blacks (22.4%) and, particularly, whites (10.4%).

The U.S. Mexican elderly growth is already making a mark on the nation's minority elderly population. For example, even though the black population was 3.5 times larger than the U.S. Mexican elderly population in 2000, the absolute change of 680,101 in the U.S. Mexican elderly population between 2000 and 2010 was larger than that of black elderly (622,290). As we will see below, the two groups of Mexican elderly (in the United States and in Mexico) are projected to grow more rapidly than those of the U.S. white and black populations in the coming decades.

Socioeconomic Characteristics

We now examine four socioeconomic indicators (education, work, median income, and poverty) among the four groups. In some cases, however, it is difficult to compare Mexico's elderly population with the three U.S. elderly populations due to variations in the measurement of particular indicators or to the lack of such information.

Table 12.2 presents the socioeconomic characteristics across the four groups of interest. Among the three groups of U.S. elderly, there is a major gap in educational

Table 12.2 Socioeconomic characteristics for selected groups of elderly, 2010. (Sources: 2010 American Community Survey Public-Use File (Ruggles et al. 2010) and 2010 Mexico Population and Housing Census Public-Use File (Minnesota Population Center 2013))

Socioeconomic characteristics	U.S.			Mexico
	White	Black	Mexican	
Pct. by educational attainment				
0–8 years/less than primary[a]	7.1	17.5	51.3	64.9
9–11 years/primary[a]	10.0	20.0	12.4	25.7
High school graduate/secondary[a]	36.4	30.9	18.8	5.5
Some college/university[a]	23.6	19.3	11.9	3.9
College graduate	22.9	12.3	5.7	–
Pct. employed				
Male	19.2	16.8	18.6	40.7
Female	11.7	11.3	9.1	10.9
Median personal income				
Male	28,800	18,000	14,400	–
Female	15,100	12,100	8.800	–
Pct. poverty/below 125% poverty line				
Male	8.5	21.6	24.9	–
Female	14.1	30.1	28.8	–

[a]The educational attainment level presented first corresponds to the U.S. groups and the one presented second corresponds to Mexico

attainment across the groups. In particular, white elderly have the highest level of education with more than four-fifths (82.9%) holding at least a high school diploma; close to one-fourth of white elderly are college graduates. In contrast, the U.S. Mexican population has the lowest level of education. Indeed, nearly two-thirds (63.7%) of U.S. Mexicans do not have a high school diploma; more than half (51.3%) of all Mexicans have less than 9 years of schooling. Black elderly fall between U.S. whites and Mexicans with respect to educational attainment. As in the case of U.S. Mexicans, Mexico's elderly population has a very low level of education. Nearly two-thirds (64.9%) have completed less than a primary education with an additional one-fourth (25.7%) completing a primary education. Obviously, education represents an important resource that allows people to use their advantageous educational position to gain more favorable aging experiences as the cumulative advantage/disadvantage perspective suggests (Ross and Wu 1996). Accordingly, U.S. Mexican elderly and Mexico's elderly population face significant challenges due to their low levels of education.

There are significant variations across groups with respect to the percentage of the elderly that are working. Mexico's elderly men are far more likely than any group to still be employed with approximately two-fifths (40.7%) holding a job (Table 12.2). There are relatively minor differences in the prevalence of employment among the three U.S. groups of men (ranging from 16.8% among blacks to 19.2% among whites) and among the four groups of women (ranging from 9.1% among U.S. Mexicans to 11.7% among whites).

Consistent with the major educational differences among elderly in the United States across racial and ethnic groups, there are significant differences across groups in median personal income. The median income of elderly white men is twice that of U.S. Mexican elderly men and 1.6 times higher than that of black elderly men (Table 12.2). While the median income of U.S. elderly women is much lower than that of elderly men, the median income of white elderly women ($ 15,100) surpasses that of U.S. Mexican elderly men ($ 14,400). The median income of elderly white women is 1.72 times higher than that of U.S. Mexican elderly women and 1.25 times higher than that of black elderly women. Unfortunately, we do not have comparable data to assess the income levels of Mexico's elderly.

Finally, there are also major distinctions in the level of poverty and near poverty across the three U.S. groups. Note that we pool here the poor and the "near poor" given that Social Security benefits are regularly adjusted to keep up with inflation resulting in many elderly being pulled slightly above the poverty level. The inclusion of the poor and the near-poor to assess poverty, thus, is appropriate here. Across the three racial and ethnic groups, poverty and near-poverty is higher among elderly women than elderly men (Table 12.2). Nonetheless, white men (8.5%) and white women (14.1%) have far lower poverty and near-poverty rates compared to black and Mexican men and women. Black and Mexican elderly women are the most likely to be poor or near-poor. Minority elderly men and women face significant health and related problems in their older years due to the lack of socioeconomic resources. Again, unfortunately we do not have data to assess the poverty levels of Mexican elderly.

Health Characteristics

We now compare the four groups of interest on three indicators of health: lack of health insurance coverage, mortality rates, and prevalence of disability. In the United States, while white and black elderly enjoy almost universal health coverage due to their U.S. citizenship, close to 6 % of U.S. Mexican elderly do not have such coverage (Table 12.3). However, the lack of health care coverage is particularly severe in Mexico where more than one-fourth of elderly in the country do not have insurance. Certainly, the lack of health insurance coverage increases the vulnerability of elderly.

Table 12.3 Health characteristics for selected groups of elderly, 2010

Health characteristics	U.S.			Mexico
	White	Black	Mexican	
Pct. lacking health insurance	0.3	1.4	5.7	26.7
Age-specific death rates (Per 100,000)				
Male				
65–74	2,257	3,275	1,745	3,030
75–84	5,770	6,849	4,442	6,925
85 and older	15,817	14,974	10,859	16,268
Female				
65–74	1,536	2,068	1,124	2,219
75–84	4,233	4,676	3,240	5,463
85 and older	13,544	12,768	8883	15,039
Pct. with a disability				
Male				
65–69	23.6	30.3	30.0	19.0
70–74	29.4	35.9	36.7	27.4
75–79	38.9	45.4	48.2	34.9
80–84	49.7	54.3	59.2	43.5
85 and older	68.2	70.8	73.2	56.7
Female				
65–69	20.9	34.3	30.6	20.2
70–74	27.1	41.0	37.5	28.2
75–79	36.8	50.7	51.5	37.0
80–84	50.5	63.4	63.6	45.6
85 and older	74.1	79.0	80.9	60.8

[a]2010 American Community Survey (ACS) Public-Use File (Ruggles et al. 2010) and 2010 Mexico Population and Housing Census Public-Use File (Minnesota Population Center 2013)
[b]Data for the U.S. are for 2010 (Murphy et al. 2013) and data for Mexico are derived from the 2011 Mexico life table (World Health Organization 2013)
[c]The death rates for Mexico are for the year 2011

Despite the moderately high lack of health insurance among Mexicans in the United States, U.S. Mexicans have the lowest age-specific death rates across age categories and sex groups (Table 12.3). Indeed, U.S. Mexican elderly have lower death rates than do white elderly across all age and sex groups with the gap being greatest at the 85-and-older category where the death rate of Mexican men is only 69% as high as that of white men and the death rate of Mexican women is only 66% as high as that of white women. This pattern, again, is consistent with the epidemiological paradox in which Mexicans—despite their low socioeconomic characteristics, relatively high levels of lack of health insurance, and related risk factors—tend to have more favorable mortality outcomes than whites. In contrast to the case of U.S. Mexicans, Mexico's elderly have relatively high death rates that are similar to those of U.S. black elderly. In fact, Mexico's elderly women consistently have the highest death rates among the four groups of interest across the three age groups. Moreover, at the oldest age category (85 and older) blacks have a mortality crossover relative to whites with black elderly having lower death rates at the oldest age group in contrast to the opposite patterns at the younger ages. Thus, at the 85-and-older category, the death rates of Mexico's elderly surpass those of the three U.S. groups.

However, given the variations associated with the prevalence of mortality among the four groups, there are some unexpected patterns related to disability, as measured by the presence of at least one disability. For example, while Mexico's elderly have high rates of mortality, they have the lowest prevalence of disability across all age groups (Table 12.3). However, this mortality-disability paradox likely reflects differences in the way disability questions are asked in the U.S. and Mexico censuses as well as perhaps differences in how people in each country interpret the questions or define disability. Furthermore, among the three U.S. groups, white elderly have a lower incidence of disability than Mexican and black elderly. U.S. Mexican elderly have the highest rates of disability in four of the five age groups (70–74, 75–79, 80–84, and 85 and older). It is clear, then, that U.S. Mexicans tend to have a high level of longevity but they do so with relatively high rates of disabilities.

Having examined variations in the socioeconomic and health patterns of the four groups of interest, we now turn to the examination of internal variations among U.S. Mexican elderly on the basis of nativity and U.S. citizenship status and among Mexico's elderly on the basis of indigenous status. We begin with an examination of the internal variations among U.S. Mexicans.

Variations Among U.S. Mexican Elderly

A slight majority (50.7%) of U.S. Mexicans were born in the United States, with slightly more than one-fourth (26.1%) of all U.S. Mexican elderly being immigrants who have become U.S. naturalized citizens, and less than one-fourth (23.2%) being immigrants who are not U.S. citizens (Table 12.4). Note that the latter group includes people who are living in the United States legally (e.g., permanent residents)

Table 12.4 Selected characteristics for U.S. Mexican elderly by nativity and citizenship status, 2010. (Sources: 2010 American Community Survey Public-Use File (Ruggles et al. 2010))

Selected characteristics	U.S.-born	Foreign-born	
		Naturalized citizen	Not naturalized citizen
Pct. distribution	50.7	26.1	23.2
Years since first coming to U.S.			
Less than 10 years	–	3.4	16.6
10–19 years	–	3.7	19.6
20–29 years	–	8.2	15.8
30–39 years	–	20.6	19.4
40–49 years	–	29.3	15.5
50 or more years	–	34.7	13.1
Pct. by educational attainment			
0–8 years	31.2	63.2	81.7
9–11 years	16.0	10.4	6.8
High school graduate	27.3	13.0	6.5
Some college	17.6	8.5	3.3
College graduate	7.9	7.9	1.7
Pct. working			
Male	16.8	19.7	21.2
Female	10.6	8.4	6.7
Median personal income			
Male	19,200	13,500	8,400
Female	10,800	8,600	4,800
Pct. poverty/below 125 % poverty line			
Male	18.3	24.7	38.5
Female	24.7	30.0	36.6
Pct. lacking health insurance	0.9	2.4	19.9

as well as unauthorized or undocumented individuals. As a whole, Mexican immigrants who are naturalized citizens have been in the United States for a long period of time with approximately 85 % first immigrating to the United States 30 or more years ago; more than one-third first arrived in this country 50 or more years ago. In contrast, Mexican immigrants who are not naturalized citizens have been in the United States for a shorter period of time, although they are fairly evenly distributed across the six length-of-U.S.-residence categories.

The three U.S. Mexican elderly groups vary with respect to their educational attainment levels. The large majority of both foreign-born groups have less than 9 years of schooling with those who are not naturalized citizens being particularly likely to have such low level of education (81.7 %) (Table 12.4). In contrast, U.S.-born Mexicans have relatively higher rates of educational attainment with close to 53 % holding a high school diploma. Still, U.S.-born Mexican elderly have lower educational levels compared to U.S. blacks and whites (see Table 12.2).

There are also differences across the groups with respect to the percentage of elderly who are employed. Among elderly men, foreign-born men, especially those who are not naturalized citizens (21.2%), are more likely than U.S.-born Mexicans to be employed (Table 12.4). In fact, Mexican foreign-born men in the United States are even more likely to be working than white elderly men (see Table 12.2). In contrast, among women, it is U.S.-born Mexican elderly women (10.6%) who are the most likely to be working with immigrant women who are not naturalized citizens (6.7%) being the least likely.

There are significant differences across the three categories on the basis of median personal income. In particular, U.S.-born elderly have the highest income levels, followed by immigrant elderly who are naturalized citizens, and immigrant elderly who are not naturalized citizens having the lowest income levels (Table 12.4). For each dollar earned by U.S.-born Mexican elderly men, Mexican immigrant men who are not naturalized citizens attain only 50 cents; the corresponding ratio is $ 1.00 to 58 cents among women. Note, however, that while the median income of U.S.-born Mexican elderly men is somewhat higher than that of black elderly men, the median income of U.S.-born Mexican elderly women is lower than that of their black counterparts (see Table 12.2). In addition, the patterns associated with the percentage of elderly who are poor or near-poor generally resemble the results for the median income.

Moreover, while almost all of U.S.-born Mexican elderly and Mexican elderly immigrants who are naturalized citizens have health care insurance, one-fifth of Mexican elderly immigrants who are not naturalized citizens do not have insurance. This group of Mexican elderly who lack U.S. citizenship and health insurance face particularly difficult challenges in their elderly years. While people who are not U.S. citizens but who are permanent legal residents are eligible to gain health care insurance through the Affordable Care Act, people who are undocumented are not eligible for coverage through this policy. Finally, the three groups of elderly Mexicans in the U.S. do not differ greatly on disability rates (data not shown).

Variations Among Mexico's Elderly

Historically, Mexico's indigenous population has experienced significant levels of racism, discrimination, and inequality (Acharya and Codina 2012; Martínez Nova 2003; París Pombo 2002; Telles and Bailey 2013; Van Cott 2007; Vargas Becerra and Flores Dávila 2002). In many respects, the nation's indigenous population has been neglected in public policy debates. Indigenous people account for one-sixth (16.7%) of Mexico's elderly population (Table 12.5). The analysis presented next provides an assessment of inequality among Mexico's elderly population along racial lines.

While Mexico's elderly population, across the board, has low levels of education, this is particularly true among indigenous elderly. Approximately 84% of indigenous elderly have completed less than a primary education compared to 61%

Table 12.5 Selected characteristics for Mexico's elderly by race, 2010. (Source: Mexico Population and Housing Census Public-Use File (Minnesota Population Center 2013))

Selected characteristics	Indigenous	Non-indigenous
Pct. distribution	16.7	83.3
Pct. by educational attainment		
Less than primary completed	83.6	61.2
Primary completed	13.6	28.1
Secondary completed	1.8	6.3
University completed	1.1	4.4
Pct. working		
Male	50.1	38.7
Female	12.8	10.5
Pct. with a disability		
Male		
65–69	22.7	18.3
70–74	29.3	27.0
75–79	39.1	34.0
80–84	50.0	42.2
85 and older	61.1	55.8
Female		
65–69	24.2	19.5
70–74	32.7	27.3
75–79	40.3	36.4
80–84	49.2	45.0
85 and older	63.5	60.2
Pct. lacking health insurance	36.3	24.8

of non-indigenous seniors (Table 12.5). Similarly, indigenous elderly men (50.1 %) are much more likely to still be working in comparison to non-indigenous men (38.7 %). Furthermore, indigenous elderly men and women have higher prevalence of disability than their non-indigenous elderly counterparts. Moreover, indigenous elderly (36.3 %) are more likely to lack health insurance coverage compared to non-indigenous elderly (24.8 %). Thus, these data suggest that Mexico's indigenous elderly have experienced substantial levels of privation with respect to social and economic resources and access to health care at the different stages of their lives. At the older stage of their lives, these deprivations have had a negative impact on their health, a matter complicated further by the lack of health care insurance.

The analysis examining internal variations within the Mexican elderly population in each country has demonstrated that the lives of the elderly are structured by nativity and U.S. citizenship in the United States and by race in Mexico. In the United States, Mexican elderly who are not U.S. citizens do not have access to health care insurance. In Mexico, indigenous elderly also have major needs for health care insurance. These two populations of Mexican elderly living in the two countries have a bundle of traits, such as very low levels of education and income, which make them particularly vulnerable in dealing with chronic and degenerative diseases that are part of the aging process. The social, economic, and health needs of these groups of elderly—as well as elderly in general—will increase significantly in the coming decades given the expected aging of the populations of the United States and Mexico. We now turn to an examination of population projections to gain a perspective of the magnitude of these anticipated changes.

A Glimpse into the Future: Projections of the Elderly Population

Over the coming decades, the populations of the United States and Mexico will continue to age significantly. We use the latest population projections for the United States and Mexico in the analysis presented below. However, there are two caveats that we need to address. First, while the U.S. population projections are based on the 2010 census (U.S. Census Bureau 2012), those for Mexico are based on the 2000 census (U.S. Census Bureau 2003). Thus, the projections are not temporally comparable, although they represent the latest population projections available for each country. Second, the U.S. population projections do not contain information specifically for the Mexican-origin population, but for the Latino population. Nonetheless, Mexicans are the largest Latino group accounting for 63% of the total in 2010.

With these cautionary notes in mind, it is clear that the elderly population of each of the four groups of interest is expected to increase significantly. By 2050 elderly individuals are due to constitute 27.4% of the U.S. white population, 19.0% of Mexico's population, 18.5% of the U.S. black population, and 13.8% of the U.S. Latino population (Fig. 12.3). The aging of the population will be particularly

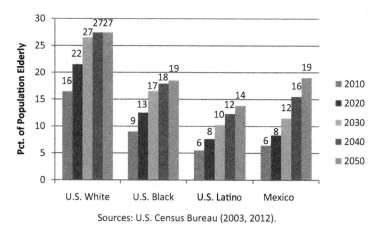

Sources: U.S. Census Bureau (2003, 2012).

Fig. 12.3 Elderly as a percentage of the projected population for selected groups, 2010–2050

Table 12.6 Projected elderly population for selected groups, 2010–2050[a]. (Sources: U.S. Census Bureau 2003, 2012)

Selected groups	2010	2020	2030	2040	2050
U.S. White	32,308,454	42,761,474	52,593,799	53,179,783	51,032,830
U.S. Black	3,400,365	5,218,318	7,483,028	8,707,857	9,602,377
U.S. Latino	2,810,207	4,830,608	8,023,014	11,694,735	15,420,507
Mexico	7,199,269	10,315,621	15,581,757	22,225,682	28,054,513

[a]The population projections for Mexico are based on the 2000 census, while those for the U.S. are based on the 2010 census

dramatic in Mexico where the percentage share of the elderly in the nation's total population is likely to triple from 6.4 % in 2010 to 19.0 % in 2050.

The population projections in absolute numbers for the elderly in the four groups are presented in Table 12.6. In terms of the absolute growth of the elderly population between 2010 and 2050, Mexico is expected to increase the most with 20.9 million more elderly during this 40-year period, followed by U.S. whites with 18.7 million, U.S. Latinos with 12.6 million, and U.S. blacks with 6.2 million. The growth of the elderly in relative terms, however, is concentrated in the U.S. Latino and Mexico's elderly populations. Indeed, the U.S. Latino elderly population is projected to increase 5.5-fold from 2.8 million in 2010 to 15.4 million in 2050, while Mexico's elderly population is expected to nearly quadruple during this period from 7.2 million in 2010 to 28.1 million in 2050.

Moreover, the centrality of these two groups in the growth of the elderly population in the United States and Mexico is consistent across the decades between 2010 and 2050. For example, the U.S. Latino elderly population is expected to increase more rapidly than the elderly of the other three groups in each decade with the percentage increase ranging from 31.9 % in 2040–2050 to 71.9 % in 2010–2020. Mexico's elderly population is expected to have the second most rapid growth of the elderly population beginning in the 2020–2030 period. The white elderly population is due to peak in absolute numbers in 2040 and decline the following decade.

Thus, over the coming decades, the United States will experience a major shift in the race/ethnic distribution of its elderly population. The population projections suggest that the share of whites among the U.S. elderly population is expected to decline from 80 % in 2010 to 61 % in 2050, while the black share of the elderly is due to grow somewhat during this period (Fig. 12.4). However, the percentage share of Latinos of the U.S. elderly population is projected to rise from 7 % in 2010 to over 18 % in 2050. In fact, the Latino elderly is expected to become the nation's largest minority elderly population sometime between 2020 and 2030.

In sum, it is clear that aging will become an increasing reality in the United States and Mexico. It is apparent also that the U.S. elderly population will become increasingly diverse with the expansion of the Latino elderly particularly noteworthy. These projected changes have major policy implications.

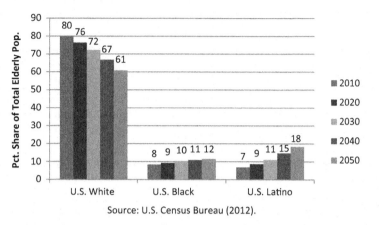

Source: U.S. Census Bureau (2012).

Fig. 12.4 Percentage share of elderly in selected groups of the projected U.S. total elderly population, 2010–2050

Conclusions

Many countries around the world are experiencing a massive aging of their populations due to low fertility rates and rising life expectancies. Due to the aging of its large baby-boomer cohort, the United States will experience a tremendous growth in its elderly population over the period between 2011 and 2029 when the cohort reaches retirement age. Mexico, too, due to a dramatic decline in its fertility rate over the last half century alongside large-scale emigration of its working-age population, will experience a significant aging of its population over the next several decades. These demographic shifts will produce major economic, political, and social challenges and will certainly place major demands on families that have been reshaped and reduced in size significantly by demographic, social, economic, and political forces. In both countries, the family structure has been impacted by reductions in fertility, postponement of marriage, increasing prevalence of divorce, rising levels of cohabitation, ascending life expectancies, and urbanization. Thus, many elderly in both countries will enter their elderly years with fewer family resources that they can tap compared to their counterparts in earlier periods.

In the case of Mexico, immigration has additionally impacted the structure and residence of the family. Working-age Mexicans have migrated to the United States for more than a century. For much of the twentieth century, migrants could readily maintain their ties to their families in Mexico through transnational movement associated with the seasonality of employment. This type of movement and the maintenance of such ties, however, have become increasingly difficult the last two decades due to heightened border security and stringent laws that criminalize immigrants (Douglas and Sáenz 2013). As such, many families have been split between the two countries, a situation that has become more severe with elevated levels of deportations during the Obama presidency. In the case of the elderly, one of the most common arrangements consists of elderly parents living in Mexico and their adult children living in the United States. Mexican immigrants who are naturalized U.S. citizens as well as those who are not U.S. citizens but who are permanent legal residents are in a more favorable position as they are relatively free to move between the two countries to care for aging parents and for the former to sponsor their movement to the United States, though this is an increasingly onerous and prolonged process. Nonetheless, for many Mexicans living in the United States and Mexico, the care of elderly is a binational situation where they must negotiate and develop strategies to provide for the healthcare of aging parents (Montes de Oca et al. 2011, 2012a, b, 2013a, b). This is an increasingly difficult task when familial resources have been weakened through international migration and recent U.S. policies that have run counter to family-friendly laws of the past which assigned importance to family reunification.

Furthermore, the extension of life among the elderly is associated with a greater prevalence of disability and chronic and degenerative disease. Hence, the social, medical, and economic needs of the elderly are likely to rise with the expansion of the elderly population. Elderly individuals will likely need greater assistance to get around and will also need more intensive medical care at a time when family resources are expected to weaken.

Moreover, as poorer populations rapidly age in the coming decades, we will see more elderly persons with limited socioeconomic resources. This will be particularly the case in Mexico and among minority populations in the United States. For minority and poor elderly in the United States and Mexico, cumulative disadvantages throughout the life course have resulted in a limited portfolio of socioeconomic resources to navigate retirement and the later stages of one's life. As shown in the analysis above, the needs of Mexican—and more broadly, Latino—elderly who are not U.S. naturalized citizens and of indigenous seniors in Mexico will be particularly acute.

Based on the demographic and socioeconomic profile in a binational setting and these trends that we have just outlined, there is an important need for the establishment of public policy to meet the needs of an elderly population that will increasingly be Mexican or Latino, increasingly have limited socioeconomic resources, and increasingly undergo aging in a binational context. First, we call for the establishment of policies and programs that provide the elderly with basic preventive and maintenance health care in order to avert the escalation of health care costs when health problems are at advanced stages. Second, we call for the establishment of collaborative binational relationships between the U.S. and Mexico governments to craft policies that are family-friendly and that recognize the realities of immigration. Such policies, for example, could allow individuals to easily travel between the two countries to provide health care assistance to aging parents and other relatives. Other policies could be designed for each government to provide its elderly citizens living abroad with affordable health insurance coverage to guarantee that their health care needs are met. However, in order for such binational arrangements to be actualized, U.S. policymakers need to recognize that the United States and Mexico are intimately connected not only economically but socially as well, a consequence of the long-standing dependence of U.S. employers on Mexican labor as well as the deep historical links between the two countries (O'Neil 2013). Discussions concerning the formation of policies to support the needs of a growing elderly population in the United States and Mexico need to begin immediately due to the demographic realities that will be unfolding in the next few decades.

This urgency is particularly apparent in the case of Mexico. For developed countries, such as the United States, the aging of the population has taken place over an extended period of time following significant reductions in fertility. As such, the institutions of these countries have had time to adjust and prepare for the aging of the population. In the United States, the Social Security system, Medicare, Medicaid, and personal retirement plans have been in place for an extended period of time and provide assistance to the elderly after they retire. Thus, such policies and programs alleviate some of the burden from the family institution in meeting the needs of the elderly. However, in many developing countries, such as Mexico and, particularly, China, the aging of the population has occurred over a relatively short period, mirroring the rapid reduction in fertility that itself took place over a relatively abbreviated time span. Given that it takes a significant amount of time for institutions to adjust, the institutions of Mexico and of other developing countries are generally not adequately prepared to deal with the needs of a rapidly growing elderly population. Policymakers in Mexico and other developing countries need to institute governmental policies and programs to address the needs of their elderly populations. Nonetheless, the United States is not an innocent bystander, for the country will see a significant growth in its Mexican elderly population, a portion of this group being Mexican citizens. The failure to develop binational policies to address the aging of Mexicans in a binational context will hurt the elderly at a time when they are vulnerable to disabilities and chronic and degenerative diseases. In addition, without such governmental intervention, the burden of caring for the elderly will fall squarely on the family.

References

Acharya, A. K., & Codina, M. R. B. (2012). Social segregation of indigenous migrants in Mexico: an overview from Monterrey. *Urbani izziv, 23*(1), 140–149.

Aoki, R. (2012). A demographic perspective on Japan's "lost decades." *Population and Development Review, 38,* 103–112.

Arias, E. (2012). United States life tables, 2008. National Vital Statistics Reports, 61, number 3. Hyattsville: National Center for Health Statistics. http://www.cdc.gov/nchs/data/nvsr/nvsr61/nvsr61_03.pdf. Accessed May 26 2014.

Balderrama, F. E. (1995). *Decade of betrayal: Mexican repatriation in the 1930s.* Albuquerque: University of New Mexico Press.

Bird, C. E., & Rieker, P. P. (1999). Gender matters: An integrated model for understanding men's and women's health. *Social Science & Medicine, 48,* 745–755.

Bonilla-Silva, E. (2013). *Racism without racists: Colorblind racism and the persistence of racial inequality in America* (4th ed.). Lanham: Rowman & Littlefield.

Britton, M. L., & Goldsmith, P. R. (2013). Keeping people in their place? Young-adult mobility and persistence of residential segregation in U.S. metropolitan areas. *Urban Studies, 50*(14), 2886–2903.

Brown, S. L., Lee, G. R., & Bulanda, J. R. (2006). Cohabitation among older adults: A national portrait. *Journal of Gerontology: Social Sciences, 61B,* S71–S79.

Carlson, E. (2008). *The lucky few: Between the greatest generation and the baby boom.* New York: Springer.

Carroll, P. J. (2003). *Felix Longoria's wake: Bereavement, racism, and the rise of Mexican American activism.* Austin: University of Texas Press.

Coale, A. J. (1972). *The growth and structure of human populations.* Princeton: Princeton University Press.

Coleman, D. (2006). Immigration and ethnic change in low-fertility countries: A third demographic transition. *Population and Development Review, 32,* 401–446.

Crystal, S., & Shea, D. (1990). Cumulative advantage, cumulative disadvantage, and inequality among elderly people. *The Gerontologist, 30*(4), 437–443.

Dannefer, D. (2003). Cumulative advantage/disadvantage and the life course: Cross-fertilizing age and social science theory. *The Journals of Gerontology, 58B,* S327–S337.

Diggs, J. (2008). Demographic transition theory of aging. In S. Loue & M. Sajatovic (Eds.), *Encyclopedia of aging and public health* (pp. 266–268). New York: Springer.

Douglas, K. M., & Sáenz, R. (2013). The criminalization of immigrants and the immigration-industrial complex. Daedalus: *Journal of the American Academy of Arts and Sciences, 142*(3), 199–227.

Dowd, J. J., & Bengston, V. L. (1978). Ageing in minority populations: An examination of the double jeopardy hypothesis. *Journal of Gerontology, 33,* 427–436.

Easterlin, R. A. (1980). *Birth and fortune: The impact of small numbers on personal welfare.* New York: Basic Books.

Elo, I. T., Turra, C. M., Kestenbaum, B., & Ferguson, B. R. (2004). Mortality among elderly Hispanics in the United States: Past evidence and new results. *Demography, 41*(1), 109–128.

Entwisle, D. R., Alexander, K. L., & Olson, L. S. (2001). Keep the faucet flowing: Summer learning and home environment. *American Educator, 47,* 10–15.

Fishman, T. C. (2010). *Shock of gray: The aging of the world's population and how it pits young against old, child against parent, worker against boss, company against rival, and nation against nation.* New York: Scribner.

Frank, R., & Heuveline, P. (2005). A crossover in Mexican and Mexican-American fertility rates: Evidence and explanations for an emerging paradox. *Demographic Research, 12,* 77–104.

Franks, M. M., Wendorf, C. A., Gonzalez, R., & Ketterer, M. (2004). Aid and influence: Health promoting exchange of older married partners. *Journal of Social and Personal Relationships, 21*(4), 431–445.

Friedlander, D., Okun, B. S., & Segal, S. (1999). The demographic transition then and now: Processes, perspectives, and analyses. *Journal of Family History, 24,* 493–533.

Gall, O. (2004). Identidad, exclusión, y racismo: Reflexiones teóricas y sobre México. *Revista Mexicana de Sociología, 66*(2), 221–259.

Gassoumis, Z. D., Wilber, K. H., Baker, L. A., & Torres-Gil, F. (2010). Latino baby boomers: A demographic and economic profile. Latinos & Economic Security Policy Brief No. 5. Los Angeles: University of California, Center for Policy Research on Aging.

Geronimus, A., Bound, J., Keene, D., & Hicken, M. (2007). Black-white differences in age trajectories of hypertension prevalence among adult women and men, 1999–2002. *Ethnicity and Disease, 17*(1), 40–48.

Goldman, N., Korenman, S., & Weinstein, R. (1995). Marital status and health among the elderly. *Social Science & Medicine, 40*(12), 1717–1730.

Grundy, E. (2001). Living arrangements and the health of older persons in developed countries. *Population Bulletin of the United Nations, 42*(43), 311–329.

Guerin-Gonzales, C. (1994). Mexican workers and the American dream: Immigration, repatriation, and California farm labor, 1900–1939. New Brunswick: Rutgers University Press.

Guralnik, J. M., Fried, L. P., & Salive, M. E. (1996). Disability as a public health outcome in the aging population. *Annual Review of Public Health, 17,* 25–46.

Hays, J. C. (2002). Living arrangements and health status in later life: A review of recent literature. *Public Health Nursing, 19*(2), 136–151.

Heaton, T. B., England, J. L., García, M., & López, G. R. (2007). The child mortality disadvantage among indigenous people in Mexico. *Population Review, 46*(1), 1–11.

Hiu, L., & Zhang, Z. (2013). Disability trends by marital status among older Americans, 1997–2010: An examination by gender and race. *Population Research and Policy Review, 32,* 103–127.

Hoffman, A. (1974). *Unwanted Mexican Americans in the great depression.* Tucson: University of Arizona Press.

House, J. S., Lantz, P. M., & Herd, P. (2005). Continuity and change in the social stratification of aging and health over the life course: Evidence from a nationally representative longitudinal study from 1986 to 2001/2002 (Americans' Changing Lives Study). *Journal of Gerontology: Social Sciences, 60*(2), 15–26.

Huber, J. C. (1998). Cumulative advantage and success-breeds-success. The value of time pattern analysis. *Journal of the American Society for Information Science, 49,* 471–476.

Hummer, R. A., Powers, D. A., Pullum, S. G., Gossman, G. L., & Frisbie, W. P. (2007). Paradox found (again): Infant mortality among the Mexican-origin population in the United States. *Demography, 44*(3), 441–457.

Kammeyer, K. C. W. (1970). A re-examination of some recent criticisms of transition theory. *The Sociological Quarterly, 11*(4), 500–510.

Katznelson, I. (2005). *When affirmative action was white: An untold history of racial inequality in twentieth-century America.* New York: W.W. Norton and Company.

Khoo, S.-E. (2012). Ethnic disparities in social and economic well-being of the immigrant aged in Australia. *Journal of Population Research, 29,* 119–140.

Koskinen, S., Joutsenniemi, K., Martelin, T., & Martikainen, P. (2007). Mortality differences according to living arrangements. *International Journal of Epidemiology, 36*(6), 1255–1264.

Lichter, D. T., Parisi, D., & Taquino, M. C. (2012). The geography of exclusion: Race, segregation, and concentrated poverty. *Social Problems, 59*(3), 364–388.

Lin, I.-F, & Brown, S. L. (2012). Unmarried boomers confront old age: A national portrait. *The Gerontologist, 52*(2), 153–165.

Manzoli, L., Villari, P., Pirone, G., & Boccia, A. (2007). Marital status and mortality in the elderly: A systematic review and meta-analysis. *Social Science & Medicine, 64*(1), 77–94.

Markides, K. S., & Coreil, J. (1986). The health of Hispanics in the Southwestern United States: An epidemiological paradox. *Public Health Reports, 101,* 253–265.

Martin, J., Hamilton, B. E., Ventura, S. J., Osterman, M. J. K., Wilson, E. C., & Matthews, T. J. (2012). Birth: Final data for 2010. *National Vital Statistics Reports, 61*(1), 1–71. http://www.cdc.gov/nchs/data/nvsr/nvsr61/nvsr61_01.pdf. Accessed May 26 2014.

Martínez Nova, C. (2003). The "culture" of exclusion: Representations of indigenous women street vendors in Tijuana, Mexico. *Bulletin of Latin American Research, 22*(3), 249–268.

Massey, D. S., & Pren, K. A. (2012). Origins of the new Latino underclass. *Race and Social Problems, 4,* 5–17.

McNicoll, G. (2012). Reflections on post-transition demography. *Population and Development Review, 38,* 3–19.

Minnesota Population Center. (2013). *Integrated public use microdata series, international: Version 6.2 [Machine-readable database].* Minneapolis: University of Minnesota Press.

Montes de Oca, V., Ramírez García, T., Sáenz, R., & Guillén, J. (2011). The linkage of life course, migration, health, and aging in Mexico and the United States: Health in adults and elderly Mexican migrants. *Journal of Aging and Health, 23*(7), 1116–1140.

Montes de Oca, V., Sáenz, R., & Molina, A. (2012a). Caring for the elderly: A binational task. In J. L. Angel, F. Torres-Gill, & K. Markides (Eds.), *Aging, health, and longevity in the Mexican-Origin population* (pp. 293–315). New York: Springer.

Montes de Oca Zavala, V., Sáenz, R., Santillanes, N., & Izazola-Conde, C. (2012b). Cuidado a la salud en la vejez y recursos familiares transnacionales en México y Estados Unidos. *Uaricha Revista de Psicología, 9*(19), 85–101.

Montes de Oca, V., García, S.J., & Sáenz, R. (2013a). Transnational aging: Disparities among aging Mexican immigrants. *Transnational Social Review: A Social Work Journal, 3*(1), 65–81.

Montes de Oca, V. Z., & Sáenz, R. (2013b). Estrategias de apoyo transnacional ante el envejecimiento en México y Estados Unidos. In V. Montes de Oca (Ed.), *Envejecimiento en América Latina y el Caribe: Enfoques en investigación y docencia de la Red Latinoamericana de Investigación en Envejecimiento* (pp. 481–528). México: Instituto de Investigaciones Sociales, Universidad Nacional Autónoma de México.

Murphy, S. L., Jiaquan, X., & Kochanek, K. D. (2013). Deaths: Final data for 2010. *National Vital Statistics Reports, 61*(4), 1–117. http://www.cdc.gov/nchs/data/nvsr/nvsr61/nvsr61_04.pdf. Accessed 17 Sept. 17 2013

Notestein, F. (1945). Population—The long view. In P. W. Schultz (Ed.), *Food for the world* (pp. 36–57). Chicago: University of Chicago Press.

O'Neil, S. (2013). *Two nations indivisible: Mexico, the United States, and the road ahead.* New York: Oxford University Press.

Palloni, A., & Arias, E. (2004). Paradox lost: Explaining the Hispanic adult mortality advantage. *Demography, 41*(3), 385–415.

París Pombo, M. D. (2002). Estudios sobre racismo en América Latina. *Politica y Cultura, 17,* 289–310.

Park, J., & Iceland, J. (2011). Residential segregation in metropolitan established immigrant gateways and new destinations, 1990–2000. *Social Science Research, 40*(3), 811–821.

Passel, J. S., Cohn, D., & Gonzalez-Barrera, A. (2012). *Net migration from Mexico falls to zero—And perhaps less.* Washington: Pew Hispanic Center.

Passel, J. S., Cohn, D., & Gonzalez-Barrera, A. (2013). *Population decline of unauthorized immigrants stalls, may have reversed.* Washington: Pew Research Hispanic Trends Project. http://www.pewhispanic.org/2013/09/23/population-decline-of-unauthorized-immigrants-stalls-may-have-reversed/. Accessed May 26 2014.

Perez, J. (2012). Residential patterns and an overview of segregation and discrimination in the Greater Washington, DC, metropolitan region. *Research in Race and Ethnic Relations, 17,* 111–131.

Peters, A., & Liefbroer, A.C. (1997). Beyond marital status: Partner history and well-being in old age. *Journal of Marriage and the Family, 59,* 687–699.

Population Reference Bureau. (2013). *2013 world population data sheet.* Washington: Population Reference Bureau.

Rosenberg, H. M., Maurer, J. D., Sorlie, P. D., Johnson, N. J., MacDorman, M. F., Hoyert, D. L., et al. (1999). *Quality of death rates by race and Hispanic origin: A summary of current research, 1999. Vital and Health Statistics, Series 2, No. 128.* Washington: National Center for Health Statistics.

Ross, C. E., & Wu, C.-L. (1996). Education, age, and the cumulative advantage in health. *Journal of Health and Social Behavior, 37*(1), 104–120.

Ruggles, S., Sobek, M., Alexander, T., Fitch, C. A., Goeken, R., Hall, P. K., et al. (2010). *Integrated public use microdata series: Version 4.0 [Machine-readable database].* Minneapolis: Minnesota Population Center [producer and distributor].

Sáenz, R., & Morales, T. (2012). The Latino paradox. In R. R Verdugo (Ed.), The demography of the Hispanic population: Selected essays (pp. 47–73). Charlotte: Information Age.

Sáenz, R., & Murga, A. L. (2011). *Latino issues: A reference handbook.* Santa Barbara: ABC–CLIO.

Saenz, R., & Siordia, C. (2012). The inter-cohort reproduction of Mexican American dropouts. *Race and Social Problems, 4*(1), 68–81.

Saenz, R., Morales, M. C., & Ayala, M. I. (2004). United States: Immigration to the melting pot of the Americas. In M. I. Toro-Morn & M. Alicea (Eds.), *Migration and immigration: A global view* (pp. 211–232). Westport: Greenwood.

Saenz, R., Douglas, K. M., Embrick, D. G., & Sjoberg, G. (2007). Pathways to downward mobility: The impact of schools, welfare, and prisons on people of color. In H. Vera & J.R. Feagin (Eds.), *Handbook of the study of racial and ethnic relations* (pp. 373–409). New York: Springer.

Seeman, T. E., & Crimmins, E. (2001). Social environment effects on health and aging integrating epidemiological and demographic approaches and perspectives. *Annals of the New York Academy of Sciences, 954*(1), 88–117.

Stokes, A., & Preston, S. H. (2012). Population change among the elderly: International patterns. *Population and Development Review, 38,* 309–321.

Teitelbaum, M. S. (1975). Relevance of demographic transition theory for developing countries. *Science, 188*(4187), 420–425.

Telles, E., & Bailey, S. (2013). Understanding Latin American beliefs about racial inequality. *American Journal of Sociology, 118*(6), 1559–1595.

Telles, E. E., & Ortiz, V. (2008). *Generations of exclusion: Mexican Americans, assimilation, and race.* New York: Sage.

Thompson, W. S. (1929). Population. *American Journal of Sociology, 34*(6), 959–975.

Ueda, R. (1994). *Postwar immigrant America: A social history.* Boston: Bedford/St. Martin's.

U.S. Census Bureau. (2003). *Mid-year population by older age groups and sex—Custom region—Mexico.* [*International data base.*]. Washington: U.S. Census Bureau. http://www.census.gov/population/international/data/idb/region.php?N=%20Results%20&T=2&A=separate&RT=0&Y=2010,2020,2030,2040,2050&R=-1&C=MX. Accessed September 17 2013.

U.S. Census Bureau. (2012). *2012 National population projections.* [*Downloadable files*]. Washington: U.S. Census Bureau. http://www.census.gov/population/projections/data/national/2012/downloadablefiles.html. Accessed September 17 2013.

U.S. Department of Homeland Security. (2013). *2012 Yearbook of immigration statistics.* Washington: U.S. Department of Homeland Security. https://www.dhs.gov/yearbook-immigration-statistics. Accessed Noveber 29 2013.

Van Cott, D. L. (2007). Latin America's indigenous peoples. *Journal of Democracy, 18*(4), 127–142.

Vargas Becerra, P. N., & Flores Dávila, J. I. (2002). Los indígenas en ciudades de México: El caso de los Mazahuas, Otomíes, Triquis, Zapotecos y Mayas. *Papeles de Población, 8*(34), 235–257.

Weeks, J. R. (2012). *Population: An introduction to concepts and issues* (11th ed.). Belmont: Wadsworth.

Williams, D. R. (1999). Race, socioeconomic status, and health: The added effects of racism and discrimination. *Annals of the New York Academy of Sciences, 896*(1), 173–188.

Williams, D. R., & Jackson, P. B. (2005). Social sources of racial disparities in health. *Health Affairs, 24*(2), 325–334.

World Health Organization. (2013). *Life expectancy: Life tables for Mexico.* Geneva: World Health Organization. http://apps.who.int/gho/data/view.main.61060?lang=En-US. Accessed September 17 2013.

Chapter 13
Access to Medical Care and Family Arrangements Among Mexican Elderly Immigrants Living in the United States

Verónica Montes-de-Oca, Telésforo Ramírez, Nadia Santillanes, San Juanita García and Rogelio Sáenz

Introduction

In Mexico, as in other Latin American countries, family, especially children, symbolizes one of the most important resources providing support for the elderly. This support manifests itself in various ways: emotionally, economically, and instrumentally (Gomes and Montes de Oca 2004). However, in some cases it can be modified as a consequence of both internal and international immigration of family members. It is well known that migrant children provide assistance to the elderly through monetary remittances, maintaining solidarity ties and communication despite the distance, although other types of support may be restricted (Montes de Oca et al. 2008). However, it has also been shown that family living arrangements serve as one strategy for Mexicans immigrants in the United States, helping meet the costs of food, education, housing and the health care of elderly adults. Given the economic crisis in the US and the demographic realities projecting Hispanics as the fastest and largest racial and ethnic group, it is critical to understand the strategies used by families in assisting elderly family members through the aging process. The Hispanic elderly are indeed a vulnerable group who face barriers in access to medical services and have dismal rates of health care coverage (Sáenz and Rubio 2007).

V. Montes-de-Oca (✉)
Universidad Nacional Autónoma de México, Mexico City, Mexico
e-mail: vmoiis@gmail.com

T. Ramírez
Researcher of CONACYT at the CRIM-UNAM

N. Santillanes
Centro de Investigaciones y Estudios Superiores en Antopologia Social (CIESAS-DF), Tlalpan, Mexico

S. J. García
The Ohio State University Department of Sociology, Columbus, USA

R. Sáenz
University of Texas, San Antonio, USA

© Springer International Publishing Switzerland 2015
W. A. Vega et al. (eds.), *Challenges of Latino Aging in the Americas,*
DOI 10.1007/978-3-319-12598-5_13

225

In the US, citizenship status is a key factor in determining labor market options and in obtaining medical care, which is essential for preventing illness and is also aligned with a positive aging experience. As people get older, their employment opportunities tend to lessen. This results in a loss in health coverage and impacts medical access which is reinstated at 65 years of age. Meanwhile, Mexican families survive by strategizing ways to care for elderly parents with the ultimate goal of overcoming barriers and lack of health care coverage (Treviño and Coutasse 2007). However, little is known about the strategies used by Mexican families in coping with caring for elderly family members in the receiving society. One way to investigate how families cope with providing care for their elderly family members is through identifying Mexican households in the US who have family members who are 60 years and older residing in the same household. In particular, we seek to arrive at an understanding of the strategies developed and used by Mexican families to address the health needs of older family members.

This chapter is organized as follows: in the first section we provide a brief description of Mexican elderly immigrants in the US. Second, we set forth the data sources and the methods used to develop the research. Third, we provide information on access to health care among the Mexican population aged 60 and over in the US. Fourth, we present information about morbidity rates of Mexican migrants in the US. Fifth, we consider the family arrangements of older Mexican immigrants in the US. On this basis, we analyze the probability of living in extended households with elderly immigrants without access to medical care. In the sixth section, we illustrate, using qualitative data, types of family arrangements which include an elderly family member residing in an extended household. Finally we present some conclusions.

Mexican Elderly Immigrants in United States

Mexican immigration to the US is a social phenomenon that began in the early twentieth century. Since then, men and women of all ages have migrated in search of work and better opportunities than those available in their communities of origin in Mexico (Massey 1987; Ramírez and Román 2007). The constant migratory flows have led to the formation of a large Mexican community in the US. It is estimated that approximately 12 million Mexicans were residing in the US in 2012. This figure increases to just over 33.7 million people when we also consider the second-generation descendants of immigrants and third-plus generations born in the US (Ramírez and Castillo 2012). The Mexican immigrant community consists mainly of people of potentially productive ages (i.e. between 18 and 59), as a feature of labor migration. Immigration policy in the US has generated a lengthening of migrants' residence, which has caused the circulatory tendency of Mexican migration to diminish and canceled plans of some migrants to return, thus leading to an increase in the numbers of elderly adult migrants.

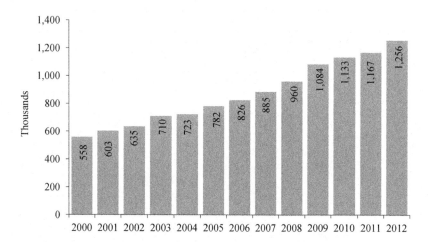

Fig. 13.1 Mexican population of 60 years and older residing in US, 2000–2012. (Source: Estimations based on the U.S. Census Bureau *Current Population Survey* (CPS), March 2000–2012)

According to Durand and Massey (2003), migratory patterns have been modified due to the crackdown in border policy. Circular migration has given way to a more prolonged residence in the country of destination, and immigrants entering the United States remain for longer periods or become permanent residents in that country. On the other hand, for a large number of legal migrants, returning home has become a vacation activity. They come and go, but reside for most of the year in the United States. Hence the migrant's average stay has become notably longer.

The data collected by the Current Population Survey (CPS) from 2012 indicates that the number of Mexicans of age 60 and over doubled from close to 558,000 people in 2000 to nearly 1.3 million in 2012, representing an increase of 125 % in just 12 years (Fig. 13.1).

According to this source, about one fifth of elderly Mexican immigrants came to the US during the years of the "Bracero Program" (1942–1964). This program hired thousands of Mexicans to work in agricultural activities in rural areas of California and other US southwestern states (Durand 2007). Two out of five men did so between 1965 and 1985. This period was known as the stage of the undocumented, due to the exponential increase in undocumented Mexican migration in those years. According to table, only 10% of older Mexicans came to the United States during the IRCA (Immigration Reform and Control Act of 1986) legalization period (1986–1993). It could be that these are people who were able to adjust their immigration status by means of the methods envisaged by the law. Finally, one fifth arrived in the last 20 years, a period called by Durand and Massey (2003) the era of clandestine migration (Fig. 13.2). Taken together, these data suggest that Mexican immigrants had greatly differing opportunities depending on the moment at which they arrived in the US, and that those differences had a socially determining impact on their old age in the US (Montes de Oca et al. 2011).

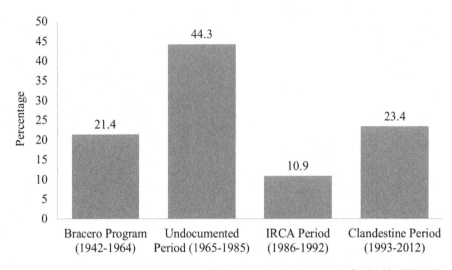

Fig. 13.2 Mexican immigrant population of 60 Years and older by period of arrival in US, 2012. (Source: Estimations based on the U.S. Census Bureau *Current Population Survey* (CPS), March 2012)

However, despite the increase in the stock of Mexican immigrant population aged 60 or over residing in the United States, little is known about the conditions in which these people live the last stage of their life cycle. To date, there continues to be a dearth of knowledge about the personal and family strategies implemented to receive medical care when needed, including physical and emotional care. In the field of health, some research suggests that Hispanic immigrants, mostly Mexicans, report a healthier profile (CONAPO 2008, 2012) and have a longer life expectancy than the native population. This is known as the "Hispanic Paradox" or "Latino epidemiological paradox" (Markides and Coreil 1986; Markides and Eschbach 2005; Palloni and Arias 2004; Smith and Bradshaw 2006), which reflects the lower mortality rates of Hispanics in the United States compared to whites and other non-Hispanics. Some researchers suggest this paradox is explained in part by the self-selectivity of migration. These researchers point to the youthfulness of migrants and suggest that the migrants who decide to migrate to the US are the healthier ones.

In addition, it is also known that as immigrants remain longer in the US and become more acculturated to the host society their health begins to deteriorate (Angel et al. 2010; Gallo et al. 2009). Why does this occur? Some scholars suggest it is due to various obstacles and barriers such as the lack of access to medical services. Some also point to psychosocial risks and the acquisition of unhealthy habits such as a poor diet, excessive alcohol, and drug use, among other unhealthy practices. Various studies have found that the state of health of immigrants deteriorates with the time spent in residence in the United States, but there is insufficient evidence to determine whether this deterioration is the result of years of arduous work and poverty, of changes in their habits and behavior that impact on health, such as diet and smoking, or a limited access to preventive medicine (CONAPO 2005).

In the absence of health insurance, immigrant families often use different strategies to help the elderly obtain medical care and other essential forms of care, in addition to being a source of emotional support especially needed to absorb the losses that come with aging. However, the role the family plays in terms of helping the elderly is determined by a variety of economic, social, and cultural factors. Thus, research has noted, for example, that the type and quality of support for the elderly may vary according to the type of family arrangement and economic conditions of its members (Gomes and Montes de Oca 2004; Montes de Oca 1999; Ramos 1994). However, there is still little empirical evidence on the link between family and the aging process among immigrant populations.

In this context, we formulate the following research questions: (1) What access do older adult Mexican immigrants have to medical services in the United States?; (2) In what types of families or households do elderly Mexican immigrants spend the last years of their lives?; and (3) What illnesses do they tend to suffer from and what family dynamics emerge in coping with these illnesses?

To answer the questions outlined above, we use quantitative and qualitative data and present a descriptive analysis of access to care and health conditions of the Mexican immigrant population aged 60 and over residing in the United States, as well as the type of family/household arrangement in which they reside. Subsequently, we present findings on the factors associated with types of family/household co-residence of elderly Mexicans living in the US.

Methods and Data

Data

We use data from the Current Population Survey (CPS March supplement 2012), which contains demographic and economic information on the Mexican immigrant population residing in the US. On the basis of the information contained in the survey it was possible to construct a series of statistical indicators related with the socio-demographic profile of elderly Mexicans, their access to medical attention, the characteristics of their households, and the time of arrival in the United States, as well as information on their situation regarding poverty and possession of US citizenship. The CPS is the official US government source for statistics on employment and unemployment, directed jointly by the US Census Bureau and the Bureau of Labor Statistics. Besides the information that is gathered monthly, the survey is used for obtaining studies of specific aspects of demographic interest.

The statistical sample is robust enough to perform an analysis of the population that concerns us. The CPS is updated every month, usually with a sample of 57,000 households, randomly selected on the basis of their residential area, in order to represent the entire population at the national, state and other specific levels. Since 2004 the sample of homes selected comprises: (1) a complete sample of the CPS for March; (2) Hispanic households, identified in November and in April (only for some groups); (3) households other than white or Hispanic, identified in the months of August, September, October, November and April; and (4) households of non-Hispanic whites with children of 18 years or less, identified in August, September, October, November and April.

With the inclusion of these populations it has been possible to improve the reliability of the sample estimations for minority groups with a certain degree of vulnerability, such as the Hispanic, among which the Mexicans are the majority, those neither Hispanic nor white, and those of whites with children under 19. For 2012, the CPS population sample was of 201,398 cases which, expanded, gives a total of 308,827,259 individuals. Of the total of the sample, 8,202 cases corresponded to individuals born in Mexico which, expanded, represented 11,877,703 Mexicans living in the country in that year. Of these, 857 cases were Mexicans aged 60 years or over, representing a total population of 1,256,064.

Statistical Methods

To analyze the factors associated with the type of residential arrangement, we carried out a multivariate analysis through the application of a binomial logistic model. This kind of statistical model enables the level of association between the variables of analysis to be determined with respect to the event that one wishes to study; it also enabled us to estimate the specific weight of each category, controlling by means of the remaining variables included in the analysis. In this case, the dependent variable is the type of living arrangement. This is a dichotomous variable that takes the value of 1 if seniors are living in an extended or composite household and the value of 0 if they reside in a nuclear, one person or co-resident household.[1] The independent variables include age, sex, marital status, US citizenship, health insurance and poverty status.

This analysis is complemented by qualitative data collected through in-depth interviews with older adult Mexican immigrants (age 50 or over) residing in Los Angeles, California, Dallas, Texas, and Chicago, Illinois, between 2010 and 2011. This data stems from a larger binational study focusing on the aging experience through a life course perspective. For this study we use interviews from transnational migrants, who regularly reside in the US but were born in the Mexican states of Zacatecas, Durango, and Guanajuato. These participants often traveled to and from Mexico to visit family and loved ones. Their life experiences allow us to further identify the barriers this population faces and the strategies families use in combating these barriers. We use pseudonyms in place of the actual names of participants

[1] Nuclear households: formed by the head of household and spouse, head of household or spouse with children, a couple living alone without children is also a nuclear household. Extended households: consisting of nuclear households plus other family members (uncles, cousins, siblings, in-laws, etc.). Composite households: consisting of nuclear or extended households and more people not related to the head of household. More specifically, this category includes (nuclear and extended household) families and other non-family people. One-person households: consisting of a single person. Co-resident households: formed by two or more persons without family relationship.

to honor agreements of confidentiality. The names of institutions, organizations, and establishments have also been omitted to ensure confidentiality. Research participants were given an informed consent form describing the scope of the research project, the procedures of the research, and their rights as research participants. The research was approved by US Institutional Review Board.

Accesses to Medical Care for Mexican Elderly in the US

According to CPS data, in 2012, 25 % of elderly women and men born in Mexico (about 300,000 people) did not have any form of health insurance. The lack of health insurance is found in all age groups, but it mainly affects older Mexican adults between 60 and 64 years, of which about 40 % are without health insurance (Table 13.1). It is likely that the low rate of health coverage for this population subgroup is related not only to the lack of economic resources, but also to immigration status (Goldring et al. 2007).

It is known that immigrants who do not have permanent residency or US citizenship face various obstacles in order to access health and employment benefits.

Table 13.1 Mexican immigrant population 60 years and older by health insurance coverage in US, 2012. (Source: Estimations based on the U.S. Census Bureau, Current Population Survey (CPS), March 2012)

Features	Health insurance coverage	
	With health insurance (%)	Without health insurance (%)
Total	75.3	24.7
Sex		
Men	75.2	24.8
Women	75.4	24.6
Age group		
60–64	59.3	*40.7*
65–69	83.2	16.8
70–74	85.6	14.4
75–79	83.9	16.1
80 or over	92.7	7.3
US citizenship		
Non-citizen	62.0	*38.0*
Citizen	86.8	13.2
Period of arrival in US		
1950–1964	*92.9*	7.1
1965–1985	*78.4*	21.6
1986–1992	63.7	*36.3*
1993–2012	58.8	*41.2*

Among these are: the failure either to have obtained a private health insurance through employment or to enjoy public insurance through various health programs provided by the state or federal government. Our data indicates that 38 % of older Mexicans who do not have citizenship do not have health insurance. This percentage without insurance is also significantly higher among those with less time living in the US, that is, immigrants who arrived in the country between 1993 and 2012; about 41 % of these do not have health insurance.

Given their relatively short time in the United States, it is likely that a significant proportion of those who have arrived in the United States since 1993 and do not have health insurance are undocumented immigrants, a group which faces various obstacles in accessing medical care. This puts them in a particularly vulnerable position. Access to medical services represents access to information, prevention of illness, and treatments serving to prolong a state of health, and thus provides opportunities for a better quality of life. It also facilitates opportunities for early detection of major chronic illnesses. Not having access to health insurance means individuals' opportunities for preventing illnesses are more limited and they may make inappropriate decisions, often resulting in catastrophic situations for the family and the elderly. Mr. Robles explains this as follows, referring to "people from Zacatecas who have diabetes mellitus and high blood pressure and other diseases":

[H]ere only the person who has health insurance and a job can get [treatment], and can go to the doctor. Here the one that suffers is someone who *doesn't* have any benefits or insurance, I mean, someone who has to deal with the illness (Mr. Robles, 67 years old, Dallas, Texas, first migration in the 1960s).

Morbidity Among Elderly Mexican Immigrants in the US

Several studies indicate that, in general, Hispanics in the US have a better health status than the native population and other immigrants from other countries and regions (Palloni and Arias 2004; Angel et al. 2012). Also, they show that Mexican immigrants have a high prevalence of chronic and communicable diseases, such as diabetes, obesity, among others (CONAPO 2008; Wong 2001).

Other studies have also pointed towards the existence of deterioration in mental health and more specifically diagnoses of depression (Gerst et al. 2010). However, research by González and colleagues (2007) points towards the complexities the immigrant paradox poses especially when examined across the life course. They conceptualize and test an acculturation-health model which takes into account "multiple points in the life course that are critical for improving health and lowering the risk of weathering effects seen in aging among minority groups". The Latino Health Paradox suggests Mexican Americans have an advantage over US born Americans (Markides and Coreil 1986). The findings show, however, that the immigrant health advantage over the US-born is not reflected among older Mexican Americans. Overall, their results suggest that a negative acculturation-health rela-

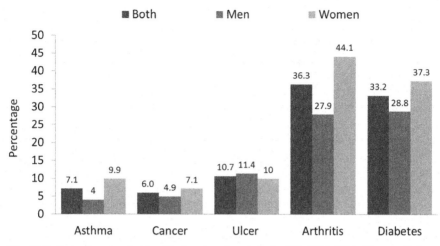

Fig. 13.3 Main diseases of older Mexican immigrants in the US, 2009–2011. (Source: Estimations based on the National Health Interview Survey (NHIS), 2009–2011)

tionship that is often mentioned regarding younger immigrant adults may indeed become a positive relationship in later life.

Data from the National Health Interview Survey (NHIS) (2009–2011), indicates that about 36 % of older adults of Mexican immigrant have been diagnosed as having arthritis by a doctor or a health specialist, and this proportion is higher among women than men (44 and 28 %, respectively). Also, older adults of Mexican inmigrants have a high prevalence of diabetes mellitus, since about one out of three have been diagnosed with this disease. Women are most affected by this disease (37 %). Finally, we note that one in ten Mexican elderly people reports having been diagnosed at least once as having ulcer problems (Fig. 13.3).

The evidence collected through our interviews with elderly immigrants in the states of California, Texas, and Illinois illustrates the feelings and concerns of elderly Mexicans. They spoke of their inability to address their health problems due to the lack of health insurance. Such is the case of Juvencio, a Mexican immigrant residing in the US for over 35 years who began to have kidney problems. However, because of being undocumented and under 65 years of age, he cannot receive health care through Medicaid.

> There was no type of insurance, I paid for the drugs, and they paid the doctor for me [visit]. Not the insurances, I didn't have to [pay], if you are already sick and they do a medical test, they no longer [want to] give you [insurance].
>
> At 66 if you are disabled, you can get it, but now the problems that all my partners and me [have] is that we do not have Medicare, there are no "insurances", that is the problem. There is no way to get the "insurance" (Juvencio, 66 years old, Chicago, first migration in 1977).

Given the absence of health insurance, many low-income immigrants frequent community health clinics, which offer primary and preventive health care to anyone, regardless of their health insurance status and immigration status. Immigrants often turn to health clinics because of their low cost and because they have doctors and staff who speak several languages, including Spanish. Also, they do not request any documents that may reveal information regarding the immigration status of people, which is the main fear among undocumented immigrants when seeking medical care (CONAPO 2012). There are preventive programs for the elderly, primary care and radiology, minor surgeries, medical examinations, rehabilitation, dentistry, among others, in these clinics in Chicago, Dallas or Los Angeles. However, there are far more barriers to getting medical care in other places. Because of these barriers, undocumented immigrants also turn to emergency rooms in search of health care.

We also found women from our study describing multi-morbidity problems, which causes them to deal with several illnesses at once. Other studies show that diabetes may be associated with high blood pressure and some forms of physical disability. This situation becomes worse if people do not have family support, medical access, and medication monitoring. For instance, two older adult women elucidate this in the following statements:

> Look, first it began with my foot, then I had high blood pressure and since then I had cholesterol and diabetes, but now, I am well controlled because I am taking my medicine. But right now what is killing me is the pain of my feet; I see people walking, and like my father-in-law used to say, I envy them. I say I am not normal anymore, like those people; but my husband helps me a lot, a lot; my son lives here (Lupita, 64 years old, California, migrated in the 1970s).
>
> Interviewer: Did they provide you insurance?
>
> M: Yes, but, the insurance stopped supporting me over a year [ago], these last days, now, I got diabetes recently. I got sick of diabetes, and none helps. I was sorry I borrow money from my children, "son, after this… I have no money to buy my medicine", one feels very bad, and I said, ay!" (Marta, 52 years old, California, migrated in the 1970's).

Family Arrangements of Elderly Mexican Immigrants in the US

Families play an important role as a source of financial and emotional support, especially when the elderly do not have resources or when illness requires care and attention. However, previous literature has found that their instrumental support is sometimes inadequate (Dietz 1995). Nonetheless, there are many cases in which immigrants decide to bring their parents and older siblings to the US, in order to seek care for and support them in the last stage of their life cycle. The type, quality, and quantity of support are closely related to the socio-demographic and economic characteristics of households.

For instance, Saturnina, an 81 year old, originally from the Mexican state of Durango, illustrates this phenomenon. When Saturnina turned 71 years old, she left Mexico to live with one of her daughters in Dallas, Texas. Her children decided to petition their mother to live with one of the siblings in the US. This decision was made collectively as a family and was prompted by a fall, which caused severe immobility for Saturnina. She is a widow and previously lived in Durango with the youngest son. In fact, this youngest son decided to join his siblings in Texas and Saturnina hesitated to leave Mexico. This resulted in her being alone for 1 year in

Mexico while all her children lived in the US. They finally convinced her to migrate to the US to be with them. As a result of her fall and severe immobility, Saturnina requires permanent medical care. She currently lives in Dallas, Texas with one of her daughters.

Despite Saturnina's health condition, she helps one of her sons who suffered from a hernia. Her time is split with living with one of her daughters and then every 15 days she attends to her son. This son is the only child who is an undocumented immigrant; this barred him from enjoying the benefits of having health insurance and receiving adequate health care. Saturnina told us that she feels it is her duty as a mother to help her son, especially as he was abandoned by his partner. She recounts her worries about her son who continues to work in arduous conditions in the building industry, which aggravates his health condition.

Although the demographic literature focusing on the aging population suggests a number of important implications impacting families, some findings point towards the importance of vertical kinship ties (i.e., up-and-down generations). This is compared to horizontal kinship ties (i.e., ties of adults to their significant partners, siblings, etc.) (Waite 2009). Our findings show the importance of vertical kinship relationships that provide intergenerational support. Additionally, family dynamics among the elderly population are usually framed as a one-dimensional process of care (i.e., the children of the elderly are viewed as the primary caregivers), Saturnina's case provides a new way of viewing these family arrangements and suggests the caregiving role can be reciprocal. In other words, although Saturnina was initially brought to the US to be cared for, she also serves as a caregiver for her son.

There are families in which family support for the elderly usually takes the form of just instrumental care for diseases, or is only economic; but there are cases where a combination of different forms of support occurs. It is well documented in Mexican and other Latin American and Caribbean countries that the more extensive and heterogeneous the support network, the higher the quality of life of older adults (Saad 2005). Care needs on account of disease, economic dependency, mental impairment and emotional care of the elderly motivate the formation of living arrangements and different forms of family support (Montes de Oca et al. 2012). Other studies have found that certain variables like marital status, age, gender and health status are linked with the living arrangements of the elderly and that such strategies are vitally important to their quality of life (Hakkert and Guzmán 2004; Pérez and Brenes 2006).

The construction of family/household arrangements, such as the extended family where the elderly can live with spouse and/or single children and/or married children and/or grandchildren and/or other family members (uncles, cousins, siblings, in-laws, etc.), or composite type (households with both relatives and non-relatives), is a strategy that has been documented both in studies on aging and in international migration (Triano 2006). In the first case, the formation of such residential arrangements is related to the needs of the elderly, who seek support both financially and emotionally (Gomes 2001; Montes de Oca 1999). In the second case, the strategy is implemented to deal with expenses for lodging, food and transportation, but also as a form of solidarity with other people of the same family or place of origin, including the older population (Ramírez 2009; Montes de Oca et al. 2008). In this regard, the 2012 statistics indicate that about 44% of older Mexican immigrants in the US reside in an extended or composite type household, and this percentage increases with age, especially when people are older. By contrast, the proportion of those living in nuclear households and co-residents tends to decrease as people advance in age. This finding is consistent with studies in Latin America and the Caribbean (Pérez and Brenes 2006).

As for the sex ratio, the data indicates older Mexican women are more likely to live in extended or composite households than men (41 and 34%, respectively) (Table 13.2). This result can be explained by the longer lifespan of Mexican immigrant women relative to men, but also because gender roles have meant that women are perpetually asked to assume the care-giving role. Thus, women are more likely than men to be incorporated into other households as helpers (Barros 2006). Furthermore, since—due to their lower mortality than that of men—women are more likely to be widowed or without a partner, marital statuses that are associated with a greater propensity to live in extended households. Nonetheless, increasing percentages of elderly people are living alone. In general terms, however, these data support the hypothesis that extended household arrangements are among the strategies undertaken by Mexican families to offset the lack of economic, social, and emotional support of elderly people facing separation, divorce, or widowhood.

Moreover, these data show that older Mexican immigrants who have not legalized their immigration status or who do not have US citizenship tend to live in extended families, as compared to those enjoying such immigration status (45 and 32%, respectively)[2] (Table 13.2). Those with citizenship usually live in one-person households or nuclear households. As we know, not having US citizenship restricts, for example, access to health insurance and access to other public benefits such as unemployment insurance, food stamps. About 46% of older Mexicans who do not have health insurance live in extended families, while a similar proportion of those who do have health insurance, either public or private, live in nuclear households. In Mexico, according to the National Survey of Demographic Dynamics (ENADID) of 2009, people 60 years and older live 59.6% in nuclear households, 27.1% in extended households and 11.9% in one-person households (Montes de Oca et al. 2014).

Hence immigrant families deploy a variety of strategies in order to reduce living expenses and deal with unexpected expenses caused by illness, accidents, etc. Obviously, this situation becomes more difficult for older low-income immigrants—persons with limited income to cover their expenses for food, clothing and health. In this regard, we note that the proportion of older adults with incomes below 150% of the Federal Poverty Line (LFP) in the US is in nuclear (38.7%), extended (32.3%), and one-person (19.7%) households.

Thus far we have built up a characterization of the types of family/household arrangements in which elderly Mexican immigrants reside in the United States.

---

[2] These households can be identified as mixed-status households, because of the different types of immigration status, some members being citizens and others not.

Table 13.2 Mexican immigrant population 60 years and older in the US by type of family/household arrangement, 2012. (Source: Estimations based on the US Census Bureau, Current Population Survey (CPS), March 2012)

Features	Type of family/household arrangement				
	Personal (%)	Nuclear (%)	Extended (%)	Composite (%)	Co-resident (%)
Total	11.9	41.9	38.2	5.6	2.4
Age group					
60–64	9.7	43.4	39.4	5.3	2.3
65–69	11.7	47.3	30.5	7.5	3.0
70–74	9.0	47.3	33.9	5.7	4.0
75–79	18.0	42.5	32.7	6.7	0.0
80–84	16.1	19.5	59.4	2.7	2.3
85 or over	18.9	16.5	62.6	0.0	1.9
Sex					
Men	9.8	47.1	34.3	6.3	2.5
Women	13.7	37.5	41.5	5.0	2.4
Marital status					
Married	2.0	63.1	30.7	4.2	0.1
Vidowed	25.0	7.8	57.9	4.4	5.0
Divorced	37.4	2.7	40.5	14.1	5.3
Separated	30.0	11.8	35.4	17.2	5.6
Never married	26.9	7.3	48.4	6.7	10.7
US citizenship					
Non-citizen	7.2	37.2	45.0	6.9	3.8
Citizen	15.9	46.0	32.3	4.5	1.3
Health insurance					
With health insurance	13.8	44.8	35.3	4.2	1.8
Without health insurance	6.1	32.9	46.8	9.8	4.3
Poverty status					
Income < 150% of FPL	19.7	38.7	32.3	6.6	2.7
Income > 150% of FPL	6.6	44.1	42.2	4.9	2.2

FPL federal poverty line in U.S

However the question arises: what factors influence or determine the type of family/household arrangements for elderly Mexican immigrants residing in the United States? In other words, how do personal and socioeconomic conditions influence the probability that an older adult will live in an extended or composite versus a nuclear or individual household?

Associated Factors of Co-residence in Extended Households

In order to identify which demographic factors influence the type of co-residence of older Mexican immigrants in the US, we used and estimated a binary logistic regression model, which not only allowed us to determine the level of association between the variables regarding the event analysis to be investigated, but also to estimate the specific value of each category, controlling by the other variables included in the model. Table 13.3 presents the results of the estimated binomial logistic regression model. In this model we see that only four of the six variables included were statistically significant: marital status, US citizenship status, health insurance coverage and income level. Regarding the marital status, the model indicates that being widowed and divorced increases the likelihood of residing in a home of extended or composite type, as opposed to the married (united), which is consistent with the descriptive analysis results. For example, being widowed increases by a factor of 3.12 the probability of residing in such homes.

Similarly, the results of the estimated binomial logistic regression model show that having US citizenship is associated inversely with the probability of living in extended or composite households. Specifically, immigration status reduces by almost half the probability of residing in this type of a family arrangement (0.448). Regarding the medical safety category, the estimated model results indicate that the failure to have health insurance, either public or private, increases the probability of older Mexicans residing in an extended or composite household, compared to people who have health insurance. This suggests that choosing to live in a family arrangement is a strategy used by many of the Mexican-origin population to facilitate the health care of their elderly family members. Finally, the model also confirms that older adults with low incomes are more likely to reside in such households (Table 13.3). This illustrates that family arrangements and solidarity among the Mexican population are resources that remain strong even in current times of vulnerability and economic crisis. More specifically, the results indicate that co-residence in extended or composite households is one of the strategies implemented by Mexican immigrants and their families to meet the needs and challenges faced during their time in the US.

Family Living Arrangements among Extended Mexican Families

The remaining task is to determine family dynamics and their impact on the quality of life of elderly women and men who reside in such families arrangements. While care tasks are easier to perform the more members there are in a household, care strategies are not without conflict especially when immigrants must work hard to cope with a difficult economic situation.

The literature notes that in other regions of the world the major support of the elderly is provided by the children who co-reside with them, and those who do not live in the same household gradually cease to support their parents in old age

Table 13.3 Associated factors of co-residence in extended households and composite. Logistic regression model. (Source: Estimations based on the US Census Bureau, Current Population Survey (CPS), March 2012)

Variables	B	Exp (β)	95.0% C.I. for Exp (β)	
			Lower	Upper
Age group				
60–64 años[a]		1.000		
65–69 años	−0.001	0.999	(0.684–1.461)	
70–74 años	−0.143	0.867	(0.536–1.401)	
75–79 años	−0.219	0.803	(0.481–1.340)	
80–84 años	0.752	2.122	(1.072–4.200)	
85 years or over	0.709	2.031	(1.000–4.125)	
Sex				
Men[a]		1.000		
Women	0.179	1.196	(0.879–1.626)	
Marital status				
Married[a]		1.000		
Widowed	1.14	3.127*	(2.051–4.768)	
Divorced	0.709	2.031*	(1.145–3.604)	
Separated	0.752	2.879	(0.981–4.588)	
Never married	0.482	2.754	(0.929–2.825)	
US citizenship				
Non-citizen[a]		1.000		
Citizen	−0.594	0.552*	(0.408–0.746)	
Health insurance				
With health insurance[a]		1.000		
Without health insurance	0.612	1.843*	(1.288–2.639)	
Poverty status				
Income < 150% of the FPL[a]		1.000		
Income > 150% of the FPL	−0.409	0.613*	(0.492–0.897)	
Constant	−0.426	0.019		
−2 Log-likelihood	1070.833[a]			
R² de Cox and Snell	0.108			
R² de Negelkerk	0.146			

FPL federal poverty line in U.S.

*$p > 0.05$

[a]Indicates the reference category used in the logistic regression model

(Knodel et al. 1992). The division of labor and follow-up in the transnational space creates tensions between siblings that often alienate families in the US.

This family conflict is illustrated by Esther's experience. She recounts sibling tensions caused by the differential assistance that parents provide for their children. Esther, a 50 year old woman from the Mexican state of Guanajuato currently residing in Chicago, tells us of the ways in which the support of her parents to two brothers in an emergency, caused sibling tension. She states the following:

> Yes, it really hurt me when it was all going on and even after. But then after it hurt because of what I'm telling you. Seeing that they [referring to her parents] would give money to my brother and my other sister, without even having a job themselves, and me I also had difficulties and emergencies, and they never gave me anything (Esther, 50 years old, Chicago, first migration in the 1970's).

Family strategies through filial obligation between parents and children follow traditional patterns among familial cultures. Hispanic caregivers are women, wives, daughters or daughters-in-law who tend to have conditions of greater poverty, unemployment or underemployment and lower level of education, with a propensity to have poorer physical and mental health than white caregivers (Aranda and Knight 1997; Scharlach et al. 2006). These characteristics of caregivers show that the care strategies of Mexican families in the US are borne most intensely by socioeconomically disadvantaged women.

For instance, Rosalba, a 54-year-old woman originally from the Mexican state of Michoacán and currently residing in Los Angeles, California, explains this in more detail. Since she was a young girl she helped her mother in supporting the family economically. Since a very young age, she began taking care of children and then helped her mother sell homemade food. This business was not enough for the family's needs—especially considering the health problems the youngest male siblings faced, with one of them dying of dengue—so this decided her to migrate with a family member to the US at 21 years of age. She married at 24 years old and formed her family home with four children and her husband. For 16 years she worked in a factory and during that time she was able to adjust her immigration status through the IRCA 1986 legislation, but she was not able to obtain citizenship because she did not pass the English exam.

At the time of our interview, Rosalba was separated from her husband and she told us that her children have formed their own homes. She lives with four nephews of whom the government granted her legal guardianship. These children are the children of one niece who has a drug addiction problem. The state had taken the children into care but Rosalba agreed to look after them out of fear of them being placed in foster homes and never seeing them again. Although the government supports her with $ 262 per child for the children's needs, Rosalba explained that it is difficult for her, especially at this stage of her life course, having to deal with children, especially with one of them who is in his teens. At the time of the interview, Rosalba had been diagnosed as having kidney problems and diabetes complications. She has been able to use Medicare to receive health care. Despite the responsibility for her own children and the four younger children, she often traveled to Mexico to help her mother. She provided both financial and emotional support to her mother who was very ill. She also managed to pay for the education of one of her brothers so that he could obtain a university education. Rosalba would love to have more support from her children, but she informed us that she has had various arguments with her own children because they tell her she should not have agreed to raise her four nephews and serve as their legal guardian.

Extended households can also be mixed homes where some members are US citizens and others who not. This may suggest that those who are citizens act to give guidance and support to those who lack institutional access to health and social services. As such, mixed-status households characterized by solidarity among family members support those disadvantaged by isolation or in the face of a catastrophic event such as widowhood or illness. These families articulate strategic ways to reduce the health risks associated with the aging process. Cross-generational and mixed social-status households can be a strategy for optimizing low-income housing in times of crisis as also happened in Mexico. The co-residence of parents, children and grandchildren is a strategy that optimizes meager incomes as in other parts of Latin America and the Caribbean (Ramos 1994).

We identified several types of mixed-status households/families in our qualitative fieldwork: (1) elderly couple and offspring who may have US citizenship or residency; (2) children and one of the older parents who may have US citizenship or residency and the other not having these statuses; (3) children who may have US citizenship or residency and both parents lacking these statuses; (4) some children with US citizenship or residency and other children not; (5) parents with US citizenship or residency and children not; along with other possible combinations. These family configurations demonstrate the complexities and critical need for further study of how legal status plays a role in understanding family relationships, household configuration and how these assist undocumented elderly family members.

Conclusions

The strategies Mexican families adopt transcend borders and perform a protective role towards members who are in a position of risk and vulnerability. For instance, extended families of Mexicans in the US play a significant role in the lives of elderly immigrants. These types of households have a greater number of older adults who arrived at different historical periods of migration between Mexico and the US. The question of who has and who does not have access to citizenship rights—including access to health care—is to a large degree contingent upon the historical period of migration during which an individual made his or her entry in the United States. This also plays a role in further understanding how time and historical context impact an individual's life trajectories and their health outcomes throughout his or her life course.

The CPS data in the analysis indicates that 38 % of elderly individuals without citizenship do not have medical care, as compared to 13 % of those who have citizenship. The data also show that 25 % of those aged 60 years and older living in the US do not have any type of health insurance. The majority in this segment are those who are not yet age 65 (60–64) and therefore do not meet the age requirement to gain access to Medicare. This exclusion prior to old age is an element that hinders disease prevention or early detection to start treatments that can improve the quality of life in old age. Faced with the aging process and the lack of health services,

Mexican immigrants resort to family arrangements, for example in extended families or composite households (44 % of the elderly live in such situations). As people age, they are more likely to be part of extended rather than nuclear households. The majority of older people who live in extended families are women, suggesting a greater role among women in providing assistance, care, and support to their children, grandchildren, and other relatives. Also the widowed, divorced or unmarried elderly tend to be more likely than married persons to live in such households, engaging in a strategy that is very common in Mexico during times of economic difficulty. Yet, it is noteworthy that nearly one-fifth (18.9 %) of persons 85 and older live in one-person households.

The interesting thing about the extended household type is that it includes vulnerable individuals without access to health care, i.e., elderly persons who are not US citizens and who do not have health insurance coverage. For example, 45 % of non-citizens live in extended households, as compared to only 32 % of US citizens living with relatives. Moreover, 47 % of seniors who do not have health insurance live in extended households, compared to only 35 % of those who do have insurance and live with relatives. As such, extended households play an important role in caring for and meeting the health care needs of extremely vulnerable family members.

In contrast, nuclear household residence is more typical of elderly people with greater stability in their lives—they tend to be married, US citizens, and better off economically. For these elderly people, social and economic stability allow for more independent living arrangements. In contrast, those with greater vulnerability must rely on the assistance and support of their families to deal with the many challenges that arise in old age with limited resources to meet them.

The qualitative data we used in this study illustrate the family dynamic, show the stress and emotions involved in view of the appearance of illness, accidents and the lack of health care access. The narratives illustrate perceptions and experiences regarding the importance of having health care and receiving preventive care. As people age, their likelihood of developing illnesses increases, and enjoying proper health care throughout their lives may prevent some illnesses from developing. Our data are also indicative of ways in which undocumented status and lack of access to health insurance can be part of a chain of disadvantages that affect the lives of Mexican immigrants residing in the US in their old age.

Some Mexican immigrants choose not to return to Mexico and opt to remain in the US so as to be close to their children, grandchildren, and great grandchildren, who provide support. Family dynamics are not easy and often create tension and conflicts within families. Finally our qualitative data shows the complexity behind the different types of extended families and the relationships between siblings, with older parents, and with other relatives. Family research among Mexican migrants is not enough to explain all the strategies between generations, citizenships status, health care access and cultural mechanisms to survive the aging process in the US.

Acknowledgements The authors would like to thank sociologist Emma Cervantes for her support during fieldwork and interview data entry. Additionally, the authors would like to thank to Christopher Follett for his translation and copy-editing of this English version.

Funding Part of the study was undertaken with the support of the Migration and Health Programme (Programa de Investigación de Migración y Salud, PIMSA) granted by the Health Initiative of the Americas of the University of California at Berkeley's School of Public Health.

Declaration of Conflicting Interests The authors declare that they have no potential conflicts of interest with respect to the research, author-ship, and/or publication of this article.

References

Angel, R., Angel, J., & Markides, K. (2003). Salud física de los mexicanos migrantes mayores en los Estados Unidos. In N. V. Salgado de Snyder & R. Wong (Eds.), *Envejeciendo en pobreza* (pp. 153–172). Mexico City: Instituto Nacional de Salud Pública.

Angel, R., Angel, J., Diaz Venegas, C., & Bonazzo, C. (2010). Shorter stay, longer life: Age at migration and mortality among the older Mexican-origin population. *Journal of Aging and Health, 22*(7), 914–931.

Angel, J., Torres-Gil, F., & Markides, K. (2012). *Aging, health, and longevity in the Mexican-origin population.* New York: Springer.

Aranda, M. P., & Knight, B. G. (1997). The influences of ethnicity and culture on the caregiver stress and coping process: A sociocultural review and analysis. *The Gerontologist, 37*(3), 342–357.

Barros, M. (2006). Las abuelas en las familias de origen mexicano en California, Estados Unidos. Un estudio de caso. In R. Esteinou (Ed.) *Fortalezas y desafíos de las familias en dos contextos: Estados Unidos de América y México* (pp. 283–313). CIESAS/ DIF, Mexico.

Consejo Nacional de Población (CONAPO). (2005). *Migración México-Estados Unidos panorama regional y estatal,* Mexico City: SEGOB.

Consejo Nacional de Población (CONAPO). (2008). *Migración y salud. Latinos en los Estados Unidos.* Mexico City:SEGOB.

Consejo Nacional de Población (CONAPO). (2009). *Migración y salud. Los hijos de migrantes mexicanos en Estados Unidos.* Mexico City: SEGOB.

Consejo Nacional de Población (CONAPO). (2010). *Migración y salud. Inmigrantes mexicanas en Estados Unidos.* Mexico City: SEGOB.

Consejo Nacional de Población (CONAPO). (2012). *Migración y salud. Jóvenes mexicanos inmigrantes en Estados Unidos.* Mexico City: SEGOB.

Dietz, T. L. (1995). Patterns of intergenerational assistance within the Mexican American family: Is the family taking care of the older generation's needs? *Journal of Family Issues, 16*(3), 344–356.

Durand, J. (2007). El programa Bracero (1942–1964) Un balance crítico. In *Migración y Desarrollo,* segundo semestre, 9, Zacatecas: Red Internacional de Migración y Desarrollo (pp. 27–43).

Durand, J. & Massey, D. (2003). *Clandestinos. Migración México-Estados Unidos en los albores del siglo XXI.* Mexico City: Miguel Ángel Porrúa and Zacatecas.

Gallo, L. C., Penedo, F. J., Espinosa de los Monteros, K., & Arguelles, W. (2009). Resiliency in the face of disadvantage: Do Hispanic cultural characteristics protect health outcomes? *Journal of Personality, 77*(6), 1707–1746. doi: 10.1111/j.1467-6494.2009.00598.x.

Gerst, K., Al-Ghatrif, M., Beard, H., Samper-Ternent, R., & Markides, K. (2010). High depressive symptomatology among older community-dwelling Mexican Americans. *The Impact of Immigration, Aging and Mental Health, 14*(3), 347–354.

Goldring, L., Beristein C., & Bernhard J. (2007). Institutionalizing precarious immigration status in Canada. Early Childhood Education Publications and Research. Paper 4. http://digitalcommons.ryerson.ca/ece/4. Accessed 16 Jan 2014.

Gomes, M. C. (2001). Desigualdad social de la vejez. Condiciones socioeconómicas de la tercera edad. *DEMOS. Carta Demográfica Sobre México, 14*, 13–15.

Gomes, C., & Montes de Oca, V. (2004). Ageing in Mexico. families, informal care and reciprocity. In P. Lloyd-Sherlock (ed.), *Living longer. Ageing, development and social protection* (pp. 230–248). London: UNRISD/ Zed Books.

González Vásquez, T., Bonilla Fernandez, P., Jauregui Ortiz, B., Yaminis Thespina J., & Salgado de Snyder, N. V. (2007). Well-being and family support among elderly rural Mexicans in the context of migration to the United States. *Journal of Aging and Health, 19*(2), 334–355.

Hakkert, R., & Guzmán, J. M. (2004). Envejecimiento demográfico y arreglos familiares de vida en América Latina. In M. Ariza & O. De Oliveira (Eds.), *Imágenes de la familia en el cambio de siglo* (pp. 479–518). Mexico City: UNAM—Instituto de Investigaciones Sociales.

Knodel, J., Chayovan, N., & Siriboon, S. (1992). The familial support system of Thai elderly: An overview. *Asia-Pacific Population Journal, 7*(3), 105–26.

Markides, K., & Coreil, J. (1986). The health of Hispanics in the southwestern United States: An epidemiological paradox. *Public Health Reports, 101*(23), 253–265.

Markides, K., & Eschbach, K. (2005). Aging, migration, and mortality: Current status of research on the Hispanic paradox. *The Journals of Gerontology Series B: Psychological Sciences and Social Sciences, 60*(Special Issue 2), S68–S75.

Massey, D. (1987). *Return to Aztlán: The social process of international migration from western Mexico*. Berkeley: University of California Press.

Montes de Oca, V. (1999). Relaciones familiares y redes sociales. In CONAPO, *Envejecimiento demográfico de México: retos y perspectivas* (pp. 289–325). Mexico City: CONAPO, Cámara de Diputados, Senado de la República.

Montes de Oca, V., Molina, A., & Avalos, R. (2008). *Migración, redes transnacionales y envejecimiento: Estudio de las redes familiares transnacionales de las personas adultas mayores guanajuatenses*. Mexico: Gobierno del Estado de Guanajuato/IISUNAM.

Montes de Oca, V., Ramírez, T., Sáenz, R., & Guillén, J. (2011). The linkage of life course, migration, health, and aging: Health in adults and elderly Mexican migrants. *Journal of Ageing and Health, 23*(7), 1116–1140.

Montes de Oca, V., Sáenz, R., & Molina, A. (2012). Caring for the elderly, a binational task. In J. Angel, M. Kyriakos, & F. Torres-Gil (Eds.), *Aging, health, and longevity in the Mexican-origin population* (pp. 293–315). New York: Springer.

Montes de Oca, V., Garay, S., Rico, B., & García, S. J. (2014). Living Arrangements and Aging in Mexico: Changes in Households, Poverty and Regions, 1992–2009. *International Journal of Social Science Studies, 2*(4). ISSN: 2324-8033.

Palloni, A., & Arias, E. (2004). Paradox lost: Explaining the Hispanic adult mortality advantage. *Demography, 41*(3), 385–415.

Pérez Amador, J., & Brenes, G. (2006). Una transición en edades avanzadas: Cambios en los arreglos residenciales de adultos mayores en siete ciudades latinoamericanas. *Estudios Demográficos y Urbanos, 21*(3), 625–661.

Ramírez, T. (2009). *El impacto de la migración masculina internacional a Estados Unidos en el trabajo femenino extradoméstico en México: Un estudio de caso en el estado de Guanajuato*. Tesis de doctorado en Estudios de Población. Mexico City: El Colegio de México.

Ramírez, T., & Castillo, M. Á. (Eds.). (2012). *El estado de la migración: México ante los recientes desafíos de la migración internacional*. Mexico City: Consejo Nacional de Población (CONAPO).

Ramírez, T., & Román, P. (2007). Remesas femeninas y hogares en el estado de Guanajuato. *Papeles de Población, 54*, 191–224.

Ramos, L. R. (1994). Family support for the elderly in Latin America: The role of the multigenerational household, in United Nations. In *Ageing and family. Proceedings of the United Nations international conference on ageing populations in the context of the family* (pp. 66–72). New York: United Nations.

Saad, P. (2005). Los adultos mayores en América Latina y el Caribe: Arreglos residenciales y transferencias informales. *Notas de Población, 80*, 127–154.

Sáenz, R., & Rubio, M. (2007). Lack of health insurance coverage and mortality among Latino elderly in the United States. In J. L. Angel & K. E. Whitfield (Eds.), *The health of aging Hispanics: The Mexican-origin population* (pp. 181–194). New York: Springer.

Scharlach, A., Kellam, R., Ong, N., Baskin, A., Goldstein, C., & Fox, P. (2006). Cultural attitudes and caregiver service use: Lessons from focus groups with racially and ethnically diverse family caregivers. *Journal of Gerontological Social Work, 47*(1/2), 133–156.

Smith, D. P., & Bradshaw, B. S. (2006). Rethinking the Hispanic paradox: Death rates and life expectancy for US Non-Hispanic white and Hispanic populations. *American. Journal of Public Health, 96*(9), 1686–1692. doi:10.2105/AJPH.2003.035378.

Treviño, F., & Coutasse, A. (2007). Disparities and access barriers to health care among Mexican Americans elders. In J. L. Angel & K. E. Whitfield (Eds.), *The health of aging Hispanics: The Mexican—origin population* (pp. 165–180). New York: Springer.

Triano, M. (2006). Reciprocidad diferida en el tiempo: Análisis de los recursos de los hogares dona y envejecidos. In M. González de la Rocha (Ed.), *Procesos domésticos y vulnerabilidad: Perspectivas antropológicas de los hogares con Oportunidades* (pp. 277–342). Mexico City: Publicaciones de la Casa Chata, CIESAS.

Visauta, B. (1998). *Análisis estadístico con SPSS para Windows. Estadística multivariante.* New York: McGraw Hill.

Waite, Linda J. (2009). The changing family and aging populations. *Population and Development Review, 35*(2), 341–346.

Wong, R. (2001) Migración internacional en la vejez. *DEMOS. Carta Demográfica Sobre México, 14*, 16–17.

Chapter 14
Intergenerational Transfers in Urban Mexico: Residential Location of Children and Their Siblings

Nekehia Quashie

Introduction

As observed in other Latin American and Caribbean countries, Mexico is undergoing a rapid demographic shift due to population aging. Although currently classified by the World Bank as a middle income country, older adults in Mexico are aging in contexts of weak and unequal access to social welfare (Aguila et al. 2011; Cotlear and Tornarolli 2011). Mexico's coverage of pension and health insurance is significantly lower than similarly developed economies in Latin America, such as Argentina, Uruguay and Brazil (Rofman and Lucchetti 2006). In contrast to developed countries such as the United States or Western Europe where older adults have stronger welfare systems and are more likely to support children than to receive support from children (McGarry and Schoeni 1997; Fritzell and Lennartsson 2005), the lack of state or market coverage in Mexico implies the family unit plays a larger role in the welfare of older adults.

Older Mexicans' economic and care needs are met through a combination of informal transfers of time and money from household members, labor income and asset accumulation (Parker and Wong 2001; DeVos et al. 2004; Gomes 2007; Wong and Higgins 2007; Aguila et al. 2011). Within the household, children are typically the providers of parental support in the absence of a spouse (Agree and Glaser 2009). Similar to other developing countries, the extent to which Mexican elderly can continue to rely on children for support is threatened by declining fertility combined with the ongoing migration of younger cohorts (Knodel et al. 2000; Izazola 2004). The issue of geographic separation and its implications for undermining or reinforcing family support is particularly relevant in Mexico given the weak and unequal coverage of social welfare for older adults. Geographic proximity provides the opportunity structure for support (Bengston and Roberts 1991). Closer proximity provides the immediacy of support exchanges in times of need but family

N. Quashie (✉)
College of Population Studies, Chulalongkorn University, Bangkok, Thailand
e-mail: Nekehia.Q@chula.ac.th

© Springer International Publishing Switzerland 2015
W. A. Vega et al. (eds.), *Challenges of Latino Aging in the Americas,*
DOI 10.1007/978-3-319-12598-5_14

members may also separate in order to better meet the needs of its members. Research in other developing countries such as China and Brazil (Bian et al. 1998; Saad 2005) shows that children living away from the parental home support their parents regardless of their precise location.

Migration is an enduring feature of Mexican society and existing research shows that family support is maintained despite distance (Donato 1994; Durand et al. 1996b; Massey et al. 2003). Moreover, the later onset of fertility decline in Mexico, relative to other countries in the region, implies larger family sizes for current cohorts of older adults (Glaser et al. 2006). Thus, older adults are likely to have children living in closer proximity to provide support. Yet we know little about how children negotiate parental care responsibilities within Mexico.

Overall, the extent to which siblings cooperate to support their parents has received little attention in research on intergenerational support in developing countries (Piotrowski 2008). Currently, the majority of existing research on shared caregiving, among siblings, to older adult parents has been conducted primarily within the United States (Horowitz 1985; Matthews and Rosner 1988; Keith 1995; Piercy 1998; Checkovich and Stern 2002), with a few exceptions in the developing countries such as Taiwan (Lin et al. 2003), Thailand (Piotrowski 2008) and Egypt (Sinunu et al. 2009). This study contributes to this literature by investigating upward transfers of financial and functional support in Mexico City, using data collected from the Mexico City sample of older adults in the 2000 Survey of Health, Well-Being and Aging of Older Adults in Latin America and the Caribbean (Pelaez et al. 2000). Specifically, I use data on the characteristics of children, as provided by the older adult parents, to investigate how children negotiate care responsibilities for their parents across geographies.

Theoretical and Empirical Review

This study views the Mexican family as a strong, stable unit such that propinquity is not always necessary for intergenerational support (Sana and Massey 2005). This does not deny that the abilities of non-coresident children to provide support may be compromised by distance; rather children adjust their support provision to match their circumstances. The modified extended family thesis (Litwak 1960) suggests that the type of support children provide will be contingent on their residential location. Children living further away from parents may provide financial support or maintain communication via telephone while children in closer proximity are more likely to fulfill daily caregiving responsibilities such as housework or related chores (DeVos et al. 2004; Knodel et al. 2010). This perspective offers the possibility to examine the negotiation of support arrangements among children depending on their location and their siblings, in relation to their parents.

On the basis that Mexican family or household members cooperate as a unit and migration is seen as a collective undertaking, it is conceivable that the location of siblings, in relation to their parents, can moderate the support provided by children living outside of the household. Siblings may work together to support their parents by strategically splitting support to reflect their circumstances such that those further away provide money while those in closer proximity provide everyday functional support that requires more time.

Coresidence provides the most immediate opportunity for support exchanges across generations (Choi 2003; Quashie and Zimmer 2013). In Mexico, intergenerational coresidence is the most common living arrangement (Ruggles and Heggeness 2008). On one hand, this may dampen support provided by non-coresident children as they may perceive fewer needs of parents or the household. On the other hand, unmeasured variables that influence a sibling's coresidence with their parents, for instance parent-child relationship quality or the economic or marital stability of the child, can also influence whether the coresident child is able to provide support and the form it may take. These factors implicate the likelihood of support provided by non-coresident children. Thus, coresiding and non-coresiding siblings may substitute their support based on the comparative advantage implied by their location. Therefore, *Hypothesis 1* proposes that children living outside of the household will be more likely to provide financial support and less likely to provide functional support, if they have a sibling within the household.

Within the literature on migration and family support, the New Home Economics of Migration is often cited as it views migration as a household-based strategy. Geographic separation may be vital to the overall well-being of household members in countries that do not have adequate systems for income-smoothing over the life course (Stark and Bloom 1985). One or more family members may migrate, either within the country or abroad, to offset the household's economic vulnerability. Economic migration can generate income, part or all of which can then be remitted to the household for consumption or investment expenditure (Stark and Lucas 1988).

As migration is embedded in a household rather than individual context, parents and other household members often support the migrant before and during their sojourn, and migrants support their parents in return (Root and De Jong 1991). This is aligned with the mutual aid model of intergenerational relations, which proposes that families operate as close-knit networks to maximize the well-being of members. Thus, parents and children provide support according to each other's needs and capacities (Lee et al. 1994). Following from this, non-coresident children may support their parents if they receive economic or other support prior to, during and in their current residence outside of the household (Lillard and Willis 1997; Menjivar 1997; Cong and Silverstein 2011).

Moreover, the new home economics perspective assumes migrants behave altruistically in maintaining their support across space. That is, migrant remittances are inextricably tied to household needs. Financial support is used for consumption, such as purchasing food, or may be used for household investments, such as savings or purchases of assets, to ensure future financial security. This is especially important for households that cannot depend on or are excluded from formal systems of social protection (Durand et al. 1996a; Massey and Espinosa 1997). These propositions are well-supported by research on remittances and remittance behavior by rural and urban migrants of low to middle income households in Latin America and the Caribbean (Itzigsohn 1995; Agarwal and Horowitz 2002).

In Mexico, however, older adults are more dependent on assets for economic security (Mason and Lee 2011). This suggests wealth is more critical than income for intergenerational transfers. As argued by Davies (2008), personal or household wealth is arguably more critical than income for overall well-being as assets provide storage of income that can be called upon during economic hardship. Furthermore, assets can be used as collateral for loans or simply liquidated for cash. The accumulation of wealth can be attributed to the migration and remittance history of older adults who returned to Mexico (Wong et al. 2007) or the past and current migration and remittances of their children. In fact, research by Torche and Spilerman (2008) identifies that circa 2000, the Gini coefficient of wealth inequality in Mexico exceeded that of income inequality, 0.70 versus 0.55.

Therefore, *Hypothesis 2* proposes that household wealth is likely to moderate patterns of financial support provided by non-coresident children. Related to the mutual aid perspective, it is also plausible that children whose parents have greater access to wealth will be more likely to reciprocate the support they received, or conversely, to provide less support if they do not perceive as much of a need for support.

Siblings may cooperate with each other, regardless of their geographic location, to improve or maintain the wealth of the household as this affects the well-being of its members, within and across generations (Heady and Wooden 2004; Torche and Spilerman 2009). This type of cooperation is likely to be more critical to households within the lower or middle of the wealth distribution, where parents' well-being is more likely to be in jeopardy and remittances provide some insurance for consumption smoothing. Conversely, children may provide financial support in order to secure membership in the household and their future inheritance (Hoddinott 1994; VanWey 2004). Furthermore, they may compete with their siblings living in closer proximity to their parents to secure this inheritance. Whether remittances are based on insurance or inheritance is influenced by motivations for migration. In this study, relative wealth is used as a proxy for household need, and the extent of cooperation among non-coresident and coresident children to meet this need.

Apart from geography, support arrangements among children depend on their abilities, gender expectations of support provision, and their competing commitments. This includes their family size, which influences the availability of siblings (Spitze and Logan 1990; Stuifbergen et al. 2008); family status, which includes their marital status and the presence of children (Sun 2002; Lin et al. 2003; Rindfuss et al. 2012); children's potential or actual resources based on their levels of education and employment status (Wolf and Soldo 1994; Lin et al. 2003; Sarkisian and Gerstel 2004); and the gender of the child and gender composition of siblings (Bialik 1992; Wolf et al. 1997; Xie and Zhu 2009; Yount et al. 2012). These factors are not always mutual exclusive.

Data and Methods

Data

Data for this study are drawn from the 2000 Survey on Health, Well-Being and Aging (SABE) in Latin America and the Caribbean, conducted between 1999 and 2000 (Pelaez et al. 2000). The SABE was a multi-center study conducted in seven urban cities of the region, inclusive of Mexico City, Mexico. Mexico's 1999 national household survey was utilized as the sampling frame. Data were collected via personal interviews and self-enumerated questionnaires in the language of each country. The final samples were all derived from multistage stratified clustered sampling with the target universe being persons aged 60 and older residing in private households, occupied by permanent dwellers. The response rate was 83.7 %. For the purposes of this study, the Mexico City sample is restricted to older adults who have at least one living child aged 15 and over and who provide information on all pertinent variables for children and parents. Children were not interviewed therefore the study relied on parents' responses about their children. Based on the sample restrictions the resulting analytical sample totals 3,779 cases of adult children, representing 957 families.

The survey covers detailed information on older adults' health conditions, access to social services, work history, household composition, personal economic status, household assets, and the exchange of support between coresident and non-coresident members. Elderly respondents were also asked detailed information on the characteristics of coresident and non-coresident household members, including their children. This included their marital, residential, education and employment status, age, gender, and parents' provisions of support to children. The analyses focus on these characteristics as they relate to children only, the information for which is all provided by the parent.

Although Mexico City is not representative of all urban areas in Mexico it is the largest city in Mexico and accounted for approximately 20 % of Mexico's total urban population in 2000 (Economic Commission for Latin America and the Caribbean (ECLAC) 2010). Additionally, in the year 2000, Mexico City had the highest median age of 31 (Rhoda and Burton 2010), which indicates that population aging is more concentrated in Mexico City than other parts of the country. Therefore, the findings from this study can serve as a base for comparison to other urban locations within the country.

Measures

Dependent Variables

The *dependent variables, receipt of financial and functional support,* were derived from respondents' answers to the following question of each child: "I would like to ask if (NAME) helps you in any way with (a) money, (b) services like transportation and housework?" The response is dichotomized as either yes, they receive money or help with services from at least one child, or not. There is no indication, however, of the time frame in which parents received support from their children. Given children serve as the unit of analysis in this study, the dependent variables in the following analyses are interpreted as children's provision of financial and functional support. As shown in Table 14.1, all non-coresident children, regardless of their specific location, are more likely to provide financial support relative to functional support. Moreover, children living abroad are most likely to provide financial support.

Independent Variables

The *residential location of the adult child* is the main independent variable, which is derived from information on the location of the child at the time of the survey. This is a categorical variable with children living in the same neighborhood as parents chosen as the reference group. The other categories of distance represent children living in the same city as parents, those living in another city but within the country, and those who lived overseas at the time of the survey. *Siblings' living arrangements* are measured by dummy variables, which represent whether any child has at least one sibling in each of the following locations; coresident with parents, same neighborhood as parents, same city as parents, another city within the country or abroad. Referring to Table 14.1, roughly half the sample of non-coresident children lives in the same city as their parents but approximately 70% have at least one sibling living in the same household as their parents. Gomes (2007) identified that in 2001, the majority of adult children with a living parent lived outside of the household, in the same city as their parents.

Covariates

Family structure is measured on the basis of the gender of the child as well as the number and gender composition of living siblings available to any child. The *number of siblings* is categorical. Children with three or more siblings were chosen as the reference group, compared to those with none, one and two siblings. According to Table 14.1, approximately 90% of children have at least one sibling. Dummy variables are created to indicate that a child has at *least one brother* and *at least one sister*, respectively. *Closest non-coresident child* was created to identify whether any non-coresident child is in fact the closest child to the parent. *Age of the adult child* is measured as a categorical variable and to the extent possible, accounts for life-course stages of children that can be correlated with their likelihoods of providing support. Children aged 35–44 years are chosen as the reference group. The majority of children are in their 30's or 40's.

Table 14.1 Distribution of children's characteristics ($n=3,779$)

Children's characteristics	Percentages/means
Provision of support	
% Provide financial support	71.2
% Provide functional support	41.3
Location of child	
% Same neighborhood	28.7
% Same city	51.4
% Another city	15.9
% Abroad	3.9
Location of siblings	
% At least 1 sibling in household	69.2
% At least 1 sibling in neighborhood	53.3
% At least 1 sibling in same city	76.9
% At least 1 sibling in diff. city	38.8
% At least 1 sibling abroad	12.2
Gender	
% Son	52.0
% Daughter	48.0
Sibship size	
3 or more siblings	89.8
No siblings	0.9
Exactly 1 sibling	3.0
Exactly 2 siblings	6.4
Gender of siblings	
At least 1 brother	93.7
At least 1 sister	93.7
Age	
35–44	42.1
15–24	3.0
25–34	27.5
45–54	22.7
55 and over	4.8
Employment status	
% Working	73.3
% Not working	26.8
Number of working siblings	
Mean (standard deviation)	4.12 (2.18)
Educational attainment	
% Elementary/middle school	34.1
% High school	30.9

Table 14.1 (continued)

Children's characteristics	Percentages/means
% College/university/professional	35.0
Marital status	
% Married	91.0
% Single	4.7
% s/d/w	4.3
Number of own children, mean (standard deviation)	2.52 (1.75)
Assistance from parent	
% Parent helps with money	20.5
% Parent helps with services	27.0
% Parent helps with things	27.9
% Parent helps with child care	20.6

Children's employment status is based on their most recent employment status within a week of the survey. It is measured by a dichotomous variable with those employed as the reference group. The *number of employed siblings* available to a child was also included as a measure of a children's potential cooperation to support their parents. As shown in Table 14.1, approximately three-quarters of children were employed within a week of the survey. Likewise, children have an average of four employed siblings. *Educational attainment* is categorical and children with elementary education are chosen as the reference group. Other categories include high school and those who complete college, university or other professional education. Children's educational attainment is substantially higher than that of their parents'; 31 % completed high school and another 35 % completed tertiary education.

Marital status of the child is assessed by a categorical variable with those married or in some form of partnership chosen as the reference group. Older adults were asked to provide information on the number of children (grandchildren) of each child. Neither the age of grandchildren nor their location at the time of the survey, that is whether they lived with their parents (the adult child), in the household with the grandparents, or elsewhere, was provided. Nevertheless, *number of own children* of each child is included as a continuous variable to account for possible competing responsibilities between generations. The majority of children are married, 91 %, and have an average of 2.5 children.

Finally, to examine and account for the role of mutual aid in motivating children's proximity to their parents and their upward transfers, *parents' provisions of support to their children* were examined by four measures: (1) financial (money); (2) functional (help with housework or transportation); (3) material support (giving food or clothing); and (4) help with child care. Parents were asked if they provided each of these forms of support to each coresident and non-coresident member. Responses were identified for each child and dichotomized to reflect that parents either provided or did not provide the respective support regardless of the location of the child. Parents were more likely to report upward flows of support relative to downward flows.

Parents' characteristics have been shown to influence children's likelihood of providing support, even when children are not in close proximity. Measures include their age; marital, health and economic status; educational attainment; gender; household composition; and household wealth (Smith 1998). In descriptive analyses, not shown, approximately 60% of respondents are women, aged 60–69 years, and married. The majority of parents, 70%, report experiencing fair to poor health. A greater proportion of the sample, 30%, experience difficulties with instrumental activities of daily living relative to difficulties with activities of daily living, 20%. Regarding economic status, 54% of the sample of older Mexicans has access to independent income through labor, pension or a combination of both sources. Despite this, half of the sample is within the lowest income quintile.

Household wealth and overall living standards were measured by a wealth index derived through principal component analysis based on respondents' indications of the quality of housing based on the type of flooring, the number of rooms, having a separate kitchen, toilets; their access to utilities such as electricity and running water; and their possession of consumer durables such as a washing machine, fan, vehicles, bicycles, television, microwave, telephone, radio, water heater, air conditioning, and other similar items (Vogel and Korinek 2012). The resulting wealth index is a combination of all assets weighted by the first principal component scores. These values were then categorized into quintiles (Zimmer 2008). In this Mexico City sample, older adults are evenly distributed among household wealth quintiles.

Analytic Strategy

Separate logistic regression models are estimated for children's provisions of financial and functional support. The theoretical focus is on the location of the adult child and the residential proximity of siblings to parents. It is likely that any child's provision of support will be correlated with their siblings' provision of support or their parents' assessment of the support received from their children. Therefore, observations within a household may be more similar than different. This violates the assumptions of independence among observations for the purpose of regression analyses. To address this issue, all analyses are adjusted for clustering to produce clustered robust standard errors of the estimated coefficients.

To assess the extent to which a non-coresident child's provision of financial and functional support is contingent on the availability of siblings in closer proximity to the parents, interactions were estimated based on the location of children and their availability of at least one sibling within the same household as the parent. Predicted probabilities are presented and discussed to facilitate easier interpretation of the interaction terms and their effects. Furthermore, as adult children's movement away from the parental home often occurs in the context of household needs, the analyses examine the influence of the relative wealth of the household in the origin in shaping patterns of support arrangements among non-coresident children with available siblings living with parents. Predicted probabilities for children's provision of financial support, specifically, are presented and discussed.

Results

Adult Children's Support Provision: Residential Location of Children and Their Siblings

Table 14.2 presents the results of logistic regression models for children's provision of financial and functional support. Net of demographic and economic circumstances of children as well as the needs and resources of their parents, children's residential location was unrelated to their provision of either financial or functional support. Children with at least one sibling residing with their parents, however, showed higher odds of providing both forms of support. Children with a sibling in the same household as their parents had twice the odds of those without a sibling living with their parents to provide money, and eight times the odds of those without a sibling in the household to help their parents with services such as household chores or transportation.

Hypothesis 1 proposed that children living further away from the household will be more likely to provide financial support and less likely to provide functional support if they have a sibling within the household. To test this hypothesis, interaction terms were included to examine differences in the effect of the presence of a coresident sibling according to a child's residential location. These results are presented in Table 14.3. Relative to children living in the same neighborhood, adult children in the same city as their parents with a sibling in the household showed lower odds of providing financial support (Odds Ratio (OR)=1.00 versus 0.74). Regarding functional support, children living in another city with a sibling in the household were less likely than those living in the same neighborhood (OR=0.28 versus 0.31), to provide functional support. To facilitate easier interpretation of the interaction terms, predicted probabilities of financial and functional support were also calculated. Figure 14.1 shows that as distance from the household increased, children with a sibling in the same household as parents had higher probabilities of providing financial rather than functional support thereby providing support for the modified extended family hypothesis. Furthermore, in results not shown, compared to non-coresident children without a sibling in the household, those with a coresiding sibling had higher probabilities of providing both forms of support.

Hypothesis 2 predicted that children's provisions of financial support will be further contingent on the relative vulnerability of the household in the origin such that non-coresident and coresident siblings may cooperate to provide support. In lieu of a three-way interaction, among children's residence outside of the household, the presence of a coresident sibling and the wealth quintile of their household, predicted

Table 14.2 Odds ratios obtained from logistic regression of financial and functional support provision by adult children to older adult parents in Mexico city ($n=3{,}779$)

Children's/parents' characteristics	Financial support	Functional support
Location of child (same neighborhood)		
Same city	0.96 (0.67, 1.38)	0.89 (0.64, 1.22)
Another city	1.11 (0.61, 2.03)	0.79 (0.46, 1.33)
Abroad	1.34 (0.68, 2.63)	0.81 (0.43, 1.55)
Location of siblings		
At least 1 sibling in household	1.85 (1.16, 2.95)**	7.83 (4.66, 13.16)***
At least 1 sibling in neighborhood	0.90 (0.59, 1.36)	1.37 (0.92, 2.04)
At least 1 sibling in same city	1.20 (0.78, 1.84)	1.61 (1.12, 2.30)**
At least 1 sibling in diff. city	1.24 (0.86, 1.78)	1.07 (0.77, 1.48)
At least 1 sibling abroad	1.02 (0.59, 1.76)	1.34 (0.79, 2.29)
Closest non-coresident child (no)		
Yes	0.93 (0.59, 1.46)	0.75 (0.49, 1.15)
Covariates		
Gender (son)		
Daughter	0.86 (0.69, 1.07)	0.97 (0.79, 1.20)
Sibship size (3 or more siblings)		
No siblings	4.86 (1.31, 18.0)*	4.79 (1.14, 20.15)*
Exactly 1 sibling	0.75 (0.32, 1.76)	2.14 (0.97, 4.73)
Exactly 2 siblings	0.92 (0.50, 1.70)	1.48 (0.80, 2.74)
Gender of siblings		
At least 1 brother	1.59 (0.99, 2.57)	1.00 (0.58, 1.72)
At least 1 sister	1.02 (0.61, 1.70)	1.94 (1.16, 3.23)**
Age (35–44)		
15–24	0.86 (0.48, 1.55)	0.68 (0.38, 1.22)
25–34	0.88 (0.69, 1.10)	1.04 (0.85, 1.27)
45–54	0.89 (0.65, 1.23)	0.96 (0.75, 1.24)
55 and over	1.14 (0.60, 2.15)	0.88 (0.55, 1.40)
Employment status (working)		
Not working	1.10 (0.85, 1.43)	0.93 (0.73, 1.19)
Number of working siblings	1.11 (0.99, 1.24)	1.05 (0.96, 1.15)
Educational attainment (primary)		
High school	0.66 (0.47, 0.94)*	1.00 (0.75, 1.33)
College/university/professional	0.75 (0.50, 1.14)	1.01 (0.72, 1.40)
Marital status (married)		
Single	1.00 (0.65, 1.58)	0.53 (0.35, 0.83)**
S/d/w	0.88 (0.54, 1.45)	0.78 (0.48, 1.28)
Number of children	0.96 (0.90, 1.03)	0.97 (0.91, 1.03)
Assistance from parent		
Parent helps with money	1.25 (0.76, 2.03)	2.12 (1.35, 3.34)***

Table 14.2 (continued)

Children's/parents' characteristics	Financial support	Functional support
Parent helps with services	3.04 (1.72, 5.38)***	1.41 (0.94, 2.12)
Parent helps with things	1.18 (0.73, 1.92)	1.15 (0.78, 1.72)
Parent helps with child care	2.08 (1.18, 3.68)**	1.25 (0.81, 1.94)
Parents' characteristics		
Age (70–74)		
60–64	0.89 (0.50, 1.55)	0.82 (0.46, 1.47)
65–69	1.17 (0.65, 2.09)	1.36 (0.77, 2.40)
75–79	3.60 (1.79, 7.25)***	0.89 (0.45, 1.79)
80–84	1.08 (0.49, 2.35)	2.51 (1.22, 5.18)**
85 and older	0.49 (0.17, 1.39)	2.23 (0.80, 6.25)
Gender (women)		
Men	0.81 (0.50, 1.32)	0.72 (0.45, 1.15)
Marital status (married)		
Widowed	1.04 (0.63, 1.69)	1.55 (0.99, 2.42)*
Never married	0.98 (0.50, 1.93)	1.65 (0.88, 3.12)
Residual household size	0.98 (0.89, 1.09)	0.94 (0.86, 1.04)
Residual household assistance (none)		
Receives at least 1 form of support	0.98 (0.61, 1.55)	1.85 (1.20, 2.86)**
Self-rated health (very good)		
Good	1.83 (0.80, 4.16)	0.59 (0.28, 1.25)
Fair	2.52 (1.09, 5.84)*	0.57 (0.28, 1.19)
Poor	2.93 (1.15, 7.52)*	0.61 (0.28, 1.36)
Disability		
Difficulty with at least 1 ADL	1.33 (0.76, 2.31)	1.44 (0.84, 2.47)
Difficulty with at least 1 IADL	0.78 (0.47, 1.28)	1.09 (0.66, 1.79)
Educational attainment (none)		
Primary	0.71 (0.43, 1.18)	0.96 (0.61, 1.50)
High-school	0.65 (0.33, 1.28)	0.49 (0.25, 0.96)*
Tertiary	0.21 (0.09, 0.50)***	0.57 (0.22, 1.44)
Work/pension status (no work/no pension)		
Pension only	0.48 (0.24, 0.95)*	1.06 (0.53, 2.14)
Work and pension	0.44 (0.19, 0.99)*	1.12 (0.49, 2.55)
Work only	0.44 (0.23, 0.83)*	1.71 (0.92, 3.17)
No info on work or pension	0.52 (0.28, 0.96)*	0.99 (0.58, 1.69)
Income quintile (I)		
III	1.10 (0.54, 2.26)	1.48 (0.72, 3.01)
IV	1.20 (0.62, 2.31)	0.60 (0.31, 1.14)
V	0.55 (0.29, 1.01)	0.50 (0.26, 0.97)*
HH wealth quintile (I)		

Table 14.2 (continued)

Children's/parents' characteristics	Financial support	Functional support
II	1.50 (0.82, 2.75)	1.78 (1.01, 3.13)*
III	2.54 (1.34, 4.79)**	1.47 (0.85, 2.55)
IV	2.02 (1.06, 3.85)*	1.64 (0.96, 2.79)
V	1.48 (0.77, 2.87)	1.46 (0.80, 2.68)
Constant	0.50 (0.11, 2.31)	0.04 (0.01, 0.17)***
Pseudo-R2	0.2174	0.2185
Wald Chi2	199.76***	200.56***
Number of clusters/families	957	957

$*p<0.05$; $**p<0.01$; $***p<0.001$; 95 % confidence intervals in parentheses

probabilities of financial support were calculated based on the interaction models presented in Table 14.3 as well as the wealth quintile of the household in the origin. Figure 14.2 shows that children with a sibling in the household, regardless of their location, were most likely to provide financial support to parents within the middle wealth quintile. In contrast, children whose parents were in the lowest wealth quintile were least likely to provide financial support.

Table 14.3 Odds ratios obtained from a logistic regression of financial and functional support provision by adult children to older adult parents in Mexico City, showing interaction effects ($n=3,779$)

Children's characteristics	Financial support	Functional support
Location of child (same neighborhood)		
Same city	1.48 (0.91, 2.41)	1.02 (0.56, 1.88)
Another city	1.43 (0.65, 3.14)	0.26 (0.10, 0.67)**
Abroad	1.77 (0.66, 4.77)	1.64 (0.58, 4.64)
Location of siblings		
At least 1 sibling in household	2.78 (1.46, 5.32)**	7.75 (4.20, 14.29)***
Interactions		
Location of child × sibling in HH		
Same city × sibling in HH	0.50 (0.27, 0.91)*	0.86 (0.47, 1.57)
Another city × sibling in HH	0.67 (0.30, 1.50)	3.53 (1.44, 8.66)**
Abroad × sibling in HH	0.65 (0.20, 2.13)	0.38 (0.12, 1.25)
Constant	0.36 (0.08, 1.71)	0.04 (0.01, 0.18)***
Likelihood ratio-test[a]	11.96**	20.31***
Pseudo-R2	0.22	0.2224
Wald Chi2	204.32***	220.19***
Number of clusters/families	957	957
Number of observations	3779	3779

$*p<0.05$; $**p<0.01$; $***p<0.001$; 95 % confidence intervals in parentheses
[a]Compared to the base models presented in Table 14.2

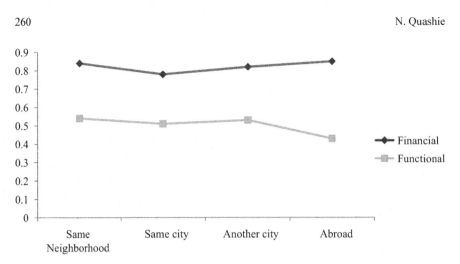

Fig. 14.1 Mean predicted probabilities of financial and functional support according to the location of the adult child and the availability of a coresident sibling ($n=3,779$)

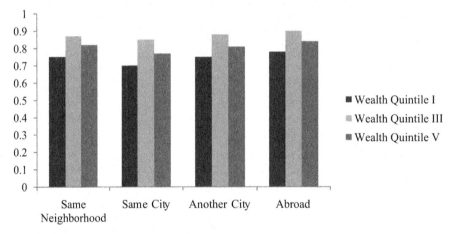

Fig. 14.2 Mean predicted probabilities of financial support by the location of the adult child, the availability of a coresident sibling and the household's location in the wealth distribution ($n=3,779$)

Discussion

Research on shared caregiving, among siblings, to their older adult parents in developing countries is nascent. Thus far, the majority of such investigations have been conducted in the United States and relatively few developing countries. Mexico is an interesting setting for this area of research as it differs from the United States, and is similar to other developing regions, on some critical dimensions which can implicate family support. First, due to its later onset of fertility decline, the average number of children to current cohorts of older adults is larger than in the United States. Second, similar to other developing countries within and outside of Latin America and the Caribbean, the family or household unit is a critical source of support for older adults (Rawlins 1999; Varley and Blasco 2000). Finally, as observed in other developing countries (Bian et al. 1998; Saad 2005) family members, especially children maintain ties across space, and continue to provide support regardless of distance to their parents. Yet, we know little about how children negotiate care arrangements among each other once they have left the parental home.

The current study contributes to this gap in the literature by examining patterns of financial and functional support provided by children to their older parents in Mexico City according to the geographic location of children and their siblings. Moreover, unlike prior research on family-based intergenerational support in Latin America and the Caribbean where upward flows of support are assessed with children as an aggregate (De Vos et al. 2004; Wong et al. 2007; Quashie and Zimmer 2013), this study examines support from the perspective of the children. Thus far, the examination of children's characteristics and their association with parental support within Latin America and the Caribbean has been conducted in Northern Brazil (Saad 2005). This type of analysis allows more direct examination of how children's personal characteristics, including the location of their siblings, condition their support.

The findings show that children living away from the parental home support their parents in Mexico City, regardless of their precise location. Net of the child's location, the presence of a coresiding sibling is associated with an increased likelihood of providing both forms of support. Interactions, however, indicate that the relationship between the presence of a coresident sibling and the likelihood of providing support differs by the location of the non-coresident child. Non-coresident children with a sibling in the household, especially those living in another city or abroad, had higher probabilities of providing financial rather than functional support. This finding suggests potential cooperation among siblings to support their parents based on the comparative advantage offered by their location, thus providing support for the modified extended family perspective.

Parents' resources also condition support arrangements among children. Non-coresident children with a sibling in the household were most likely to provide financial support to parents in the middle wealth quintiles relative to parents in households with less or more wealth. Moreover, children living abroad at the time of the survey had the highest probability of providing such support. This result may reflect differences in the patterns of intergenerational exchange, which may be related to differences in perceived need, across wealth quintiles. Descriptive analyses, not shown, indicate that the highest proportion of older adults providing and receiving financial support were found in the middle wealth quintile (quintile III). Drawing upon the migration literature, migrants are self-selected by resources in that those with access to more resources prior to migration, economic and otherwise, are more likely to emigrate (Portes and Rumbaut 2006). Migrant children who receive support will then provide support in return. Relative to households in the higher echelons, this exchange of support may be more critical for households in the middle wealth quintile in order to circumvent falling into poverty.

Although there is support for the second hypothesis that children's provision of support will be moderated by the relative wealth of the household, the degree of cooperation or competition among children is uncertain. Research on Mexican migration and remittances (Durand et al. 1996) has shown that migrants are more likely to remit if they have assets in the home country. If the provision of financial support is motivated by self-interest for future inheritance, non-coresident children may be competing with their coresident siblings for favor. At the same time, non-coresident and coresident siblings may be working together to pool incomes to maintain the economic standing of the household. Affirmative conclusions from these findings cannot be drawn, however, as the data do not give any indication of the ownership of household assets, the relative contribution of parents and children to these assets or children's intentions for migration. Therefore, children may be more entitled to and invested in the household if they have cumulatively, over time, contributed to the current economic standing of the household. Likewise, children may be more or less likely to provide financial support according to their level of contribution to the household's assets.

Limitations

Foremost, information regarding children's characteristics and their support behavior are all provided by the parents. Parents' recollections of support from their children are likely to differ. Relatedly, the estimated effects neither account for the frequency of support nor the amount of support that is given at any point in time by children and their siblings. This implicates the observed patterns of support from children in the present analysis. An additional drawback is the lack of detailed information on children's employment status, whether full time or part-time. Such data provide more holistic understanding on the degree to which children negotiate caregiving support arrangements. Additionally, information of children's income, both absolute and relative, can provide stronger assessments of cooperation among siblings. Children in better financial positions relative to their siblings may bargain to provide financial rather than functional support, even if they live within the same location and relative distance, from their parents. Finally, this sample is restricted to one urban location but wealth inequalities, family size, and migration patterns are likely to differ in rural areas and other urban locations. A more complete understanding of family support can be provided by future research that compares other urban and rural areas in Mexico.

Conclusion

This study provides several important insights for aging research in Mexico and the wider Latin America and Caribbean region. First, comprehensive assessments of intergenerational support require inclusion of children's characteristics and that of their siblings. The latter is particularly important in Mexico as families, until recently, have been larger than found in more developed societies within and outside of Latin America and the Caribbean. Additionally, given the weak institutional contexts of support for older and younger cohorts, parents are likely to need support from multiple children and children are likely to depend on each other to meet the needs of their parents. The findings also underscore the importance of examining care-giving and well-being in single-child families. As shown, children without siblings were five times as likely as those with three or more siblings to provide both financial and functional support. Although this a small proportion of the current sample of families, declining fertility increases the likelihood of single child families. This has implications for the well-being of older adults, whose sole child may live at far distances and not be able to provide support as needed. There are also implications for the well-being of the caregivers who are burdened with the sole responsibility of their parents' well-being.

Second, future research on intergenerational support in Mexico should examine the gendered dimensions of siblings' systems of support, giving attention to whether non-coresident sons' and daughters' provision of support is further conditioned by the gender composition of their siblings in closer proximity to their parents. This expands our understanding of gender roles in family care and how these may be reinforced or adjusted by geographic separation. The findings from this study show that the gender composition of siblings influences support transfers in that non-coresident children with at least one sister are more likely to provide functional support.

Finally, intergenerational support is embedded in mutual aid and exchange of resources. This reflects the sustainability of social norms regarding family support, which may be reinforced by the insecurity of social protection systems for both younger and older cohorts both within and outside of Mexico. In light of weak and unequal institutional support systems, current demographic, economic and social changes in Mexico provide unmeasured opportunity for researchers and policy makers to evaluate how children negotiate support for older adults and the geographic, ethnic, class and gender differences in these patterns.

References

Agarwal, R., & Horowitz, A. W. (2002). Are international remittances altruism or insurance? Evidence from Guyana using multiple-migrant households. *World Development, 30,* 2033–2044. doi:10.1016/S0305-750X(02)00118-3.

Agree, E. M., & Glaser, K. (2009). The demography of informal caregiving. In P. Uhlenberg (Ed.), *The international handbook of population aging* (pp. 647–668). Dordrecht: Springer.

Aguila, E., Diaz, C., Fu, M. M., Kapteyn, A., & Pierson, A. (2011). *Living longer in Mexico: Income security and health*. Washington, DC: AARP.

Bengston, V. L., & Roberts, R. E. L. (1991). Intergenerational solidarity in aging families: An example of formal theory construction. *Journal of Marriage and Family, 53,* 853–870. doi:10.2307/352993.

Bialik, R. (1992). Family care of the elderly in Mexico. In J. I. Kosberg (Ed.), *Family care of the elderly: Social and cultural changes* (pp. 31–46). London: Sage.

Bian, F., Logan, J. R., & Bian, Y. (1998). Intergenerational relations in urban China: Proximity, contact, and help to parents. *Demography, 35*(1), 115–124.

Checkovich, T., & Stern, S. (2002). Shared caregiving responsibilities of adult siblings with elderly parents. *Journal of Human Resources, 37,* 441–478. doi:10.2307/3069678.

Choi, N. G. (2003). Coresidence between unmarried aging parents and their adult children: Who moves in with whom and why? *Research on Aging, 25,* 384–404. doi:10.1177/01640275030 25004003.

Cong, Z., & Silverstein, M. (2011). Intergenerational exchange between parents and migrant and non-migrant sons in rural China. *Journal of Marriage and Family, 73,* 93–104. doi:10.1111/j.1741-3737.2010.00791.x.

Cotlear, D., & Tornarolli, L. (2011). Poverty, aging, and the life cycle in Latin America. In D. Cotlear (Ed.), *Population aging: Is Latin American ready?* (pp. 79–134). Washington, DC: The International Bank for Reconstruction and Development.

Davies, J. B. (2008). An overview of personal wealth. In J. B. Davies (Ed.), *Personal wealth from a global perspective* (pp. 1–23). Oxford: Oxford University Press.

De Vos, S., Solis, P., & Montes de Oca, V. (2004). Receipt of assistance and extended family residence among elderly men in Mexico. *International Journal of Aging and Human Development, 58*(1), 1–27.

Donato, K. M. (1994). US policy and Mexican migration to the United States, 1942–1992. *Social Science Quarterly, 75*(4), 705–729.

Durand, J., Kandel, W., Parrado, E. A., & Massey, D. S. (1996a). International migration and development in Mexican communities. *Demography, 33*(2), 249–264.

Durand, J., Parrado, E. A., & Massey, D. S. (1996b). Migradollars and development: A reconsideration of the Mexican case. *International Migration Review, 30,* 423–444. doi:10.2307/2547388.

Economic Commission for Latin America and the Caribbean. (2010). *Statistical yearbook for Latin America and the Caribbean, 2010.* Santiago: Economic Commission for Latin America and the Caribbean.

Fritzell, J., & Lennartsson, C. (2005). Financial transfers between generations in Sweden. *Ageing and Society, 25,* 397–414. doi:10.1017/S0144686X04003150.

Glaser, K., Agree, E. M., Costenbader, E., Camargo, A., Trench, B., Natividad, J., et al. (2006). Fertility decline, family structure, and support for older persons in Latin America and Asia. *Journal of Aging and Health, 18,* 259–291. doi:10.1177/0898264305285668.

Gomes, C. (2007). Intergenerational exchanges in Mexico: Types and intensity of support. *Current Sociology, 55*(4), 545–560.

Heady, B., & Wooden, M. (2004). The effects of wealth and income on subjective well-being and ill-being. *Economic Record, 80*(1), S24–S33.

Hoddinott, J. (1994). A model of migration and remittances applied to Western Kenya. *Oxford Economic Papers, 46*(3), 459–476.

Horowitz, A. (1985). Sons and daughters as caregivers to older parents: Differences in role performance and consequences. *The Gerontologist, 25,* 612–617. doi:10.1093/geront/25.6.612.

Itzigsohn, J. (1995). Migrant remittances, labor markets, and household strategies: A comparative analysis of low income household strategies in the Caribbean basin. *Social Forces, 74,* 633–655. doi:10.2307/2580495.

Izazola, H. (2004). Migration to and from Mexico City, 1995–2000. *Environment and Urbanization, 16*(1), 211–230.

Keith, C. (1995). Family caregiving systems: Models, resources and values. *Journal of Marriage and Family, 57,* 179–189. doi:10.2307/353826.

Knodel, J., Friedman, J., Si Anh, T., & Cuong, B. T. (2000). Intergenerational exchanges in Vietnam: Family size, sex composition and the location of children. *Population Studies, 54,* 89–104. doi:10.1080/713779067.

Knodel, J., Kespichayawattana, J., Saengtienchai, C., & Wiwatwanich, S. (2010). How left behind are rural parents of migrant children? Evidence from Thailand. *Ageing and Society, 30,* 811–841. doi:10.1017/S0144686X09990699.

Lee, Y., Parish, W. L., & Willis, R. J. (1994). Sons, daughters and intergenerational support in Taiwan. *American Journal of Sociology, 99*(4), 1010–1041.

Lillard, L. A., & Willis, R. J. (1997). Motives for intergenerational transfers; evidence from Malaysia. *Demography, 34,* 115–134. doi:10.2307/2061663.

Lin, I. F., Goldman, N., Weinstein, M., Lin, Y. H., Gorrindo, T., & Seeman, T. (2003). Gender differences in adult children's support of their parents in Taiwan. *Journal of Marriage and Family, 65*(1), 184–200.

Litwak, E. (1960). Geographic-mobility and extended family cohesion. *American Sociological Review, 25*(3), 385–394.

Mason, A., & Lee, R. (2011). Population aging and the generational economy: Key findings. In R. Lee & A. Mason (Eds.), *Population aging and the generational economy: A global perspective* (pp. 3–31). Cheltenham: Edward Elgar Publishers.

Massey, D. S., & Espinosa, K. E. (1997). What's driving Mexico-U.S. migration? A theoretical, empirical, and policy analysis. *The American Journal of Sociology, 102*(4), 939–99.

Massey, D. S., Durand, J., & Malone, N. J. (2003). *Beyond smoke and mirrors: Mexican immigration in the era of economic integration.* New York: Russell Sage Foundation.

Matthews, S. H., & Rosner, T. T. (1988). Shared filial responsibility: The family as the primary caregiver. *Journal of Marriage and Family, 50*(1), 185–195.

McGarry, K., & Schoeni, R. F. (1997). Transfer behavior within the family: Results from the asset and health dynamics study. *The Journals of Gerontology Series B: Psychological Sciences and Social Sciences, 52*(Special Issue), 82–92.

Menjivar, C. (1997). Immigrant kinship networks: Vietnamese, Salvadoreans and Mexicans in comparative perspective. *Journal of Comparative Family Studies, 28*(1), 1–24.

Parker, S. W., & Wong, R. (2001). Welfare of male and female elderly in Mexico: A comparison. In E. G. Katz & M. C. Correia (Eds.), *The economics of gender in Mexico: Work, family, state, and market* (pp. 249–290). Washington, DC: The World Bank.

Pelaez, M., Palloni, A., Albala, C., Alfonso, J. C., Ham-Chande, R., Hennis, A., et al. (2000). *SABE—survey on health, well-being, and aging in Latin America and the Caribbean, 2000. ICPSR03546-v1.* Ann Arbor: Inter-university Consortium for Political and Social Research (distributor), 2005. http://doi.org/10.3886/ICPSR03546.v1. Accessed 20 Sept 2010.

Piercy, K. W. (1998). Theorizing about family caregiving: The role of responsibility. *Journal of Marriage and Family, 60,* 109–118. doi:10.2307/353445.

Piotrowski, M. (2008). Sibling influences on migrant remittances, evidence from Nang Rong, Thailand. *Journal of Population Ageing, 1,* 193–224. doi:10.1007/s12062-009-9011-7.

Portes, A., & Rumbaut, R. (2006). *Immigrant America: A portrait* (3rd ed.). Berkeley: University of California Press.

Quashie, N., & Zimmer, Z. (2013). Residential proximity of nearest child and older adults' receipts of informal support transfers in Barbados. *Ageing and Society, 33,* 320–341. doi:10.1017/S0144686X1100122X.

Rawlins, J. M. (1999). Confronting ageing as a Caribbean reality. *Journal of Sociology & Social Welfare, 26*(1), 143–153.

Rhoda, R., & Burton, T. (2010). *Geo-Mexico: The geography and dynamics of modern Mexico.* British Columbia: Sombrero Books.

Rindfuss, R. R., Piotrowski, M., Entwisle, B., Edmeades, J., & Faust, K. (2012). Migrant remittances and the web of family obligations: Ongoing support among spatially extended kin in North-East Thailand, 1984–1994. *Population Studies, 66,* 7–104. doi:10.1080/00324728.2011.644429.

Rofman, R., & Lucchetti, L. (2006). *Pension systems in Latin America: Concepts and measurements of coverage.* Washington, DC: The World Bank.

Root, B. D., & De Jong, G. F. (1991). Family migration in a developing country. *Population Studies, 2,* 221–233. doi:10.1080/0032472031000145406.

Ruggles, S., & Heggeness, M. (2008). Intergenerational coresidence in developing countries. *Population and Development Review, 34,* 235–281. doi:10.1111/j.1728-4457.2008.00219.x.

Saad, P. M. (2005). Informal support transfers of the elderly in Brazil and Latin America. In A. A. Camarano (Ed.), *Sixty plus: The elderly Brazilians and their new social roles* (pp. 169–210). Rio de Janeiro: Institute of Applied Economic Research.

Sana, M., & Massey, D. S. (2005). Household composition, family migration and community context: Migrant remittances in four countries. *Social Science Quarterly, 86,* 509–528. doi:10.1111/j.0038-4941.2005.00315.x.

Sarkisian, N., & Gerstel, N. (2004). Explaining the gender gap in help to parents: The importance of employment. *Journal of Marriage and Family, 66,* 431–451. doi:10.1111/j.1741-3737.2004.00030.x.

Sinunu, M., Yount, K. M., & El Afify, N. A. W. (2009). Informal and formal long-term care for frail older adults in Cairo, Egypt: Family caregiving decisions in a context of social change. *Journal of Cross-Cultural Gerontology, 24,* 63–76. doi:10.1007/s10823-008-9074-6.

Smith, G. C. (1998). Residential separation and patterns of interactions between elderly parents and their adult children. *Progress in Human Geography, 22,* 368–384. doi:10.1191/030913298673626843.

Spitze, G., & Logan, J. (1990). Sons, daughters, and intergenerational social support. *Journal of Marriage and Family, 52,* 420–430. doi:10.2307/353036.

Stark, O., & Bloom, D. E. (1985). The new economic of labor migration. *The American Economic Review, 75*(2), 173–178.

Stark, O., & Lucas, R. E. B. (1988). Migration, remittances and the family. *Economic Development and Cultural Change, 36*(3), 465–481.

Stuifbergen, M. C., Van Delden, J. J. M., & Dykstra, P. A. (2008). The implications of today's family structures for support giving to older parents. *Ageing and Society, 28,* 413–434. doi:10.1017/S0144686X07006666.

Sun, R. (2002). Old age support in contemporary urban china from both parents' and children's perspectives. *Research on Aging, 24,* 337–359. doi:10.1177/0164027502243003.

Torche, F., & Spilerman, S. (2008). Household wealth in Latin America. In J. B. Davies (Ed.), *Personal wealth from a global perspective* (pp. 150–176). Oxford: Oxford University Press.

Torche, F., & Spilerman, S. (2009). Intergenerational influences of wealth in Mexico. *Latin American Research Review, 44,* 75–101. doi:10.1353/lar.0.0089.

VanWey, L. K. (2004). Altruistic and contractual remittances between male and female migrants and households in rural Thailand. *Demography, 41*(4), 739–756.

Varley, A., & Blasco, M. (2000). Intact or in tatters? Family care of older women and men in urban Mexico. *Gender and Development, 8,* 47–55. doi:10.1080/741923623.

Vogel, A., & Korinek, K. (2012). Passing by the girls? Remittance allocation for educational expenditures and social security in Nepal's households 2003–2004. *International Migration Review, 46,* 61–100. doi:10.1111/j.1747-7379.2012.00881.x.

Wolf, D. A., & Soldo, B. J. (1994). Married women's allocation of time to employment and care of elderly parents. *The Journal of Human Resources, 29*(4), 1259–1276.

Wolf, D. A., Freedman, V. A., & Soldo, B. J. (1997). The division of family labor: Care for elderly parents. *Journal of Gerontology Series B: Psychological Sciences and Social Sciences, 52,* 109–109. doi:10.1093/geronb/52B.Special_Issue.102.

Wong, R., & Higgins, M. (2007). Dynamics of intergenerational assistance in middle-and old-age in Mexico. In J. K. Angel & K. E. Whitfield (Eds.), *The health of aging Hispanics* (pp. 99–120). New York: Springer.

Wong, R., Palloni, A., & Soldo, B. J. (2007). Wealth in middle and old age in Mexico: The role of international migration. International Migration Review, *41,* 127–151. doi:10.1111/j.1747-7379.2007.00059.x.

Xie, Y., & Zhu, H. (2009). Do sons or daughters give more money to parents in urban China? *Journal of Marriage and Family, 71,* 174–186. doi:10.1111/j.1741-3737.2008.00588.x.

Yount, K. M., Cunningham, S. A., Engelman, M., & Agree, E. M. (2012). Gender and material transfers between older parents and children in Ismalia, Egypt. *Journal of Marriage and Family, 74,* 116–131. doi:10.1111/j.1741-3737.2011.00881.x.

Zimmer, Z. (2008). Poverty, wealth inequality and health among older adults in rural Cambodia. *Social Science & Medicine, 66,* 57–71. doi:10.1016/j.socscimed.2007.08.032.

Chapter 15
Texas Self-Help Informal Settlement and *Colonia* Housing Conditions, Aging, and Health Status

Peter M. Ward

Introduction: *Colonias* and Aging

Two major datasets and studies offer excellent health and morbidity panel data for Mexican and Mexican-American populations among the elderly. First, in Mexico the Mexican Health and Aging Study (MHAS) contains panel data from 2001 for a representative sample of people born in 1951 or earlier, and which has been extended with follow-up sampling for 2003, 2012 and (proposed) 2015 (http://www.mhasweb.org). The second, Hispanic Established Population for Epidemiological Study of the Elderly (HEPESE), a multiple-wave study on Mexican American populations aged 65 or over at the time of the baseline survey in 1993–1994. However, neither these databases, nor most other major surveys (including the US Census) provide detailed data about housing conditions that can be readily linked to epidemiological outcomes. Thus it is often difficult to analyze the ways in which impaired health and mobility may be associated to specific housing conditions and household residential arrangements.

This paper will explore some of these issues in the specific context of self-built or self-managed housing in two informal homestead subdivisions (IfHSs) located in the rural hinterland several miles outside of the city of San Marcos in Guadalupe County, Central Texas. Akin to their more widely known counterpart *colonias* in the Texas border region, the populations that live in these settlements are poor or very poor home owners who acquire or produce their housing informally by purchasing un-serviced lots in rural areas outside city boundaries and either self-build a home, or place a manufactured home on site, often extending the dwelling into a hybrid structure over time (Ward 1999). *Colonias* in the border are almost exclusively Hispanic (Mexican origin) and low income, as are the residents of IfHSs elsewhere in Texas (although in these cases a minority is also likely to be Caucasian). By residential standards lots are quite large—half to one acre or more—and the yard, even if unimproved, offers an important amenity space for the household (Sullivan and Ward 2012).

P. M. Ward (✉)
The LBJ School of Public Affairs and The Department of Sociology,
The University of Texas, Austin, TX, USA

© Springer International Publishing Switzerland 2015
W. A. Vega et al. (eds.), *Challenges of Latino Aging in the Americas,*
DOI 10.1007/978-3-319-12598-5_15

An earlier contribution to this conference series describes how even poor housing conditions in *colonias* and IfHSs offer pragmatic opportunities for home ownership among Hispanic populations in the US, as well as the flexibility to adjust household dynamics to lot and housing arrangements throughout the life course (Ward 2007). Moreover, in that paper it was proposed that these neighborhoods and housing arrangements may, in the future, play an increasingly important role as spaces in which elderly parents and grandparents are accommodated, both as part of family reunification among first generation Mexican households, and as Mexican American households and families discover that they cannot afford offsite formal residential care for their aging parents. Most alternative dwelling arrangements such as rental apartments or formal housing estates do not readily allow for family extension and shared home care space for elderly kin. However, an advantage of *colonias* and IfHSs is that the large lot size, low density settlement, and greater likelihood that neighbors will be physically around to watch out for one another, offers flexible and self-managed housing arrangements in which to care for sick, immobile, or an Alzheimer-suffering elderly family member (Varley 2008). However, at the same time, the dwelling structures and often deteriorated housing conditions can also produce poor health outcomes, especially for the elderly, and this current chapter offers some preliminary observations drawing primarily upon survey data collected for two self-help neighborhoods.

Colonias, Housing Conditions and (Ill)Health

Much of the early (late 1980s) concerns about *colonias* focused both upon the appalling housing conditions as well as the extreme health hazards associated with low-income settlements that had little or no services of water, drainage, power supply, etc. Rates of shigellosis and hepatitis A were twice that of the national average, occasional cholera outbreaks occurred, and gastroenteritis and respiratory disease were rife (Ward 1999; Anders et al. 2008, 2010). Moreover, health problems were exacerbated by the relative isolation of *colonias* to health care providers and institutions (Mier et al. 2008). State intervention and legislation to improve and extend infrastructure to *colonias* from the early and mid-1990s onwards began to alleviate some of the exposure to these health hazards and extreme housing conditions, most notably through providing access to water supplies, regulating drainage and septic fields, by providing some levels of consumer protection to would be home owners, and by limiting the development of new *colonias*—at least in the border region (Ward 1999). In 2006, Senate Bill 827 established a three-level color-coded classification to identify *colonias* exposed to severe public health risks throughout the six Texas counties with the largest colonia populations, all located along the border. "Red" *colonias* are those that pose the greatest health and safety risks because they lack piped water and wastewater disposal; "yellow" *colonias* are those with adequate water and wastewater systems but which, due to a lack of road paving, drainage, or solid waste disposal, still pose certain health risks; and "green" *colonias* being those with adequate infrastructure that pose minimal health and safety risks (Durst and Zhang 2013; Durst 2014).

As is the case in Mexico among low-income populations who aspire to owning their home, for the most part the actual process of creating shelter is left to the families themselves, either through self-building or through the self-management of home sites.[1] In Texas the latter is most usually achieved by moving a second hand or a new manufactured (trailer) home onto a home site which used has minimal or absent service infrastructure. While legislative controls were imposed in 1995 upon any new *colonia* development in the border region, IfHSs continue to proliferate elsewhere, and although the median income is generally $ 10,000 higher (around $ 25,000 household per year), IfHSs remain the primary route to home ownership among Mexican and Mexican American populations whose incomes are too low or irregular to qualify for home mortgages through the formal sector. The same situation applies, of course, to lower-income Caucasian households that also choose to live in these settlements, albeit as a relatively small minority group compared to Hispanics. In Texas most *colonias* were developed and platted in the 1980s, with many households taking up occupancy around that time and through the 1990s, although a significant proportion of lots remain vacant to this day (Ward and Carew 2001; Drust et al. 2012). Many IfHSs also date to that period, although generally they have been far less visible to policy makers who were mostly concerned with the 350–400,000 people estimated to be living in some 1400 border colonias at that time.[2]

The relatively long period since these neighborhoods were first established has several important implications for the following discussion of housing conditions, aging, and health outcomes. First, as in Mexico and other developing countries, self-help housing shows a trajectory of expansion and consolidation over time, and is invariably in synch with the family building and life course. Moreover, both settlements and dwelling structures are improved or "consolidated" over time. Survey work across two time horizons (2002–2010) in a number of *colonias* outside of Rio Grande City show significant improvements in community level infrastructure, median home values of $ 47,500 in 2011, and median household expenditures on improvements of $ 9,500 over that time period (Durst and Ward 2014). The large majority of the dwellings that were restudied after 10 years showed significant improvement and consolidation, although less than two-fifths of owners regarded their houses as "finished", and most expected to continue to make improvements or add extensions. Second, most household heads were usually aged in their early thirties at the time they first came to reside in the colonia, and were then most likely to be nuclear families with young children. Third, once the family had established itself

---

[1] This contribution will focus exclusively on the USA. However, many of the insights relating to informality, self-help housing, caring for the elderly, household and life course dynamics, transgenerational asset transfers and inheritance come from Mexico (see Ward et al. 2014; www.lahn.utexas.org, Varley 2008).

[2] Today the population is estimated to be closer to half a million, as vacant lots in existing colonias have been settled. Nor is this a solely Texas phenomenon: *colonias* are found extensively in New Mexico and Arizona (Donelson and Esparza 2010), not to mention the extensive IfHSs found outside of the border in Texas and elsewhere (Ward and Peters 2007).

in the neighborhood, there is relatively little outward mobility of home owners, in part because of the ongoing "use-value" to the household, and partly because the lack of formal financing stymies the market, making sale of one's home to other low-income buyers extremely difficult. As in Mexico and elsewhere, once established, self-builder home owners have little opportunity for spatial mobility and, de facto, "a home is forever" (Gilbert 1999; Ward 2012).

These last two factors have led to an aging population of resident owners today, many of whom after 30 years or more of residence are now entering late middle age. Most of their children have left home to establish their own households elsewhere, such that dwelling densities have declined, and the proportion of elderly-couple households are on the rise (Ward and Durst 2014). Although adult children and grandchildren occasionally share the lot with their aging parents, more often they live nearby, leaving their parents as vestige households living on what are now very low (largely retirement) incomes. For these elderly families, home improvement and upgrades, retrofits, and routine maintenance are less affordable, such that housing conditions are more likely to be deteriorating. Americans are estimated to spend 80 % of their time indoors—this is even higher among the elderly—and low-income populations spend a higher proportion of their income on heating and A/C than do middle and upper-income households, such that the elderly are especially likely to be "fuel poor" (Olmedo 2014). Faced with conditions of deteriorating housing, a lack of access to home weatherization improvements such as better dwelling insulation, many of the elderly couples living in *colonias* and IfHSs are less able to control the internal micro climate within their home, resulting in likely exacerbation of their poor health status.

The following two figures show a fairly typical example of self-managed and self-built housing in Texas *colonias*. This particular case comes from a *colonia* outside of Rio Grande City and formed one of several intensive case studies undertaken in 2011 where several researchers conducted multiple interviews with family members in order to better understand the process of build-out in relation to family expansion, home improvement financing, and the use of space inside the home, as well as outside in the yard. The lot was purchased in 1987 for $ 1,700 and the family (then the couple and three children) began living in the 45' × 10' trailer. Figure 15.1 shows the build-out process which started with the trailer home that was later added-to (1993) through self-building and contracted labor in a manner that connected the two sections of the home although there is a 6" gap between the two sections (making it difficult to achieve an effective energy "envelope"). The septic tank was later moved from the far rear corner of the lot to a new tank closer to the house, and a covered workspace was added (Fig. 15.2).

The family of five lived in the trailer until the eldest son left in 1988 when he got married, which allowed changes in the room allocation and use. The family began working on the extension and when the daughter became pregnant in 1991, she got married and her husband and child remained living in the home, again with some rotation in the room use as the mother and father moved into the extension. Meanwhile the couple's youngest son continued to live in the trailer section. Around the same time the daughter purchased the lot opposite and gradually began to build on

3D Build-out Diagram

Fig. 15.1 Trailer and home extension build out, Las Lomas Colonia, Starr County

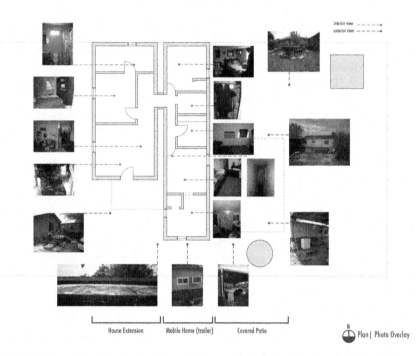

Fig. 15.2 Trailer and home extension plans and room photographs. Las Lomas Colonia, Starr County

the new site while continuing to live with her family (husband and by 2000 three young children) in her parents' home. Relatively recently (2010) she and her family moved out, leaving the elderly couple and their single remaining (now adult) son in the dwelling, although he too expects to move out in the short term which will then essentially close out the need to use the trailer section. Having suffered serious overcrowding during most of the previous 20 years, downsizing has today left several rooms unused or partially used as storage space.

Fig. 15.3 Owner's home in
B&H Colonia (Starr County)
on steeply inclined lot, showing daughter's house below
(*left*) and steep steps and
unfenced hazard (*front and
right of photo*.)

In 2011 the parents are aged 67 and 64, and the mother in particular has serious health problems (diabetes, poor eyesight from a detached retina, and suffers from depression), and relies heavily upon care and support from her daughter now living opposite. The father has mobility problems and carries a walking stick. The roof leaks, and the trailer and extension have only partial window unit AC in some bedrooms, with stand-alone floor fans elsewhere. The AC and electricity costs run at around $ 150 per month. Several features are worth noting in this case—all quite typical. First is the way in which self-help allows for home extension and considerable flexibility in the allocation and use of bedrooms tied to family expansion. Second is the poor quality of the housing itself: small rooms, the lack of privacy, and problems of inadequate heating and cooling. Third is the reliance upon close kin for accommodation and support—in the past to the children and grandchildren, and increasingly from the siblings towards their aging parents. Co-residence or spatial propinquity among close family members are important attributes of *colonia* and IfHSs, especially as parents age. Fourth is the advantage of relatively large lot size which in several cases studied allowed for a second home to be built on the same site. In this particular case it was not necessary since the daughter's family lived opposite. However the yard offers additional space as well as amenity opportunities outside of the home.

In another case not described here (Fig. 15.3) the large lot stretched up a steep slope. The family house at the top was originally accessed from a vacant lot at the rear, but this has since been occupied so the elderly owners now have a steep climb from the street and from yard. Both front and side of the lot falls away with a drop of 8–10 ft., creating a hazard for the young grandchildren in particular since there are no barriers. The daughter has built her own family home in the lower part of the lot and will ultimately inherit the whole site. Now daughter and parents are considering doing a swap, to mitigate the need for the elderly parents to climb such steep stairs to their home. Once again, one sees the importance that ample space, flexible home construction over time, and co-residence can play in managing the home environment to meet the needs of the elderly.

Housing Conditions in Redwood and Rancho Vista, Guadalupe County

The data discussed in the following part of this chapter come from several recent surveys undertaken at the LBJ School of Public Affairs about the nature of self-help housing improvement processes from which we can begin to think about the possible interrelationship between the dwelling environment and health and physical mobility status. It draws primarily upon a 2010 survey of Housing Conditions, Sustainability and Self-help in two IfHSs in Guadalupe County, Central Texas (LBJ School of Public Affairs 2010). The study was undertaken through the Community Development Clinic of the Law School at the request of community residents who were interested in seeking funding for home and weatherization improvements from the American Reinvestment and Recovery Act (Table 15.1).

Two other studies also inform this discussion. The first is a 2010–2011 research project sponsored by the Ford Foundation which examined *colonia* population density and housing quality changes between a baseline survey undertaken in 2001–2002 and a follow-up restudy of the same settlements and households in 2010 (Durst et al. 2012; Durst and Ward 2014) The second is a recently completed major property title study conducted on behalf of the Texas Department of Housing and Community Affairs (TDHCA) in 2011–2012, which examined the changing use of Contracts for Deed since major legislation was enacted in 1995 and 2001 (Ward et al. 2012). While random surveys were undertaken for that TDHCA study across eight counties (two of which were in Central Texas and focused upon IfHSs), only the data gathered in six border counties are amenable to extrapolation for *colonias* in those counties. The reports and databases for all of these studies may be accessed at www.lahn.utexas.org (Texas Housing Studies button).

The housing conditions and health problems in the Redwood and Rancho Vista subdivisions are the main focus of this paper, and are described more fully below. Given the close links and good relations that the Law School had with the community leaders in these two adjacent neighborhoods, it was felt that a mail-out survey, properly explained and primed, might be expected to work reasonably in these

Table 15.1 Comparative data for Redwood and survey settlements

Variable	Redwood CDP	UT survey data Redwood and Rancho Vista	Guadalupe county	Guadalupe county 2000	Guadalupe county 2010
Hispanic/Latino	84%	88%	–	33%	36%
Households with disability	22%	57%	–	19%	–
Median household income	$ 31,100	$ 43,300	–	$ 44K	$ 61.3K
Education <12 grade, no diploma	46%	43%	–	22%	16%
Mobile home	80%	65%	66% est.	22%	18%
% Lots with extensions or additional dwelling units	–	40%	–	–	–
Median value of home	$ 44.5K	$ 53,200	–	$ 91.4K	$ 142.4K

settlements. But as a failsafe, and in order to offer some triangulation about the relative effectiveness of each technique, we also conducted random face-to-face interviewer applied surveys with every fifth dwelling unit, dropping-off mail back surveys at the intervening lots. This provided a total of 133 completed questionnaires almost equally split across the two settlements. Seventy percent of these were mail backs representing a 15% response rate—more than double the response achieved in most mail out surveys.[3] The survey gathered information about household structure, access to housing, length of residence on site, housing type and construction, a rating of perceived housing problems, and self-reporting on health problems and personal mobility.[4]

Demographic Data Profiles and Housing Conditions

Household sizes in the survey neighborhoods (Rancho Vista and Redwood combined) vary a little but are typically between 2 and 5 household members with relatively few outliers. The average size is 3.94, almost matching exactly the 2000 Census average household size of 3.98 for Redwood CDP (Census Defined Place). Twenty-seven percent of households had only one (8%) or two (19.2%) members, and a further 15% had three members. These three categories almost certainly concentrate the more elderly households across our sample. Fourteen percent of the lots contain two housing units, with most of the persons in the second home being

---

[3] Postal (mail) survey response rates are sometimes notoriously low (Babbie 2002, p. 259) but may also be raised very substantially depending upon: the nature of the survey (postal or e-mail—the latter becoming increasingly widely used (Gosling et al. 2004)) and whether it targets a relatively captive and receptive constituency (for example a home institution's students or faculty). Other features that shape response rates are: if respondents are recompensed in cash or kind; the number of "tickler" reminders sent after the first call; instrument readability and design, etc. The sort of housing mail-back surveys that we are describing here rarely get above 6–7% response, although Babbie (ibid) and others argue that only when one gets above a 50% response rate can one have confidence in the material. Moreover, here it should be remembered that we are dealing primarily with a very low-income Spanish-speaking population many of whom have limited education and literacy, and previous experience had also suggested that a 9–10% response rate would provide valuable insights, albeit not for extrapolation purposes (Ward and Carew 2001). Our mail survey invitation to respondents allowed for in-house consultation and while the unit of analysis was the household and the dwelling, it could be completed by a surrogate adult on behalf of the owner. Thus we were gratified to achieve a 15% return which we considered relatively high, and was largely the result of having the community leaders' support and participation, and a constituency that was engaged and interested in improving housing and community infrastructure. Naturally this can create problems of biased responses, which is another reason why we also sought to conduct parallel random face to face interviews using the same instrument. In the event the differences between the two types of survey were relatively small (see LBJ School of Public Affairs 2010).

[4] The instrument was two sided: odd pages written in Spanish, evens in English. It was self-administered except for the face-to-face interviews. All the IRB protocols were observed, and a letter was retained or left with each respondent (www.lahn.utexas.org [Texas Housing Studies, Sustainability Project—Instruments]).

Fig. 15.4 Rancho Vista and Redwood IfHSs in Redwood CDP

related to the primary householder's dwelling. Almost half (46%) of the households have one member employed, with a further 30% having two members contributing to the household income, although in neither case did the survey record whether this was part- or full-time employment. Sixteen percent of households were not in paid employment—again these are most likely retired householders although the survey did not gather systematic data about age of household members, nor of the owner(s). Including benefits, one-third (33.3%) reported a monthly household income of between $ 2,000 and $ 3,000 (only 6.5% earned above $ 3,000), almost another third (31.7%) reported earnings between $ 1,000 and $ 2,000; while 28% stated a monthly household income at under $ 1,000.

Most are long term residents: 29.5% have lived in the neighborhood for over 20 years, 31.3% for 15–19 years, and only 19% for less than 9 years. This emphasizes the relative population stability and the likelihood that this is a middle aged and aging population. Figure 15.4 shows the two neighborhoods and their location in Redwood CDP. Rancho Vista has the more regular grid layout at the top of the photo while Redwood actually comprises a series of small subdivisions some of which comprise only one or two streets or a single cul-de-sac. Typical house types

Fig. 15.5 Typical manufactured home (trailer) with
"skirt". Rancho Vista. (Note
also the self constructed
porch and steps rather than
a ramp)

Fig. 15.6 Older—style
manufactured (trailer) with
self-built extensions, porch
and false roof for shade,
Rancho Vista

are shown in Figs. 15.5, 15.6, 15.7, 15.8, 15.9 selected here to show the range of
dwellings that will be discussed further below. Manufactured homes predominate
(Table 15.2), and self-building entirely from scratch is far less common in central
Texas than in border *colonias*, although many trailer homes have self-built do-it-
yourself (DIY) extensions and modifications (porches, carports, false roofing) for
shade (Figs. 15.6 and 15.7).

Manufactured homes are those that are built off site and moved onto the lot. They
are usually trailers and may be single or double-wide. They all share the feature of
being on a wheel-base chassis although the wheels may be removed when placed on
site. Generally they are supported on brick piers, and often have a "skirt" around the
base to disguise the wheel base and to keep out animals and pests (Fig. 15.5). Older
trailers/manufactured homes are easily distinguished by their unique snout-nosed

Fig. 15.7 Two homes on site (probably two related households). Front is manufactured home; rear is custom home. Rancho Vista

Fig. 15.8 Self-built/modular home in Rancho Vista (Note "camper" alongside, home extensions at rear, and "new" verandah at the front)

Fig. 15.9 Manufactured home (trailer in front) with custom home under construction at the rear. Redwood

shape at one end (Fig. 15.6). These older trailer homes often predate the 1986 Housing and Urban Development (HUD) codes for manufactured home production. In particular, some of the doublewides are quite spacious and come in sections with a porch or front extension and peaked roofs. More recent manufactured homes are built to higher code specifications and to higher energy efficiency.

Table 15.2 Different types of self managed home development in Redwood and Rancho Vista

Dimension/Item	Rancho Vista		Redwood		Combined	
	%	N	%	N	%	N
Housing structure type		65		66		131
Manufactured home (trailer)	69.2	45	68.2	45	68.7	90
Single-wide	53.8	35	45.5	30	49.6	65
Double-wide	15.4	10	22.7	15	19.1	25
Constructed home	21.5	14	15.2	10	18.3	24
Contractor built on-site	1.5	1	7.6	5	4.6	6
Self-built on-site	20.0	13	7.6	5	13.7	18
Modular manufactured home (not on wheel base) assembled on site	4.5	3	1.5	1	3.1	4
Camper (recreational vehicle)	0.0	0	1.5	1	0.8	1
Other	4.6	3	13.6	9	9.2	12

The (approximate) average age of the primary dwelling units was 22 years, indicating that many houses are relatively old and in need of repair or weatherization. Only 1/3 of the homes are less than 15 years old. Further, 21 % of homes are 30 years or older. One quarter of respondents (25.6 %) report the presence of additional units on lot used either by themselves or by members of their household. In addition, 39.5 % have extended or added to their primary housing unit, 2/5 of which are trailers. Self-built extensions are especially common. Where they exist, most of the additional secondary home units on the lot have 1–2 usable rooms (77 %). The age of the extensions and additional units varies: just over one quarter of the units are under 5 years old (29 %); while 1/3 are over 30 years old (33 %). Extension and addition types and purposes vary, and include uses such as storage, porches, garages, and outhouses. However, the majority of reported additions are for the purpose of living space: bedrooms, family rooms, etc. Indeed, three-quarters of the respondents (78 %) indicate that their addition or extension is for the purpose of sleeping or the primary residence of another household.

The most common power source is electricity (62 %) and mixed electric and propane (tanks). Eighty-six percent use electricity to heat water, although one-fifth have serious water heating issues. One-half of the homes have partial AC, and only a minority reported full AC (31 %). Many have a system of partial AC (a window unit in the bedroom for example), and use ceiling or floor fans elsewhere—as described in the case study earlier. Just over one third reported problems with air conditioning, including the high costs.

A central part of the survey was to analyze the housing problems as they were perceived and experienced by the residents themselves. The unit of analysis was the household, and survey participants were asked to rate 24 housing dimensions on an ordinal scale where: (1) was a constant or severe problem; (2) an occasional problem; (3) satisfactory or broadly okay; and (4), good, i.e. not a problem. They were also allowed to answer "(5) not relevant or no opinion". Additionally, respondents were asked in an open-ended question to list the five most severe problems in their

Table 15.3 Comparison of the percent of respondents that answered "constant or occasional problem" vs. "satisfactory or good, not a problem." (Source: LBJ School of Public Affairs (2010: 64))

	% Constant or occasional problem	% Satisfactory or good, not a problem	% Problem index
1 Doors do not shut properly	71.9	24.2	47.7
2 Unit is too hot in summer	68.5	29.0	39.5
3 Unit is too cold in winter	63.9	33.6	30.3
4 Poor insulation	61.9	33.0	28.9
5 Pest infestation	55.7	36.5	19.2
6 Problems with septic tank	55.0	40.8	14.2
7 Poor venting from bathroom	53.8	42.0	11.8
8 Roof leaks	533	45.0	8.3
9 Poor flooring	52.6	42.9	9.7
10 Poor venting from kitchen	52.6	42.4	10.2
11 Unstable foundation	52.1	42.7	9.4
12 Windows do not close properly	51.7	45.9	5.8
13 Electrical wiring and/or outlets	50.8	45.7	5.1
14 Plumbing leaks	48.6	46.8	1.8
15 Mold	47 0	46.1	0.9
16 Poor venting from toilets	46.5	50.0	−3.5
17 Humidity/condensation problems	46.1	50.4	−4.3
18 Lack of privacy (poor sound proofing)	45.2	49.6	−4.4
19 Poor air quality	44.0	47.4	−3.4
20 House shakes when wind blows	43.2	51.6	−8.4
21 Steps to the front door	36.6	56.2	−19.6
22 Insufficient hot/warm water	30.9	62.7	−31.8
23 Missing shingles	27.8	40.8	−13.0
24 Inadequate number of electrical outlets	27.2	66.7	−39.5

homes. These ratings provided us with an inventory of the range of problems that residents face, as well as the severity of each problem.

Table 15.3 shows a general (aggregate) percentage index that accounts for how the 133 respondents rated the 24 housing dimensions and the top (most frequent) problem noted by residents was that doors do not close properly, thus creating drafts and security concerns. Seventy-two percent of respondents report this to be a constant or occasional problem. The second, third and fourth top problems are somewhat related—that the dwelling unit is too hot during the summer or too cold during the winter (64 %); and that it is poorly insulated (62 %). Between 50 and 56 % of the households rated the following housing dimensions as a constant or occasional problem of the dwelling units: pest infestation, septic tanks, bathroom venting, roof leaks, flooring, kitchen venting, foundation, windows closing properly, and electrical wiring (Table 15.3).

Household quartiles were also constructed based upon the severity of problems recorded, and a total of 18 % of households fell into the 1st quartile, meaning that almost one in five homes identified extensive and serious housing problems since they ranked between 19 and 24 of the housing dimensions as a constant or occasional problem. Another 24% (almost one in four) households fell within the 2nd quartile with substantial housing problems. Combined, this translates to two out of five homes (42%) reporting substantial to extensive housing troubles related to their dwelling structures. Quartiles 3 and 4 reported "modest problems" (21.4%) and few housing problems (36.4%) respectively. In another study using these same data Sullivan and Olmedo (2014) report a statistical analysis using an ordered logit model in which these same housing quartiles became the dependent variable, and a series of independent variables are used to ascertain the level of association. The independent variables were: household income, age of the dwelling unit, house value, constructions skills of a household member, wastewater problems, AC problems, type of garbage collection service, and self-reported severe health or mobility problems. Their analysis shows that the number and extent of housing problems increases with age of the dwelling structure, household reporting of problems with the septic tank system and with the source of air cooling, and if the household has a member with health issues or disabilities. On the other hand, the number of housing problems is estimated to decline (negative coefficients) as the value of the home increases, and especially if a member of the household has previous construction experience such that they can more readily overcome structural and other physical dwelling problems. The linkage between having a poorer air cooling and many other consequent problems is less easy to discern, but they argue that it seems likely to relate to inadequate insulation and the costs of effective AC, both of which are a surrogate for poor quality homes and fittings, and are likely to aggravate health problems associated with higher summer ambient temperatures and poor air quality. Ensuring effective cooling can be both difficult and expensive, and if the home contains persons with serious health problems or disabilities then those individuals are more likely to be adversely affected by those conditions unless that household has someone to provide the necessary resources to deal with dwelling dilapidation (Sullivan and Olmedo 2014; Olmedo and Sullivan forthcoming).

Health Problems and Disabilities and their Relation to the Dwelling Unit

More than half of the surveyed population (57%) indicated that they have at least one member of their household with some sort of severe health problem or disability. The most frequently reported health problem among respondents is diabetes; 29% of the population reporting that a member of their household was afflicted by diabetes (Fig. 15.10). This statistic, striking in itself, is actually an underrepresentation of the incidence of diabetes within these communities. Because residents were only asked if they or a member of their household was affected, it probably undercounts the true number of total members of the household with diabetes. Oftentimes respondents answered "yes" (that someone in their household is affected) and then

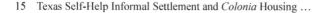
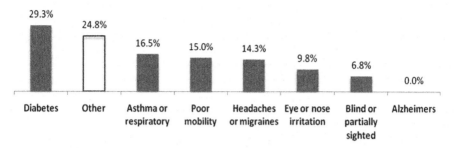

Fig. 15.10 Percentage of households reporting a member with a serious health problem or disability

listed that both they and their spouse or a parent suffered, or that diabetes was an issue for several members. It is clear that the high prevalence of diabetes in these communities is also associated with very-low reported incomes and high poverty levels of this particular population—both factors that have been shown to contribute to diabetes (Black 2002; Chaufan et al. 2011).

The next most reported category for health and illness was the "other" category (25%) which is mainly related to cardiovascular disease or orthopedic problems (e.g., issues with back or joints). There were several other health problems that affect at least one member in about 15% of the households: poor mobility (15%), asthma or respiratory problems (16.5%), and migraines or headaches (14%).

Because almost 70% of our respondents live in manufactured housing, we were also interested to see if this group of residents reported different rates of health problems. We restricted the groups by those that live in manufactured housing vs. those that do not. Results showed no difference in the rate of reporting health problems: 58% of manufactured home residents report some health problem vs. 57% of non-manufactured home residents. Comparing the same groups (manufactured-vs-non-manufactured homes) within each category of poor mobility, Alzheimer's, diabetes, asthma, migraines, sight problems, eyes/nose irritation, and other problems no significant differences were found in the rate of reporting of these health issues. Based on the data, living in a manufactured home does not, of itself, appear to necessarily contribute to negative health outcomes.

The Links Between Housing Problems and Health Outcomes

When we asked the residents themselves how the aforementioned health and mobility issues were affected by their housing situations, many listed poor access (for those with limited mobility), and generalized resulting stress. However, the housing condition they cited most often as contributing to illness and poor health was poor indoor air quality which would include mold, noxious odors, humidity, dust, and poor air circulation generally. This response is noteworthy given a large body of

Table 15.4 Percentage of households that answered "constant" or "occasional" problem to selected housing dimensions by population sub-groups reporting illness

	Asthma and respiratory problems		Migraines and headaches		Eyes and nose irritation	
	% Asthma population	% Total population	% Migraine population	% Total population	% Eye/nose population	% Total population
	(N=15)	(N=133)	(N=19)	(N=133)	(N=13)	(N=133)
Poor air quality	81.0	44.0	56.3	44.0	76.9	44.0
Humidity and condensation	68.0	52.5	64.7	52.5	76.9	52.5
Mold	81.8	47.0	76.5	47.0	92.3	47.0
Poor venting from kitchen	81.0	52.5	82.4	52.5	69.2	52.5
Poor venting from bathrooms	85.7	53.8	61.1	53.8	83.3	53.8
Poor venting from toilets	86.4	46.5	58.8	46.5	76.9	46.5
Poor insulation	90.5	61.9	76.5	61.9	76.9	61.9
Doors do not shut properly causing drafts	86.4	71.9	89.5	71.9	100.0	71.9

research that links health outcomes such as asthma and lung cancer to the quality of indoor home environments in inner city environments among minority populations (McCormack et al. 2009). Americans tend to spend between 80 and 90 % of their time indoors, and children and the elderly especially spend a disproportionate amount of time indoors (Diette et al. 2007). However, as the Environmental Protection Agency advises, indoor air is often more polluted than outdoor air for a variety of reasons.[5]

The relationship between negative health outcomes and the condition of the physical house becomes clearer when one analyzes reported housing problems within the groups that also report health problems. For instance, selecting only those cases that report having a member of their household affected by asthma, and then examining the housing problems that they listed as being a constant or occasional problem, this group was found—across the board—to be more likely to list mold, poor air quality, humidity and condensation, poor venting from the kitchen or bathroom or toilet, or drafts from doors as problems. The same is true for those that report problems with migraines or headaches and eyes or nose irritations.

Table 15.4 summarizes these findings by comparing the total population (N=133) against the subgroup affected by severe health problems on the rate of reporting problems with certain housing dimensions. Although the sample size (N) is not large for these illnesses, it is probably sufficient to allow us to draw some

---

[5] http://www.epa.gov/iaq/ia-intro.html. "An Introduction to Indoor Air Quality".

preliminary conclusions. The table shows that those reporting asthma or respiratory problems, migraines or headaches, and eye or nose irritation are more likely than the total population to rate these housing issues as problematic. Across the board, residents affected by illness are more likely to report housing issues related to indoor air quality. For instance, residents affected by asthma are almost twice as likely to report poor indoor air quality as a problem than the general population.

Elderly populations also have high exposure to hazards in the home (Carrillo Zuniga et al. 2011) such as slippery surfaces, loose mats, lack of proximity of toilets to the bedroom, etc. Their study also revealed that households with an elderly person present were often unprotected by smoke detectors, and that 25 % of dwellings in such households were exposed to between 4 and 6 hazards.

Concluding Remarks

It should be a relatively easy next step to examine some of the major data sets relating to health among Hispanics in Texas and to generate an accurate profile of the health problems that confront aging and Mexican origin populations. Tying those data to housing conditions will be more of a challenge, however. The data presented, if not demonstrating a causal link, do offer at least a prima facie case for links between housing quality, poor health, and chronic illness. That in itself is not especially surprising. What is interesting, however, is the way in which the housing conditions described in this study do appear to have a segmented and adverse effect upon elderly Hispanic populations—not only because of their current prominence in *colonias* and IfHSs, but also because of their expected demographic increase over the next 10–20 years. Poor elderly Hispanics should be a target group for attention in these self-built and self-managed neighborhoods, not least given their vulnerability to chronic illnesses such as diabetes; and to heart and respiratory ailments, all of which are exacerbated by excessive heat or cold within the home, and impaired physical mobility problems around the house. The additional challenge that they face is that even within the low income self-built and self-managed housing contexts examined here, the elderly are often the least able to rectify or address their housing problems, given their age, infirmity and much lower income stream late in life.

Acknowledgements The author wishes to thank the anonymous reviewer for a helpful critique, and Mr. Noah Durst, Dr. Lissette Aliaga and Ms Edna Ledesma for their assistance in fieldwork and in preparing Figs. 15.1 and 15.2 of this chapter. The Policy Research Institute of the LBJ School of Public Affairs is thanked for funding support to develop the intensive case study instrument (see also Ward et al. 2014), and for fieldwork support in Rio Grande City. The author was the director of the 2010 survey of Housing Conditions, Sustainability and Self-help in two IfHSs in Guadalupe County, Central Texas (see LBJ School of Public Affairs 2010).

References

Anders, R. L., Olson, T., Wiebe, J., Bean, N. H., DiGregorio, R., Guillermina, M., & Ortiz, M. (2008). Diabetes prevalence and treatment adherence in residents living in a colonia located on the West Texas, USA/Mexico border. *Nursing & Health Sciences, 10*(3), 195–202.

Anders, R. L., Olson, T., Robinson, K., Wiebe, J., DiGregorio, R., Guillermina, M., Albrechtsen, J., Bean, N. H., & Ortiz, M. (2010). A health survey of a colonia located on the US/Mexico border. *Journal of Immigrant and Minority Health, 12*, 361–369.

Babbie, E. R. (2002). The basics of social research, Wadsworth.

Black, S. A. (2002). Diabetes, diversity, and disparity: What do we do with the evidence?. *American Journal of Public Health, 92*(4), 543–548. http://www.ncbi.nlm.nih.gov/pmc/articles/PMC1447113/. Accessed 15 Nov 2014.

Carrillo Zuniga, G., Mier, N., Seol, Y.-H., Villarreal, E., Garza, N. I., & Zuniga, M. (2011). Home hazards assessment among Elderly in South Texas Colonias. *Texas Public Health Journal, 63*(4), 14–17.

Chaufan, C., Davis, M., & Constantino, S. (2011). The twin epidemics of poverty and diabetes: Understanding diabetes disparities in a low-income Latino and immigrant neighborhood. *Journal of Community Health, 36*(6), 1032–1043

Diette, G. B., Hansel, N. N., Buckley, T. J., Curtin-Brosnan, J., Eggleston, P. A., Matsui, E. C., et al. (2007). Home indoor pollutant exposures among inner-city children with and without asthma. *Environmental Health Perspectives, 115*, 1665–1669.

Donelson, A. J., & Esparza, A. X. (2010). *The Colonias reader: Economy, housing, and public health in U.S.-Mexico border Colonias*. Tucson: The University of Arizona Press.

Durst, N. J. (2012). Second-generation policy priorities for Colonias and informal settlements in Texas. *Housing Policy Debate*. Published online May 1: http://www.tandfonline.com/doi/full/10.1080/10511482.2013.879603#.U2zsv6ROVdg. Accessed 5 May 2014.

Durst, N. J., & Zhang, W. (2013). Health risks and disparities in Texas Colonias: Policy analysis and potential solutions. Unpublished paper.

Durst, N. J., & Ward, P. M. (2014). Measuring self-help home improvements in Texas *Colonias*: A ten year snapshot study. *Urban Studies, 51*, 2143–2159

Durst, N., Ward, P. M., Olmedo, C., & Rojas, D. (2012). Report #2, documenting a decade of change in Starr county Colonias: Survey design and results. In *Housing sustainability, self-help and upgrading in Texas Colonias: A longitudinal perspective—2002 plus 10*. Final report of a study commissioned by the Ford Foundation in 2010. www.lahn.utexas.org.

Environmental Protection Agency (EPA). An Introduction to Indoor Air Quality. http://www.epa.gov/iaq/ia-intro.html. Accessed 15 Nov 2014.

Gilbert, A. (1999). A home is forever? Residential mobility and homeownership in self help settlements. *Environment and Planning, 31*, 1073–1091.

Gosling, S., Vazire, S., Srivastava, S., & Oliver, J. (2004). Should we trust web-based studies? A comparative analysis of six preconceptions about internet questionnaires. *American Pyschologist, 59*, 93–104.

LBJ School of Public Affairs. (2010). Housing conditions, sustainability and self help in Rancho Vista and Redwood informal homestead subdivisions in central Texas. Final report and database for the community residents and for the community development clinic of the UT law school. http://www.lahn.utexas.org/Texas%20Colonias/TexasColonias2.html. Accessed 15 Nov 2014.

McCormack, M. C., Breysse, P. N., Matsui, E. C., Hansel, N. N., Williams, D. A., Curtis Brosnan, J., Eggleston, P, Diette, G. B., et al. (2009). In home particulate concentration and childhood asthma morbidity. *Environmental Health Perspectives, 117*(2), 294–298.

Mier, N., Ory, M. G., Zhan, D., Conklin, M., Sharkey, J. R., & Burdine, J. N. (2008). Health-related quality of life among Mexican American living in colonias at the Texas-México border. *Social Science & Medicine, 66*, 1760–1771.

Olmedo, C. & Sullivan, E. (Forthcoming). Assessing housing's contribution to poor health in Texas informal communities, from border colonias and model subdivisions to metro-centered informal subdivisions. *Urban Affairs*.

Olmedo, C. (2014). Fuel Poverty and Housing Conditions in Texas Colonias. Presentation at the 44th Urban Affairs Association Conference in San Antonio, Texas, March 20, 2014.

Sullivan, E., & Ward, P. M. (2012). Sustainable housing applications and policies for low-income self-build and housing rehab. *Habitat International, 36*(2), 312–323.

Sullivan, E., & Olmedo, C. (2014). Informality on the urban periphery: Housing conditions and self-help strategies in Texas informal subdivisions. *Urban Studies* (in press).

Varley, A. (2008). A place like this? Stories of dementia, home and the self. *Environment and Planning D: Society and Space, 26*(1), 47–67

Ward, P. M. (1999). *Colonias and public policy in Texas and Mexico: Urbanization by stealth*. Austin: University of Texas Press.

Ward, P. M. (2007). Colonias, informal homestead subdivisions, and self-help care for the elderly among mexican populations in the United States. In J. L. Angel & K. E. Whitfield (Eds.), *The health of aging Hispanics* (pp. 141–162). New York: Springer.

Ward, P. M. (2012). 'A patrimony for the children': Low-income homeownership and housing (im)mobility in Latin Amercian cities. *Annals of the Association of American Geographers, 102*(6), 1489–1510.

Ward, P., & Carew, J., (2001). Tracking absentee lot owners in Texas colonias: A methodology. *Journal of Land Use Policy, 18*(2), 73–86.

Ward, P. M., & Peters, P. A. (2007). Self-help housing and informal homesteading in Peri-urban America: Settlement identification using digital imagery and GIS. *Habitat International, 31*(2), 205–218.

Ward, P. M., Way, H. K., & Wood, L. (2012). The contract for deed prevalence project: A final report to the Texas department of housing and community affairs (TDHCA). http://www.lahn.utexas.org/Texas%20Colonias/TDHCA.html. Accessed 15 Nov 2014.

Ward, P. M., Jiménez, E., & Virgilio, M. D. (2012). *Housing policy in Latin American cities: A new generation of strategies and approaches for UN-Habitat III in 2016*. New York: Routledge.

Chapter 16
Achieving Sobriety Among Latino Older Adults

Erick G. Guerrero, Tenie Khachikian, Yinfei Kong and William A. Vega

Achieving Sobriety Among Latino Older Adults: A Challenge in the Americas

Substance abuse has emerged as a major health issue in the Americas, and the prevalence of use and related burden on health services in the United States is one of the highest in the world. The proportionate increases in illicit substance use in Latin America, especially in Mexico which experienced an 87% increase in illicit drug use (0.8–1.5%) in the past decade, are an ascending concern (Villatoro et al. 2012). Increases in substance use are influenced by circular migration, drug trafficking patterns, and globalization. The scope and interrelationships of these influences are complex and beyond the goal of this report. A self-evident implication is the growing need in the United States and Latin America for effective prevention and treatment services for socially marginal populations. As part of an expanding body of research on this topic, this study supplies information about Latinos who received treatment in Southern California drug treatment centers, and describes how treatment provider and patient characteristics are associated with attaining sobriety.

During the past 10 years, Latinos have been identified as the fastest-growing population to entering substance abuse treatment (SAT) in the United States, comprising 12% of the total receiving treatment (Guerrero 2010). Latinos in the United States also report higher rates of substance use disorders (9.7%) compared to Whites (8.9%) and African Americans (8.2%; Substance Abuse and Mental Health Services Administration (SAMHSA) 2012). These statistics represent a general view of US Latinos as a homogenous group. However, growing research has suggested that it is necessary to consider Latino subgroups, including nation of origin, ethnicity, race, region of residence, sex and sexual orientation, co-occurring

E. G. Guerrero (✉) · T. Khachikian · Y. Kong · W. A. Vega
School of Social Work, University of Southern California, 655 West 34th Street, Los Angeles, CA 90089, USA
e-mail: erickgue@usc.edu

© Springer International Publishing Switzerland 2015
W. A. Vega et al. (eds.), *Challenges of Latino Aging in the Americas*,
DOI 10.1007/978-3-319-12598-5_16

medical conditions, and age, to better identify risk of drug use and response to treatment (Guerrero 2013b).

Substance use, which in this report refers to both licit and illicit drugs, and alcohol is by far the most commonly used and abused substance in the Americas, especially in older people. The latest comparative research on drug use showed that Mexicans, the largest group of Latinos residing in the United States, report the highest rate of binge drinking (26%) among all racial and ethnic groups (Chartier and Caetano 2010). This in addition, national statistics have suggested that Latino women report lower lifetime prevalence of substance use on average compared to Latino men in the general population (SAMHSA 2011). However, in serving time in correctional facilities, Latino women report high rates of substance abuse as the reason for detention, and also report more mental health problems (Amaro et al. 2005, 2007). This suggests that, in contrast to men, Latino women are less likely to have a drug problem, and have a narrower substance use profile when they enter treatment, reporting higher rates of substance use severity in combination with other emotional and health conditions (e.g., depressive and posttraumatic stress disorder symptoms) on average than men (Amaro et al. 1999, 2007). This high severity and comorbidity has been associated with greater barriers to enter treatment for women, mainly due to personal stigma and the limited number of gender-appropriate programs (Amaro et al. 1999, 2005). Yet, emerging research has shown that Latino women with more combined drug and health problems are more likely to achieve sobriety when provided with comprehensive and gender appropriate services (e.g., child care, transportation, etc.) compared with Latino men (Amaro et al. 1999, 2007; Grella et al. 2004; Grella and Stein 2006; Guerrero et al. 2014; Marsh 2012). Research on sex disparities in treatment outcomes is growing, yet few studies have focused on Latino age groups despite growing concern about substance use among older adults of differing sexes.

National statistics have shown that Latinos aged 50 or older have an increased rate of alcohol and drug use during the previous 10 years (Andrews 2008; Gfroerer et al. 2003), and the rate of substance use disorder in this population is expected to increase from 2.8 million in 2006 to 5.7 million in 2020 (Han et al. 2009). In 2011, SAMHSA estimated 4.8 million adults aged 50 or older used illicit drugs during the previous year, with marijuana and prescription drugs being most commonly used. Statistics have shown that at least 3.4% of Latino older adults report heavy alcohol use and 14.1% report binge drinking (Andrews 2008; SAMHSA 2005a, b). These statistics show the need for treatment for this population. However, there has been limited research on Latino older adults' access and response to SAT (Andrews 2008; Bartels et al. 2005; Han et al. 2009; Korper and Raskin 2002; Suro and Singer 2002).

Program factors are generally believed to improve the effect of SAT interventions on racial and ethnic minority groups (Amaro et al. 2006; Guerrero 2013; Guerrero et al. 2013b; Vega and Lopez 2001). For instance, emerging evidence has suggested that counselors' Spanish language proficiency is associated with higher treatment access and retention in publicly funded SAT (Guerrero et al. 2013a). The extant literature also has shown that programmatic efforts to decrease wait time to enter treatment and increase retention in care are associated with positive treatment outcomes among all ethnic groups (Marsh et al. 2009; Zhang et al. 2003). Yet little is known about the extent to which program factors, such as counselors' Spanish speaking proficiency, and efforts to decrease wait time and increase retention help Latino older adults achieve sobriety.

In the general population entering SAT, wait time to enter treatment (Appel et al. 2004; Claus and Kindleberger 2002; Hadland et al. 2009) and treatment retention, or time spent in treatment, are considered robust predictors of reduced post treatment substance use (Simpson et al. 1997; Zhang et al. 2003). These measures have been recognized as proxies of program performance and standards of care (Garnick et al. 2009; McCarty et al. 2007). It is feasible to use wait time, retention, and quality of communication via counselors' Spanish language proficiency to examine program effects on client odds of sobriety and recovery from substance use, particularly for the majority of bilingual Latinos in Los Angeles County, who are at high risk of treatment dropout.

Sobriety Among Latino Older Adults

Examining sobriety as an outcome is problematic because of the varying meanings of sobriety for individuals based on primary drug of choice (e.g., alcohol vs. heroin) and the different types of drug treatment approaches (outpatient vs. hospitalization). Sobriety may be more challenging for individuals addicted to heroin compared to Latino older adults with a history of marijuana use. However, most programs in the United States follow an abstinence or sobriety orientation as the main goal of treatment (D'Aunno 2006). By controlling for treatment modality and drug of choice, it is possible to accurately examine sobriety rates in this population.

Achieving sobriety among older adults is critical because their use of substances may cause psychological, physical, and social effects that may in turn decrease their functioning and increase their risk of harm. One of the major challenges of the treatment system is to engage this population. Emerging studies have suggested that Latino older adults are difficult to engage in treatment because they experience more language and cultural barriers than any other age group (Andrews 2008; Kouyoumdjian et al. 2003). Providing appropriate and efficient treatment for Latino older adults generally requires on-demand access to culturally competent, age-specific, and integrated care models due to their stage of life (Abramson et al. 2002). Although admissions of individuals aged 50 or older have nearly doubled during the previous 8 years (Andrews 2008), there is limited knowledge about the quality of care, namely staff proficiency of Spanish language provision and programmatic efforts to enhance access and retention in this population to improve the likelihood of sobriety.

It is important to explore sobriety among Latino older adults because of the significant implications of drug use for their overall health and their risk of multiple health problems, functional impairments, and disabilities (Andrews 2008; Feidler et al. 2002). Latinos will comprise 25 % of the U.S. population by 2050, with at least half of them speaking Spanish at home (Guzmán 2001), and the health care service delivery system is poised to examine the extent to which provider linguistic competence can be fully attained and determine its effects on treatment outcomes in this population (Kouyoumdjian et al. 2003). It is therefore critical to deliver health care in Spanish to immigrants and monolingual Spanish-speaking older adults (Alegría et al. 2007; Iwashita et al. 2008; Santiago-Rivera et al. 2009) and rigorously test the extent to which counselors' Spanish language proficiency and treatment wait time and retention are associated with increased odds of achieving sobriety at treatment discharge among Latinos served by the publicly funded SAT system.

Methods

Sampling Frame and Data Collection

This study used a fully concatenated program and client data set collected in 2010–2011. The sampling frame included all 408 nonprofit SAT programs funded by the Department of Public Health in Los Angeles County, California. Client data were drawn from the Los Angeles County Participant Reporting System. These system wide evaluation data, collected by each provider on an ongoing basis, reflect the treatment experiences and immediate outcomes of a racially and ethnically diverse client population in the largest treatment system in the United States. Client data represented 15,100 client treatment episodes collected from July 1, 2010, to December 30, 2011.

Data were also collected from a random sample of 147 publicly funded and nonprofit programs from the 350 programs located in communities with a population of 40 % or more Latino and African American residents in Los Angeles County. The clinical supervisor was the key informant for program survey measures, and additional sources of data were used to cross-validate survey measures during follow-up site visits with 91 % of the sample. Consistent information from at least two of three sources of data was necessary for inclusion of each program in the analytic sample—i.e., (1) a review of program characteristics and service delivery information reported to the funding organization (L.A. County Department of Public Health); (2) qualitative reports from one counselor per program; and (3) a review of printed material available at each provider site (e.g., brochures, group activities, posted signs). More information on the study design and sampling frame is available elsewhere (Guerrero, 2013).

Analytic Sample

Of the 147 participating programs, 50 eligible programs with Spanish-speaking counselors ($n=103$) were randomly selected to complete the Spanish proficiency test. Only one counselor was selected from each program, ideally the counselor with the longest tenure in the program. The final analytic sample included 41 programs with a senior counselor who completed the Spanish proficiency test. Counselors reported on 1,658 client treatment episodes in 2010–2011. We excluded clients younger than 18 to conduct comparisons among Latino adults.

Dependent Variable

We examined one dependent variable: client sobriety at discharge. This variable was measured at discharge for all clients. Counselors assessed the extent to which clients reported and were observed maintaining sobriety from alcohol and other drugs.

Explanatory Variables

This study relied on four primary variables of interest: (1) age (18–28, 29–49, 50 or older), (2) counselors' Spanish language proficiency; (3) client-reported wait time to treatment; and (4) client retention in treatment. Counselors' Spanish language proficiency was tested using the Language Proficiency Assessment (LPA) of the American Council on the Teaching of Foreign Languages. The LPA in Spanish was conducted via phone as a 30-min computer-generated and recorded interaction featuring contextual questions within a structured conversation. Counselors responded to questions and recordings, which were later evaluated by two expert raters. The validity and reliability of a rater-based assessment is a function of raters applying a shared mental model structured in the test. The interrater reliability of the phone-based test was 0.94 and internal consistency was $\alpha=0.96$ (Surface et al. 2009; Thompson et al. 2007).

Client wait time was measured at intake as days spent on a waiting list before starting treatment (78 % of clients reported no wait). Retention reflected the number of days between admission and discharge dates as noted by counselors. As analytic measures, wait time and retention have been successfully used in several analyses (Friedmann et al. 2003; Guerrero et al. 2012; Guerrero et al. 2013a).

We also accounted for other variables as controls. Client variables included gender, Medi-Cal eligibility, history of mental health issues, homelessness status, primary substance, prior treatment episodes, and social support, referring to participation in recovery groups. Program characteristics included acceptance of Medi-Cal reimbursement, acceptance of private insurance reimbursement, professional accreditation (the Joint Commission), program type (outpatient, methadone, or residential), and whether the program was part of a larger parent organization.

Data Analysis

The initial analysis examined disparities among client characteristics based on age group. We used Stata version 12 to conduct analysis of variance and global chi-square tests to identify disparities. Stata 12 was also used to conduct multilevel logistic regressions to examine the extent to which program and client characteristics were associated with odds of sobriety among Latino clients, with a focus on clients in the 50 or older subgroup.

Maximum likelihood estimation in multivariate logistic regressions was used to effectively respond to missing data (Allison 2002). Using this approach with the current rate of missing data (highest rate was 8 %) is considered the most adequate way to obtain unbiased estimation parameters (Allison 2002).

Results

Table 16.1 shows differences among program and client characteristics across age groups. Older clients (50 or older) were more likely than members of other age groups to participate in programs with higher acceptance rates of Medi-Cal (Medicaid in California) and private insurance and accreditation by the Joint Commission. Older adults were less likely to receive outpatient services and more likely to receive methadone and residential treatment.

Individual differences across age groups were also statistically significant. Compared with younger groups, the older adult group had higher Medi-Cal eligibility, disproportionately fewer women, more members with a history of mental health issues, and higher rates of alcohol as primary substance of abuse. Compared to younger groups, Latino older adults reported two to three times as many treatment episodes. Finally, compared to younger groups, Latino older adults reported longer treatment duration and reduced wait time, but these differences were only marginally statistically significant ($p < 0.10$).

Table 16.2 summarizes results of a multilevel logistic regression model of sobriety among Latinos. Age was not associated with odds of sobriety ($p > 0.05$). However, treatment retention was associated with increased odds of sobriety (OR = 1.006, 95 % CI = 1.004, 1.009). Women had higher odds of sobriety in relation to men (OR = 1.496, 95 % CI = 1.064, 2.103). Furthermore, Latinos reporting cocaine or crack as their primary drug were more than twice as likely to be sober at discharge than Latinos using heroin as primary drug (OR = 2.468, 95 % CI = 1.180,

Table 16.1 Program and client characteristics by age group

Variable	18–28 ($n=655$) M (SD) or %	29–49 ($n=884$) M (SD) or %	50+ ($n=181$) M (SD) or %
Program characteristics			
Spanish language proficiency	8.80 (1.02)	8.74 (1.00)	8.77 (0.84)
Acceptance of Medi-cal**	67.3	64.9	80.1
Acceptance of private insurance**	26.0	28.3	45.1
Accreditation**	12.2	13.7	24.3
Client characteristics			
Outpatient**	94.0	91.3	64.6
Methadone**	2.6	7.2	31.5
Residential*	3.4	1.5	3.9
Treatment duration (days)	90.00 (87.16)	95.43 (95.79)	106 (110.09)
Wait time to enter treatment (days)	2.17 (6.31)	2.09 (6.59)	1.07 (5.04)
Medi-cal eligibility**	33.4	29.3	47.5
Female**	47.0	33.3	22.7
History of mental health issues**	10.5	16.5	25.4
Homeless	12.1	13.8	14.4
Alcohol as primary substance**	12.4	12.4	28.2
Primary treatment episodes**	0.96 (1.67)		2.98 (5.93)

$*p<0.05; **p<0.001$

5.163). Marijuana and hashish were also associated with increased odds of sobriety compared with heroin (OR = 2.013, 95 % CI = 1.015, 3.990). Odds of sobriety at discharge decreased by 48.8 % for each additional day clients used their primary drug (OR = 0.512, 95 % CI = 0.390, 0.672). Clients with a history of mental illness had significantly lower odds of sobriety compared to those without such a history (OR = 0.606, 95 % CI = 0.387, 0.950).

Clients attending programs whose counselors reported higher scores in Spanish proficiency had lower odds of sobriety (OR = 0.744, 95 % CI = 0.561, 0.986). Finally, the odds of sobriety for clients attending programs with parent organizations were lower than for programs without parent organizations (OR = 0.495, 95 % CI = 0.295, 0.830).

Discussion

This study provided preliminary evidence of differences in treatment outcomes among Latinos based on age. Although differences in wait time or retention were not statistically significant, the comparative analysis revealed important differences in the type of programs serving Latino older adults and client characteristics. Latino

Table 16.2 Logistic regression on sobriety among Latinos at discharge from substance abuse treatment

Variable	OR	SE	95% CI
Client characteristics			
Retention	1.006***	0.001	1.004, 1.009
Wait time to enter treatment	0.994	0.010	0.975, 1.015
Age[a]			
29–49	0.806	0.117	0.607, 1.071
50+	0.750	0.201	0.444, 1.269
Female	1.496*	0.260	1.064, 2.103
Education[b]			
High school	0.839	0.136	0.611, 1.153
College	0.762	0.140	0.532, 1.094
Postgraduate	0.401	0.346	0.074, 2.170
Employed at admission	1.359	0.243	0.957, 1.929
History of crime	1.361	0.271	0.921, 2.010
Primary substance[c]			
Alcohol	2.048	0.752	0.997, 4.207
Cocaine/crack	2.468*	0.929	1.180, 5.163
Marijuana/hashish	2.013*	0.703	1.015, 3.990
Methamphetamine	1.597	0.561	0.802, 3.180
Other	1.315	0.675	0.481, 3.594
Days of primary drug use	0.512***	0.071	0.390, 0.672
Age at primary drug initiation	0.931	0.078	0.790, 1.096
Days of injecting drug use	0.778	0.412	0.276, 2.195
Prior treatment episodes	0.989	0.019	0.953, 1.027
Medication[d]			
Methadone	0.645	0.353	0.221, 1.883
Other	1.168	0.403	0.594, 2.299
History of mental illness	0.606*	0.139	0.387, 0.950
Social support	1.016	0.010	0.996, 1.036
Homeless	0.999	0.201	0.673, 1.482
Program characteristics			
Accreditation	1.293	0.470	0.634, 2.637
Spanish language proficiency	0.744*	0.107	0.561, 0.986
Parent organization	0.495**	0.131	0.295, 0.830
Number of programs	41		
Number of clients	1658		

*$p < 0.05$; **$p < 0.01$; ***$p < 0.001$

[a]18–28 was reference group

[b]Less than high school was reference group

[c]Heroin was reference group

[d]None was reference group

older adults were more likely than other age groups to enter publicly funded programs that accepted public and private insurance. In addition, younger groups were primarily served by outpatient services (more than 90%), whereas only 65% of Latino older adults received this type of treatment. Older adults were more likely to receive methadone or residential treatment. This finding suggests that publicly funded SAT programs mainly treat an older adult population with higher substance use severity and use of harder drugs (i.e., opiates) than the overall Latino population, indicating that Latino older adults also have poorer health status.

Retention in treatment was the only program factor associated with higher odds of sobriety for this sample, whereas Spanish language proficiency was associated with lower odds of sobriety. The association between Spanish language proficiency of counselors and lower sobriety rates of clients is counterintuitive. Albeit conjectural, programs with counselors with higher proficiency may have been smaller programs located in smaller communities with fewer resources (ancillary and social services) to support recovery. Therefore, language proficiency may have been associated with under resourced programs rather than quality of care in this analysis.

Other studies have showed that although Latinos benefit from culturally and linguistically responsive practices, other resources such as public health insurance (see Guerrero 2013) and comprehensive services (Guerrero et al. 2014; Marsh et al. 2009) play an important role in clients achieving their goals in treatment and successfully completing treatment.

Limitations

Several limitations associated with study data must be acknowledged. First, all measures were derived from cross-sectional data, preventing analysis of causality or directionality. However, the large sample of clients provided robust estimates. Second, client measures were not directly related to their counselor's Spanish language proficiency. Most programs reported several Spanish-speaking counselors; we selected the primary counselor assigned to Spanish-speaking clients. Some bilingual Latino clients may have received services in English or from different Spanish-speaking counselors. Spanish language proficiency was tested as an indicator of program quality; however, the single measure was not a proxy of each program's linguistic competence.

Another limitation was the subjective measure of sobriety, as reported and assessed by counselors at discharge. A urine drug test would have been more objective. However, by using auxiliary survey measures such as successful treatment completion and drug use prior to discharge, we determined that the sobriety measure was consistent with reports on other outcome measures capturing a similar concept. We also accounted for treatment modality to control for different sobriety expectations—for instance, sobriety from alcohol in outpatient treatment versus methadone maintenance for opioid treatment. Finally, analysis of the data only allowed findings regarding service delivery and client outcomes to be generalized to a subsample of programs with a Spanish-speaking counselor serving communities with a population of 40% or more Latino residents, or approximately 7.7 million residents in L.A. County, California.

Conclusion

This study explored age differences among Latinos attending publicly funded SAT and their relation to sobriety. Although few significant differences were found among age subgroups, considering these differences in future research is critical to examine other factors that may help older adults with addictions engage in and benefit from treatment to achieve sobriety.

We accounted for critical individual and program factors that have been associated with differences in treatment response for Latino older adults. This approach accurately captured the real differences within Latinos that are becoming known as distinguishing factors in response to treatment in this population. Gender (male) and history of mental health issues were associated with lower odds of sobriety as expected. Older adults entering treatment are less likely to be women and more likely to report mental health issues than other age groups. These findings have clear implications for research, policy, and program design.

Implications for the Americas

These preliminary findings highlight the importance of conducting further research to explore the treatment experiences of Latino older adults in publicly funded programs in the United States, Mexico, and elsewhere in Latin America. These nations are linked by drug control problems, continuing migratory processes, accelerating growth of aging populations, and the need to create sustainable pension and health care systems. Specifically Mexico and the United States, linked by a long common border and high human transit rates, are facing the requirement for major expansions in their respective healthcare delivery systems nationally, and penetrating into very low income populations with health services including drug treatment programs, and have been historically underserved or unserved. In particular, it is critical to examine program capacity to deliver not only culturally and linguistically appropriate care but also comprehensive and integrated substance abuse, mental health, and medical care. Primary care doctors are the first point of service for individuals with drug and associated medical conditions in both nations, thus primary care doctors should be trained in therapies and motivated as key participants in care systems designed to receive patients with substance use problems and addictions, and to become partners in sustaining these patients in treatment to achieve sobriety.

Substance use addictions are chronic conditions, and often last into later life as they are commonly used to self-medicate depression, pain, and to coping with functional limitations. Recovery is attainable at any age and requires provider skills and coordination of services. As the new health care system seeks to improve standards of care and reduce disparities in health (Andrulis et al. 2010), it is critical that health care policies support program designs that consider gender and comorbidities in the delivery of SAT. Policies should emphasize program performance benchmarks using retention, a factor associated with rates of sobriety.

A limitation of this report is the inability to make direct generalizations to other geographic areas in the United States or Mexico. No doubt many people think about Los Angeles as a border city because the County is host to literally millions of Mexican immigrants and their children, despite being 120 miles north of Tijuana, Baja California Norte. While geopolitical differences are undeniable across nations, we are comfortable in setting forth our findings as instructive for future research regarding patient and treatment system profiles, and the way characteristics of both may interplay in shaping treatment effectiveness. Providers of SAT in L.A. County have a unique opportunity to develop a culturally and linguistically responsive system of integrated care to engage this vulnerable, low-income, and mainly bilingual–bicultural older adult population. The results of these efforts may ultimately have far reaching impact on investigations and health care systems in other regional, national, international settings.

References

Abramson, T. A., Trejo, L., & Lai, D. W. (2002). Culture and mental health: Providing appropriate services for a diverse older population. *Generations, 26,* 21–27.

Alegría, M., Mulvaney-Day, N., Torres, M., Polo, A., Cao, Z., & Canino, G. (2007). Prevalence of psychiatric disorders across Latino subgroups in the United States. *American Journal of Public Health, 97,* 68–75. doi:10.2105/AJPH.2006.087205.

Allison, P. D. (2002). *Missing data.* Thousand Oaks: Sage.

Amaro, H., Nieves, R., Johannes, S. W., & Cabeza, N. M. L. (1999). Substance abuse treatment: Critical issues and challenges in the treatment of Latina women. *Hispanic Journal of Behavioral Sciences, 21,* 266–282. doi:10.1177/0739986399213005.

Amaro, H., McGraw, S., Larson, M. J., Lopez, L., Nieves, R., & Marshall, B. (2005). Boston consortium of services for families in recovery: A trauma-informed intervention model for women's alcohol and drug addiction treatment. *Alcoholism Treatment Quarterly, 22*(3–4), 95–119. doi:10.1300/J020v22n03_06.

Amaro, H., Arévalo, S., Gonzalez, G., Szapocznik, J., & Iguchi, M. Y. (2006). Needs and scientific opportunities for research on substance abuse treatment among Hispanic adults. *Drug and Alcohol Dependence, 84,* S64–S75. doi:10.1016/j.drugalcdep.2006.05.008.

Amaro, H., Dai, J., Arévalo, S., Acevedo, A., Matsumoto, A., Nieves, R., & Prado, G. (2007). Effects of integrated trauma treatment on outcomes in a racially/ethnically diverse sample of women in urban community-based substance abuse treatment. *Journal of Urban Health, 84,* 508–522. doi:10.1007/s11524-007-9160-z.

Andrews, C. (2008). An exploratory study of substance abuse among Latino older adults. *Journal of Gerontological Social Work, 51,* 87–108. doi:10.1080/01634370801967570.

Andrulis, D. P., Siddiqui, N. J., Purtle, J. P., & Duchon, L. (2010). *Patient protection and affordable care act of 2010: Advancing health equity for racially and ethnically diverse populations.* Washington, DC: Joint Center for Political and Economic Studies.

Appel, P. W., Ellison, A. A., Jansky, H. K., & Oldak, R. (2004). Barriers to enrollment in drug abuse treatment and suggestions for reducing them: Opinions of drug injecting street outreach clients and other system stakeholders. *American Journal of Drug and Alcohol Abuse, 30*, 129–153. doi:10.1081/ADA-120029870.

Bartels, S. J., Blow, F. C., Brockmann, L. M., & Van Citters, A. D. (2005). *Substance abuse and mental health among older Americans: The state of the knowledge and future directions.* Rockville: Substance Abuse and Mental Health Services Administration, Older American Substance Abuse and Mental Health Technical Assistance Center.

Chartier, K., & Caetano, R. (2010). Ethnicity and health disparities in alcohol research. *Alcohol Research & Health, 33,* 152–160.

Claus, R. E., & Kindleberger, L. R. (2002). Engaging substance abusers after centralized assessment: Predictors of treatment entry and dropout. *Journal of Psychoactive Drugs, 34*, 25–31. doi:10.1080/02791072.2002.10399933.

D'Aunno, T. (2006). The role of organization and management in substance abuse treatment: Review and roadmap. *Journal of Substance Abuse Treatment, 31*, 221–233. doi:10.1016/j.jsat.2006.06.016.

Feidler, K., Leary, S., Pertica, S., & Strohl, J. (2002). *Substance abuse among aging adults: A literature review.* Rockville: Substance Abuse and Mental Health Services Administration, Center for Substance Abuse Treatment.

Friedmann, P. D., Lemon, S. C., Stein, M. D., & D'Aunno T. A. (2003). Community referral sources and entry of treatment-naïve clients into outpatient addiction treatment. *American Journal of Drug and Alcohol Abuse, 29*, 105–115. doi:10.1081/ADA-120018841.

Garnick, D. W., Lee, M. T., Horgan, C. M., Acevedo, A., & Washington Circle Public Sector Workgroup. (2009). Adapting Washington circle performance measures for public sector substance abuse treatment systems. *Journal of Substance Abuse Treatment, 36*, 265–277. doi:10.1016/j.jsat.2008.06.008.

Gfroerer, J., Penne, M., Pemberton, M., & Folsom, R. (2003). Substance abuse treatment need among older adults in 2020: The impact of the aging baby-boom cohort. *Drug and Alcohol Dependence, 69*, 127–135. doi:10.1016/S0376-8716(02)00307-1.

Grella, C. E., & Stein, J. A. (2006). Impact of program services on treatment outcomes of patients with comorbid mental and substance use disorders. *Psychiatric Services, 57*, 1007–1015. doi:10.1176/appi.ps.57.7.1007.

Grella, C. E., Gil-Rivas, V., & Cooper, L. (2004). Perceptions of mental health and substance abuse program administrators and staff on service delivery to persons with co-occurring substance abuse and mental health disorders. *Journal of Behavioral Health Services & Research, 31*, 38–49. doi:10.1007/BF02287337.

Guerrero, E. G. (2010). Managerial capacity and adoption of culturally competent practices in outpatient substance abuse treatment organizations. *Journal of Substance Abuse Treatment, 39*(4), 329–339. doi:10.1016/j.jsat.2010.07.004.

Guerrero, E. G. (2013). Enhancing access and retention in substance abuse treatment: The role of medicaid payment acceptance and cultural competence. *Drug and Alcohol Dependence, 132*, 555–561. doi:10.1016/j.drugalcdep.2013.04.005.

Guerrero, E. G., Cepeda, A., Duan, L., & Kim, T. (2012). Disparities in completion of substance abuse treatment among Latino subgroups in Los Angeles county, CA. *Addictive Behaviors, 37*, 1162–1166. doi:10.1016/j.addbeh.2012.05.006.

Guerrero, E. G., Khachikian, T., Kim, T., Kong, Y., & Vega, W. A. (2013a). Spanish language proficiency among providers and Latino clients' engagement in substance abuse treatment. *Addictive Behaviors, 38*, 2893–2897. doi:10.1016/j.addbeh.2013.08.022.

Guerrero, E. G., Marsh, J. C., Khachikian, T., Amaro, H., & Vega, W. A. (2013b). Disparities in Latino substance use, service use, and treatment: Implications for culturally and evidence-based interventions under health care reform. *Drug and Alcohol Dependence, 3*, 805–813. doi:10.1016/j.drugalcdep.2013.07.027.

Guerrero, E. G., Marsh, J. C., Cao, D., Shin, H.-C., & Andrews, C. (2014). Gender disparities in utilization and outcome of comprehensive substance abuse treatment among racial/ethnic groups. *Journal of Substance Abuse Treatment, 46,* 584–591. doi:10.1016/j.jsat.2013.12.008.

Guzmán, B. (2001). *The Hispanic population: Census 2000 brief.* Washington, DC: U.S. Census Bureau.

Hadland, S. E., Kerr, T., Li, K., Montaner, J. S., & Wood, E. (2009). Access to drug and alcohol treatment among a cohort of street-involved youth. *Drug and Alcohol Dependence, 101,* 1–7. doi:10.1016/j.drugalcdep.2008.10.012.

Han, B., Gfroerer, J. C., Colliver, J. D., & Penne, M. A. (2009). Substance use disorder among older adults in the United States in 2020. *Addiction, 104,* 88–96. doi:10.1111/j.1360-0443.2008.02411.x.

Iwashita, N., Brown, A., McNamara, T., & O'Hagan, S. (2008). Assessed levels of second language speaking proficiency: How distinct? *Applied Linguistics, 29,* 24–49. doi:10.1093/applin/amm017.

Korper, S. P., & Raskin, I. E. (2002). The impact of substance use and abuse by the elderly: The next 20 to 30 years. In S. P. Korper & C. L. Council (Eds.), *Substance use by older adults: Estimates of future impact on the treatment system* (*Analytic Series A-21, DHHS Publication No. SMA 03-3763*). Rockville: Substance Abuse and Mental Health Services Administration, Office of Applied Studies.

Kouyoumdjian, H., Zamboanga, B. L., & Hansen, D. J. (2003). Barriers to community mental health services for Latinos: Treatment considerations. *Clinical Psychology: Science and Practice, 10,* 394–422. doi:10.1093/clipsy.bpg041.

Marsh, J. C. (2012). Gender differences among Latino subsample in a national sample of substance abuse treatment programs. *Journal of Substance Abuse Treatment, 43,* e8. doi:10.1016/j.jsat.2012.08.050.

Marsh, J. C., Cao, D., Guerrero, E., & Shin, H.-C. (2009). Need-service matching in substance abuse treatment: Racial/ethnic differences. *Evaluation and Program Planning, 32,* 43–51. doi:10.1016/j.evalprogplan.2008.09.003.

McCarty, D., Gustafson, D. H., Wisdom, J. P., Ford, J., Choi, D., Molfenter, T., Capoccia, V., & Cotter, F., (2007). The network for the improvement of addiction treatment (NIATx): Enhancing access and retention. *Drug and Alcohol Dependence, 88,* 138–145. doi:10.1016/j.drugalcdep.2006.10.009.

Santiago-Rivera, A. L., Altarriba, J., Poll, N., Gonzalez-Miller, N., & Cragun, C. (2009). Therapists' views on working with bilingual Spanish–English speaking clients: A qualitative investigation. *Professional Psychology: Research and Practice, 40,* 436–443. doi:10.1037/a0015933.

Simpson, D. D., Joe, G. W., & Brown, B. S. (1997). Treatment retention and follow-up outcomes in the drug abuse treatment outcome study (DATOS). *Psychology of Addictive Behaviors, 11,* 294–307. doi:10.1037/0893-164X.11.4.294.

Substance Abuse and Mental Health Services Administration. (2005a). *Older adults in substance abuse treatment: Update.* Rockville: Substance Abuse and Mental Health Services Administration, Office of Applied Studies.

Substance Abuse and Mental Health Services Administration. (2005b). *Substance use among older adults: 2002 and 2003 update.* Rockville: Substance Abuse and Mental Health Services Administration, Office of Applied Studies.

Substance Abuse and Mental Health Services Administration. (2011). Illicit drug use among older adults. http://oas.samhsa.gov/2k11/013/WEB_SR_013_HTML.pdf. Accessed 12 Nov 2013.

Substance Abuse and Mental Health Services Administration. (2012). *Results from the 2011 national survey on drug use and health: Summary of national findings.* Rockville: Substance Abuse and Mental Health Services Administration, Center for Behavioral Health Statistics and Quality.

Surface, E. A., Harman, R. P., Watson, A. M., & Thompson, L. F. (2009). *Are human- and computer administered interviews comparable?* Paper presented at the 24th annual meeting of the Society for Industrial and Organizational Psychology, New Orleans, LA.

Suro, R., & Singer, A. (2002). *Latino growth in metropolitan America: Changing patterns, new locations*. Washington, DC: Brookings Institution.

Thompson, L. F., Surface, E. A., & Whelan, T. J. (2007). *Examinees' reactions to computer-based versus telephonic oral proficiency interviews*. Paper presented at the 22nd annual conference of the Society for Industrial and Organizational Psychology, New York, NY.

Vega, W. A., & Lopez, S. R. (2001). Priority issues in Latino mental health services research. *Mental Health Services Research, 3,* 189–200. doi:10.1023/A:1013125030718.

Villatoro, J., Medina-Mora, M. E., Fleiz Bautista, C, Moreno Lopez, M., Bustos Gamino, M., Buenabad, N. (2012). El consume de drogas en Mexico: Resultados de la Encuesta Nacional de Adicciones, 2011. *Salud Mental, 35,* 447–457.

Zhang, Z., Friedmann, P. D., & Gerstein, D. R. (2003). Does retention matter? Treatment duration and improvement in drug use. *Addiction, 98,* 673–684. doi:10.1046/j.1360-0443.2003.00354.x.

Chapter 17
The Profile of Mexican Elder Migration Flow into the US (2004–2013)

Silvia Mejía-Arango, Roberto Ham-Chande and Marie-Laure Coubès

Introduction

Demographic aging and international migration have been dominant processes in sociodemographic change that represent outcomes of economic change and societal modernization. Although later life mobility occurs at a lower rate compared to mobility in other age groups, it brings forward several implications (1) The number of aged migrants has grown and will increase during the coming decades; (2) cultural differences with the host population represent limitations for integration processes (3) The perceived limited dimension of later-life migration has caused a tendency to minimize it, and it has affected the implementation of adequate policies.

Elderly migrants are a heterogeneous group with different migrant life trajectories and motivations to migrate. This group includes those who have grown old in the host country, those who are already elderly when they emigrate or return to their country of origin, as well as those with immigrant background (second and third generation). Motivations to migrate may also generate different categories which include forced migration, migrating for retirement, and migrating to rejoin their family.

Later-life migration varies between countries having Europe as the most advanced (Torres 2001; Bolzman et al. 2006; Warnes and Williams 2006). Reports, particularly from northern European countries like Sweden and Switzerland, describe increasing transnational lifestyles affecting elderly migrants who have established in both, the adopted country and the country of origin, and in some cases even in a third country (Casado-Diaz et al. 2004; Duncombe et al. 2003). This condition differs substantially from what happens in countries with other socioeconomic and political conditions like Mexico.

S. Mejía-Arango (✉) · R. Ham-Chande · M.-L. Coubès
El Colegio de la Frontera Norte, Tijuana, Baja California, Mexico
e-mail: smejia@colef.mx

© Springer International Publishing Switzerland 2015
W. A. Vega et al. (eds.), *Challenges of Latino Aging in the Americas,*
DOI 10.1007/978-3-319-12598-5_17

Theoretical Framework

The life course model has been one of the theoretical models central to migration research. This model is based on the idea that mobility rises in response to particular life course events such as retirement, onset of disability, income decline as well as widowhood. These events represent relevant life course events for later-life migration which vary along with different antecedents and long-term consequences in the temporal, geographical and socio-political context where they occur (Walters 2002).The above mentioned life course events may trigger migration depending on characteristics such as migration history, access to family networks in the host country, geographic proximity and possession of legal documents.

Warnes and colleagues studying European later-life migration found several typologies developed within the life-course model (Warnes and Williams 2006; Longino et al. 2002). The researchers propose a typology of life-course events associated to health status and disability that trigger migration including a healthy and disability-free status. A first type refers to retirement of healthy individuals. International retirement migration has received great attention from European and US researchers who have related this type of mobility pattern with long-distance migration of healthy seniors searching for alluring weather and leisure amenities. Walters (2000) introduced a category labeled *amenity migration* referring to migrants who are relatively affluent, healthier, more active among older people, less family-oriented and more concerned with quality of life (Murray et al. 2003; Haas and Serow 2002).

A second category deals with moderate illness, low income and widowhood leading to the migration of functionally independent elders, as proposed by Warner. This category corresponds to the type of mobility labeled by Walters (2000) as *assistance migration* which generally promotes short-distance migration for seniors to the homes of adult children for informal care, preservation of income and strengthening of kinship ties.

Finally, there is migration in severe disability, generally associated with relocation of elders to nursing homes or to adult children's dwellings who provide informal health care. It points to co-residence as a strategy for reducing living costs and coping with moderate or severe disability (Bures 2009).

Among elder migrants are two relevant groups: the most enterprising, affluent older people, and the most disadvantaged, deprived, socially-isolated and socially-excluded older people (Warnes and Williams 2006; Keister and Deeb-Sossa 2001). Heterogeneity of later-life migration reveals the importance of geographical and socio-political contexts. For the second group type migration represents the best way to preserve wellbeing. Many of these migrants come from rural areas and had relatively little education and few have formal or technical job skills. Most of them have adult children and have formed social networks at the destination which constitute an important antecedent when they seek informal family and community support. (Silveira and Allebeck 2001). The need for reciprocal relationships makes family reunification the main reason for migrating.

The environment of insecurity (EOI) conceptualization is another framework relevant to explain the new migration patterns in Europe during the last two decades that has also been used by some researcher to understand migration patterns in other countries with increasing violence. The EOI is not just a combination of push factors but an opportunity framework. The environment of insecurity is elaborated in terms of general political aspects, socio-economic indicators, legal framework, and demographic implications. Relative poverty is one of the most important aspects referred within this conceptualization. However, the basic function of migration here is not that poverty promotes migration but poverty within an environment of insecurity can facilitate migration through an opportunity framework created by these circumstances. Bohra-Mishra and Massey (2011) find that violence has a non-linear effect on migration, low to moderate levels of violence reduce the odds of migrating, while at high levels of violence, the odds of movement increase.

Mexican Elder Migration Studies

Studies that analyze migration in Mexican elder population basically considered two issues: (1) the effect migration has in household composition and, (2) the effect remittances have on elder's wellbeing that stay in Mexico. The first shows that migration restricts the availability of relatives, increasing the odds of elder's loosing family support. Remittances represent a financial contribution with a positive effect extending the period of independent living (Lopez-Cordova et al. 2008).

Other studies have analyzed Mexican elder population with migration history (Wong and Gonzalez 2010; Ross et al. 2006; Palloni and Arias 2004). Disability, wealth and access to health care in returned migrants are some of the aspects studied in the Mexican Health and Aging Study (MHAS). Other studies have revolved around health, family and other aspects related to the wellbeing of elders of Mexican origin who have grown old in the USA (Ottenbacher et al. 2009; Angel et al. 2008).

Despite abundant scientific information on the above issues, no studies have been found on the analysis of the profile of Mexican elders during migration process.

In the present study we sought to describe migration flow of elder subjects (60 and over) who move from south to northern border of Mexico with the intention to cross to the USA during the last 10 year period (2004–2013) and identify demographic and migration characteristics that describe the profile of these elder subjects dividing time in two periods: 2004–2008 and 2009–2013, to observe recent changes with increasing migration.

Methods

All subjects were participants of the Border Survey of Mexican Migration [*Encuesta sobre Migración en la Frontera Norte de México* (EMIF)]. This survey has been operating since 1993, as a continuous research program tracking original data on the number of people crossing the U.S.-Mexico border, whether legally or illegally. It's conducted in 12 Mexican border cities and at airports in the interior of Mexico by

El Colegio de la Frontera Norte (COLEF), a government-funded social science research institution and with the financial support of the Ministry of Foreign Affairs, Ministry of Labor and Social Welfare, National Institute of Migration/Migratory Politic Department, National Population Council and National Council of Science and Technology. The objective of the Border Survey of Mexican Migration is to provide unbiased estimates of the size, composition, and characteristics of Mexican labor flows taking place across the U.S.-Mexico border (http://www.migrationmonitor.com). The data created by EMIF relate to events rather than individuals since a migrant can contribute to one or more separate migrations during a quarter or a year. For more information about EMIF survey go to http://www.colef.mx/emif.

All subjects who participated in each of the yearly flows from 2004 to 2013 of the EMIF survey who met the following criteria were included: (1) subjects who migrated from south to northern border, (2) age 60 and more, (3) subjects who responded yes to the question if they intend to cross the U.S.-Mexico border. The total number of subjects analyzed was 2193.

Variables

The 10 year migration flow (2004–2013) was divided in two groups: the flow from 2004 to 2008 ($n = 662$) and from 2009 to 2013 ($n = 1,531$). Socio-demographic variables analyzed included: age, sex, education, marital status and locality of origin (urban/non-urban). Migration variables considered for the analyses were: reasons to cross the border, time staying in the US, have legal documents to cross the border, travel alone, or not, speak English, state of origin and state of destiny.

Statistical Analysis

Descriptive statistics were performed to describe and compare socio demographic and migration characteristics of subjects in each of the two time periods. Significance level was set up at < 0.05. A first logistic regression model was then conducted to determine the association between the two time periods as dependent variable, using as reference the period from 2004 to 2008 and the demographic characteristics as predictor variables. The second logistic regression model included migration variables adjusted for demographic variables that entered in the first model.

Results

In the last 10 years, Mexican migration to the US has changed. After a 5 year period of constant increase associated with the labor demand of Mexican workers in construction jobs, migration flow has been descending drastically since 2008. The

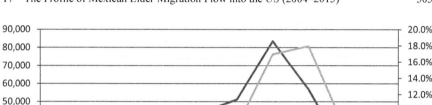

Fig. 17.1 South-North migration flow who intend to cross, total and percent 2004–2013. (Source: El Colegio de la Frontera Norte, Secretaría del Trabajo y Previsión Social, Consejo Nacional de Población, Instituto Nacional de Migración, Secretaría de Relaciones Exteriores, Encuesta sobre Migración en la Frontera Norte de México, http://www.colef.mx/emif)

economic crisis and the security enforcement laws for migrants are two important factors explaining the reduction of labor migration flow of Mexicans to the US until the present (Coubès 2011).

Figure 17.1 shows that Mexican elders 60 years and more constitute a small proportion of the total migration flow to the US. However, this group showed an opposite behavior of that observed in the general migrant population, increasing it's flow since 2008. As a result of these opposite tendencies, elder migration reached a point where it represented 17 and 18 % of the total flow in 2010 and 2011. Although it decreased during 2013 the flow of elder population still represents 10 % of the total flow. Comparing the 2 year periods allows seeing if there has been a change in the profile of the elders with increasing migration.

Demographic characteristics in each time period are presented in Table 17.1. The proportion of elder women during the second migration flow period (2009–2013) increased while men ratio lowered; subjects in the oldest age groups were higher; a decrease in married subjects reflected a moderate increase in widowed and single or divorced elders, low schooling (< 7 years) lowered while subjects with 7 or more years of education were higher and finally, subjects from urban locality increased significantly during this period.

Migration characteristics are shown in Table 17.2. When subjects were asked about the reasons they have to cross the border, meeting with family and friends was the most frequent answer during both migration flow periods, however in the last 5 years "other and stay living" category has increased significantly. On the contrary, crossing the border to work showed an important reduction. The time elders planned to stay in the US has also changed towards an increase, particularly evident

Table 17.1 Distributions and bivariate associations in demographic characteristics of Mexican elder migrants who intend to cross the U.S.-Mexico border during two periods in the last 10 years. (Source: El Colegio de la Frontera Norte, Secretaría del Trabajo y Previsión Social, Consejo Nacional de Población, Instituto Nacional de Migración, Secretaría de Relaciones Exteriores, Encuesta sobre Migración en la Frontera Norte de México, http://www.colef.mx/emif)

	2004–2008[a]	2009–2013[a]	(p value)[b]
Sex (%)			0.0001
Masculine	51.8	42.4	
Feminine	48.4	57.6	
Age %			0.0001
60–64	40.6	34.4	
65–69	31.3	26.6	
70–74	20.3	19.9	
75–79	5.6	13.5	
80>	2.2	5.7	
Marital status %			0.02
Married	66	60.5	
Widowed	25	27	
Divorced/single	8.9	12.4	
Education %			0.016
No schooling	20.2	19.5	
1–6 years	64.7	60.2	
7> years	15.1	20.3	
Locality of origin			0.0001
Urban	36.2	52.6	
Non-urban	63.8	47.4	
Total (intend to cross)	30.2%	69.8%	
N	662	1,531	

[a]weighted percentages
[b]p-value is based on chi square test

in those who want to stay for more than a year. In the same direction, an increase was observed in the proportion of subjects who travel alone, have legal documents and speak English. For the state of origin and destiny, no differences were observed between both migration flow periods. The geographical pattern of migration from origin to destination observed in the elderly is concentrated around what has been called the historic migration region or west-central states of Mexico and include almost exclusively the border states of the US as destiny (Massey et al. 2010).

Results of logistic regression models for the second migration flow period are shown in Table 17.3. Demographic variables in model 1 included sex, age and locality of origin, while education and marital status were not significant predictors. Subjects in the second migration flow were more likely to be women, age 75 and

Table 17.2 Distributions and bivariate associations in migration characteristics of Mexican elders who intend to cross the U.S.-Mexico border during two periods in the last 10 years. (Source: El Colegio de la Frontera Norte, Secretaría del Trabajo y Previsión Social, Consejo Nacional de Población, Instituto Nacional de Migración, Secretaría de Relaciones Exteriores, Encuesta sobre Migración en la Frontera Norte de México, http://www.colef.mx/emif)

	2004–2008[a]	2009–2013[a]	(p value)[b]
Reasons to cross and stay %			0.0001
Meet with family/friends	69.8	76.6	
Work/business	20	11.5	
Travel/shopping	9.2	8	
Other + stay living	0.9	3.8	
Time staying in the USA %			0.0001
<1 month	30.6	19.6	
1–3 months	31.9	38.4	
4–12 months	15.2	7.5	
>1 year	22.2	34.4	
Travels alone (yes) %	62.5	66.8	0.05
Has legal documents (yes) %	94.2	96.1	0.04
Speak English (yes) %	7.3	13.5	0.0001
State of origin %			0.0001
Historic region (West-Central)	62	62.9	
Border region	19	19	
Central region	17.6	16.4	
Southeast region	1.3	1.7	
State of destiny %			0.0001
Borderlands	95.2	94.6	
Northwest	2.7	3.7	
Other[c]	2.2	1.6	
N	662	1,531	

[a]weighted percentages
[b]p-value is based on chi square test
[c]Other: includes Great lakes, Northeast, Southeast, Deep South, Plains, see Massey et al.

over and came from an urban locality compared with the first period. All migration characteristics included in model 2 were significant predictors: Crossing the border to work was less likely during the last 5 years while crossing for other reasons and to stay there living was three times more likely. The duration of time staying in the US has also changed in the last years, elders stay in the US for longer periods, staying for more than 1 year is also three times more likely. The odds of having legal documents to cross, traveling alone and speaking English range from 1.54 to 2.00 times during the second period.

Table 17.3 Multivariate regression models for the likelihood of 60> migration flow during 2009–2013 according to demographic and migration variables

	Model 1[a]			Model 2[a]		
	OR	Sig.	(95 % CI)	OR	Sig.	(95 % CI)
Sex						
Masculine	1			1		
Feminine	1.37	0.001	(1.13–1.65)	1.600	0.0001	(1.22–2.08)
Age						
60–64	1					
65–69	0.95	0.64	(0.76–1.19)	0.908	0.49	(0.68–1.19)
70–74	1.12	0.35	(0.86–1.46)	1.554	0.010	(1.11–2.17)
75–79	2.96	0.0001	(2.02–4.32)	3.010	0.0001	(1.84–4.90)
80>	3.02	0.0001	(1.72–5.31)	4.064	0.0001	(1.96–8.44)
Locality of origin						
Non-urban	1			1		
Urban	1.94	0.0001	(1.60–2.35)	2.09	0.0001	(1.65–2.66)
Reasons to cross						
Meet with family/friends				1		
Travel/shopping				1.24	0.32	(0.81–1.89)
Work/business				0.55	0.002	(0.38–0.80)
Other/stay living				3.05	0.011	(0.29–7.22)
Time staying in the USA						
<1 month				1		
1–3 months				2.11	0.0001	(1.53–2.89)
4–12 months				1.60	0.04	(1.02–2.54)
>1 year				3.36	0.0001	(2.34–4.82)
Has legal documents						
No				1		
Yes				2.21	0.002	(1.32–3.66)
Travels alones						
No				1		
Yes				1.56	0.0001	(1.20–1.99)
Speaks English						
Yes				1		
No				1.54	0.03	(1.04–2.27)

[a]Model run with population weight

Conclusions

In the present study we found a significant increase of elder migration flow to the US during the last 5 years, a result contrary with the decrease in global migration. The general drop in migration from Mexico to the USA has been explained by the effects of the economic crisis and the security enforcement laws for migrants after 2008. The same reasons may have implications for the increasing elder flow during the last 5 years assuming that migrant children may be restricted from traveling back home and care for the elderly parent; this may act as an especially hard constraint for unauthorized migrants. Insecurity at their states of origin can also be a meaningful explanation for the increase of elderly migration during the last 5 years. Mexico has been experiencing a significant rise in violence during the last years (Alvarado and Massey 2010) affecting elder subjects who may have a worse perception of insecurity in their environments, as well as their children living in the US.

The demographic and migration profile of elders in this study showed that migrating during the last 5 years is more likely in women, in old ages, from urban areas, migrating for reunification reason, having legal documents, traveling alone and planning to stay for more than 1 year.

From the typology of amenity and assistance migration proposed by Walters (2000), elder migration in Mexico can be related to the search for better quality of life, preservation of income and strengthening of kinship ties. However, the environment of insecurity (EOI) conceptualization (Sirkeci 2005) used to explained new migration patterns in Europe can be useful to understand some of the results in the present study and provide a framework where the environment of insecurity model considers not only the lack of human security that elders have in their places of origin, but also other indicators as poverty, debilitating social networks, family oriented tradition which together with the presence of social networks at destination and having the legal documentation to stay in the US provide an opportunity framework that call attention to a new model for understanding migration in old ages.

An important limitation of our study relates to the fact that all participants in the EMIF survey included in the present study were migrants that arrive to the northern border and who express their intention to cross. Although an intention not necessarily transforms in a real behavior, we considered it's a valid way to measure real border crosses if elders leave their places of origin and arrive to the northern border. Another limitation is the absence of information on health status. The EMIF survey has added a health module since 2012 which will be extremely useful to characterize elder subjects crossing the border.

In conclusion, the study of Mexican elders that arrive to the northern border with the intention to cross the border, suggests that there's an increasing trend of later-life migration in Mexico that has been mostly ignored in the literature but must be considered a new issue in the study of Mexican-American elder population.

References

Alvarado, E., & Massey, D. (2010). In search of peace: Structural adjustment, violence, and international migration. *The Annals of the American Academy of Political and Social Science, 630,* 137–161.

Angel, R., Angel, J., & Hill, T. (2008). A comparison of the health of older Hispanics in the United States and Mexico: Methodological challenges. *Journal of Aging and Health, 20,* 3–31.

Bolzman, C., Fibbi, R., & Vial, M. (2006). What to do after retirement? Elderly migrants and the question of return. *Journal of Ethnic and Migration Studies, 32*(8), 1359–1375.

Bohra-Mishra, P., & Massey, D. S. (2011). Individual decisions to migrate during civil conflict. *Demography, 48*(2), 401–424.

Bures, R. M. (2009). Moving the nest: The impact of co-residential children on mobility in later midlife. *Journal of Family Issues, 30*(6), 837–851.

Casado-Diaz, M. A., Kaiser, C., & Warnes, A. M. (2004). Northern European retired residents in nine southern European areas: Characteristics, motivations and adjustment. *Ageing and Society, 24*(3), 353–381.

Coubès, M.-L. (2011). Tendencias recientes de los flujos de migración de México hacia Estados Unidos: algunas hipótesis explicativas. El Colegio de la Frontera Norte. Reporte tecnico.

Duncombe, W., Robbins, M., & Wolf, D. A. (2003). Place characteristics and residential location choice among the retirement-age population. *The Journals of Gerontology Series B: Psychological Sciences and Social Sciences, 58*(4), S244–S252.

Haas, W. H., & Serow, W. J. (2002). The baby boom, amenity retirement migration, and retirement communities: Will the golden age of retirement continue? *Research on Aging,* 24(1), 150–164.

Keister, L. A., & Deeb-Sossa, N. (2001). Are baby boomers richer than their parents? Intergenerational patterns of wealth ownership in the United States. *Journal of Marriage and Family, 63*(2), 569–579.

Longino, C. F., Perzynski, A. T., & Stoller, E. P. (2002). Pandora's briefcase: Unpacking the retirement migration decision. *Research on Aging, 24*(1), 29–49.

Lopez-Cordova, E., Tokamn, A., & Verhoogen, E. A. (2008). Globalization, migration, and development: The role of Mexican migrant remittances. *Economia, 6*(1), 217–256.

Massey, D., Rugh, J., & Pren, K. (2010) The geography of undocumented Mexican migration. *Mexican Studies, 26*(1), 129–152.

Murray, M., Pullman, D., & Rodgers, T. H. (2003) Social representations of health and illness among 'baby–boomers' in eastern Canada. *Journal of Health Psychology, 8,* 485–99.

Ottenbacher, K. J., Graham, J. E., Snih, S. A., Raji, M., Samper-Ternent, R., Ostir, G. V., & Kyriakos, M. (2009). Mexican Americans and frailty: Findings from the Hispanic established populations epidemiologic studies of the elderly. *American Journal of Public Health, 99,* 673–679.

Palloni, A., & Arias, E. (2004). Paradox lost: Explaining the Hispanic adult mortality advantage. *Demography, 41,* 385–415.

Ross, S., Pagan, J., & Polsky, D. (2006). Access to health care for migrant returning to Mexico. *Journal of Health Care for the poor and undeserved, 17,* 374–385.

Silveira, E., & Allebeck, P. (2001) Migration, ageing and mental health: An ethnographic study on perceptions of life satisfaction, anxiety and depression in older Somali men in east London. *International Journal of Social Welfare, 10,* 309–320.

Sirkeci, I. (2005). War in Iraq: Environment of insecurity and international migration. *International Migration, 43,* 197–214

Torres, S. (2001). Understandings of successful aging in the context of migration: The case of Iranian immigrants to Sweden'. *Ageing & Society, 21,* 333–355

Walters, W. H. (2000). Types and patterns of later-life migration. *Geografiska Annaler, 3,* 129–147.

Walters, W. H. (2002). Later-life migration in the United States: A review of recent research. *Journal of Planning Literature, 17,* 37–66.

Warnes A., & Williams, A. M. (2006). Older migrants in Europe: An innovative focus for migration studies', *Journal of Ethnic and Migration Studies, 32*(8), 1257–1281.

Wong, R., & Gonzalez, C. (2010) Old-age disability and wealth among returned migrants from the USA. *Journal of Aging and Health, 22,* 932–954.

Part IV
Cost and Coverage: Fiscal Impacts

Chapter 18
Overview: Policy Responses, Political Realities and Cross-National Variations: Responding to Health and Social Needs of the Latino Century

Fernando M. Torres-Gil

The demographic imperative facing the America's and the new emerging population of older Latinos has a profound relevance for how we pay, for whom we provide benefits and for how nation's respond to these new realities. This volume has shed important intellectual and analytical light on the demographic trends, economic, social and cultural factors, methodological needs and the bi-national and transnational migration issues facing aging in the America's. This last section brings together the mezzo aspects of health care financing and programs in the U.S. and its impact on Latino elders and the macro aspects of how governments and nation's vary in their response to aging and the new realities of longevity.

Within the mezzo sphere lie important questions about how U.S. entitlement programs and recent health care policy developments address health care disparities faced by Hispanics in general and Latino elders, in particular. If we, as scholars and advocates, are to influence how governments ought to serve the growing numbers of Latino elders, in the U.S. and throughout the Americas, it behooves us to first ask: what is occurring within current social policies designed to provide health care to older persons in general and what structural and policy changes might be evolving that impinge on the ability of Latino elders to receive access to medical, health and long-term care services commensurate with their particular needs?

Two papers in this section provide intriguing insights to these mezzo concerns. Brown, Wilson and Angel address "head-on" the existing system of Medicaid benefits for U.S. beneficiaries (eligibility is determined via income testing and though designed for poor and disabled often becomes nursing home coverage for older persons with declining assets) and the Affordable Care Act (ACA), designed to expand health care coverage through the private sector, subsidies for small businesses, an individual mandate to obtain health insurance and expanding state Medicaid programs for low-income individuals. Brown et al. raise an intriguing concern: what happens when states are given the option (through a recent U.S. Supreme Court decision) to "opt-out" of expanding Medicaid and thus reject federal subsidies. Their

F. M. Torres-Gil (✉)
University of California, Los Angeles, CA, USA
e-mail: torres@luskin.ucla.edu

© Springer International Publishing Switzerland 2015
W. A. Vega et al. (eds.), *Challenges of Latino Aging in the Americas*,
DOI 10.1007/978-3-319-12598-5_18

paper examines the relationship and causality of opting out of Medicaid expansion and seek to understand the impact of such a decision on the ability of the ACA to address the health disparities facing Latino's and Latino elders. Perhaps not surprisingly, this study demonstrated that in fact, this opt-out feature harms the ability of the ACA to achieve its stated goals and the paper points out the irony that those states resisting Medicaid expansion are in fact the states with the highest number of uninsured Latinos and where medical and health care needs (as well as long-term care) are the greatest.

The greater irony perhaps in the ideological debates about the role of government (too much, not enough), the role of the private sector (more or less) and the public's mercurial relationship with taxation (disenchantment with paying taxes) is an alternative policy response by conservative and free market advocates and those states that opt out of the ACA Medicaid expansion: encourage individuals to save for their health needs and provide financial incentives/disincentives to overutilization of medical services. Thus, the next paper in this section sheds an important indicator about this "individual responsibility" to pay for medical care. Odufuwa, Berrens and Valdez examine the "racial and ethnic disparities in willingness to pay for improved health." The salience of this paper lie with understanding the economic behavior of populations and the "willingness to pay" (WTP). Comparing the proclivity for WTP by non-Hispanic whites and minorities, the analyses found that in general, minorities evidence a modest WTP to consider paying more yet when examined within the terms of absolute added dollars and sacrificing current consumption of goods and services, the authors found that minorities are less willing than whites to pay increasing dollars for improved health care, even with the presence of risky behaviors (alcohol and tobacco use). The behavioral reasons for a WTP is unclear but the authors surmise that other economic and social mechanisms play a role in the WTP. Regardless of the ultimate reasons, this study provides important clues that the use of tax policies such as health savings accounts, high deductibles and co-pays and otherwise relying on private pay approaches may not work effectively with minority and diverse populations.

The economic complexity of aging in the America's is further amplified in the Gassoumis, Wilbur and Torres-Gil treatise on the "Economic Security of Latino Baby Boomers." A new line of inquiry is presented via this paper—the subgroup that represents the nexus of baby boomers and Latinos: Aging Latino Boomers. The approximately 8 million in this subgroup may well be the "canary in the mineshaft" since aging Latino Boomers are both the next generation of Hispanic elders in the U.S. and represent the new governing elites in the Hispanic community (the political, economic, academic, cultural leadership). Gassoumis et al. assess the nature of the racial and ethnic structural economic disparities facing this emerging subgroup and ascertain the root factors that may enable or inhibit their economic, social and health advancement. Utilizing comparative sampling among Baby Boomers and Silent Generation cohorts and comparing to Asians populations, the authors surmise that of all the challenges facing the next generation of Latino elders; education and naturalization go to the "head-of-the-class." In their estimation, the two most likely avenues for redressing the health and economic disparities among Latino Baby

Boomers—improving educational outcomes and promoting naturalization and a "path to citizenship"—are most likely to promote long-term positive outcomes.

Yet, the future of aging in the America's is not confined to the northern hemisphere or to the United States. Ultimately the commonalities facing aging in South America, Central America and North America are universal yet the responses vary dramatically. Two papers lend themselves to important insights about the role of government and public policy and the bi-national opportunities for reshaping the hemisphere's response to it's aging. Gutiérrez Robledo, Ortega and Campos provide a comprehensive overview of the "Present State of Elder Care in Mexico." This seminal piece brings together the past, present and future endeavors of our neighbor to the south as it grapples with a demographic transition little-noted; rapid aging and declining fertility levels. The demographic dividend of a large youthful and working age-population is about to transform to a large elder population, decline in the youth cohort and a shift from traditional reliance on family for social, health and long-term care of the elderly. Gutiérrez Robledo et al. characterize Mexico's response to this new reality as "untimely and inadequate" and with a reliance on a health care and social system of benefits and services leaving large segments of its population vulnerable to the vagaries of longevity. Yet, Mexico is making incremental advances in expanding social insurance and public benefits and the creation of a "National Institute of Geriatrics." The authors raise an intriguing possibility of a "Gray Vote" that may empower future generations of Mexican elders to exercise their potential political influence as they demonstrated in a 2000, Mexico City (el Distrito Federal) law creating a non-contributory pension for the 18.2 % of D.F. elders.

The issue of retirement security in the Americas receives important investigation by Angel and Perieira on "Pension Reform, Civil Society and Old-Age Security in Latin America." Although all nations in the Americas face a common demographic reality of longevity, declining fertility rates and changes in family roles and care giving, each, in turn, addresses these realities in different ways. Argentina, Chile, Mexico and Uruguay present contrasting models in their efforts to mitigate the vulnerabilities of old age. Chile, for example, is well known for its privatization of a pension system based on social insurance. The Pinochet government experimented with an ideological model based on private investments and individual responsibility. While this approach had initial success, the continued vulnerabilities faced by retirees led to a more balanced approach of public subsidies, personal savings and individual retirement accounts. Argentina continues to rely on public pensions while encouraging private accounts and personal savings. Mexico has long utilized public pensions for select constituencies (e.g. oil workers and government employees) but with large segments of Mexico elders without a safety net, it is creating a mix of public subsidies and expanded pensions. Uruguay appears to focus on individual retirement accounts. What these four nations have in common are attempts to incrementally provide a minimal measure of retirement supports while the social and economic disparities faced by growing numbers of older persons continue to test the ability of the public and private sectors to provide a comprehensive safety net. Yet, each nation, provides important experiments and lessons that can serve as models for the entire hemisphere including as the authors suggest, an expanded role for civil society and non-governmental and faith based organizations.

Finally, in "A Politics of Aging in a Majority-Minority Nation," Demko and Torres-Gil examine a development unique to the United States but evidenced in other parts of the globe: the interconnection of two profound trends: the aging of a society while its mix of racial and ethnic group changes. The advent of longevity and growth of older populations while the previous minority groups become a new majority, creates a new dynamic that both complicate and create new opportunities for addressing aging and diversity. Other nations throughout Asia and Europe are facing similar developments but the United States presents an important model given the propensity of older persons to be an important political and electoral force while minority and racial groups, including Latino's evolve their own political and electoral influence. Where and how this new politics of aging and diversity will influence the public and private sectors response to longevity is uncertain but the authors suggest that the potential for intergenerational and interethnic/interracial tensions and relationships can both complicate and create opportunities for new alliances based on age, race and ethnicity.

The papers in this sections raise perhaps more questions than answers but they present an important contribution in outlining the frameworks by which we can better understand the twenty-first century reality that the Americas are aging and undergoing profound changes related to financing, public policy and politics: such changes will have an immeasurable impact on the health, long-term care and social conditions facing new generation of elders throughout the Americas.

Chapter 19
Profile of Pre-retirement Age Hispanics in the Medicaid and Private Insurance Markets: Expanding vs. Non-expanding States

Henry S. Brown, Kimberly J. Wilson and Jacqueline L. Angel

When the Supreme Court upheld the Patient Protection and Affordable Care Act (ACA), it allowed states to opt out of expanding Medicaid without losing existing Medicaid funds. This decision means that uninsured persons who would newly qualify for Medicaid under the ACA will not obtain Medicaid if they live in non-expanding states. The decision also hinders those buying private insurance in the exchanges. At least initially, sign-ups under the state-run exchanges were stronger, perhaps due to the superior design of the state-run sites, or the higher use of navigators. Hispanics, despite being in an economic position to receive subsidies and/or Medicaid, are lagging in Medicaid and private enrollment (Alonso-Zaldivar 2014; Chang 2014; Medrano 2014; A. P. Press 2014a, b; Seipel 2014). Therefore, the ability of states to both stop the expansion of Medicaid and to complicate the expansion of private insurance would mean few people will obtain insurance and hence will not be able to afford health care. The vast majority of eligible people were uninsured and lacked access to care prior to ACA. Non-expansion could therefore mute the disparity-reducing impact of ACA. Given well-documented race/ethnic health disparities in the U.S., a natural question is whether non-expansion affects aging Hispanics more than Non-Hispanic whites. In other words, to what extent does non-expansion appreciably mute the disparity-reducing effects of ACA, especially as they pertain to aging Hispanics?

Aging Hispanics are potentially greatly influenced by a state's decision to expand. Our study focuses on Hispanics aged 51–64. First, uninsured people up to 133 % of poverty are eligible to obtain Medicaid in expansion states. Second, private insurance in the exchanges is highly subsidized, and given the low incomes of Hispanics, even those nearing retirement are under the 400 % of poverty threshold for subsidy. Further, ACA's community rating ensures that Hispanics will not be

H. S. Brown (✉) · K. J. Wilson
School of Public Health Austin Regional Campus, The University of Texas Health Science Center, Houston, USA
e-mail: Henry.S.Brown@uth.tmc.edu

J. L. Angel
The University of Texas, Austin, USA

© Springer International Publishing Switzerland 2015
W. A. Vega et al. (eds.), *Challenges of Latino Aging in the Americas*,
DOI 10.1007/978-3-319-12598-5_19

penalized with higher premiums for having a chronic illness, like diabetes, which are common among aging Hispanics (Cantu et al. 2013). Finally, although legal residents must wait 5 years to be eligible for Medicaid, they are eligible for the exchanges immediately. This is an important advantage for aging Hispanics who are foreign-born.

Non-expansion may greatly affect whether aging Hispanics obtain health insurance coverage. Of the top ten states in terms of the share of Hispanics, only Texas and Florida did not expand. In 2012, 19,769,438 Hispanics lived in non-expansion states, or 37.2 % of all Hispanics (American Factfinder 2012). While over 58.2 % of all Hispanics live in expansion states, these states started with policies much closer to ACA than non-expansion states (Some states are undecided). On average, non-expansion states set the Medicaid maximum income threshold for working parents at 62 % of poverty level versus 117 % of poverty level for those in expanding states (author's calculations (Heberlein et al. 2012)). This difference means that ACA would help more uninsured persons obtain Medicaid in non-expanding states than in expanding states. Therefore, non-expansion could be muting disparity reduction greatly.

As noted earlier, persons in non-expanding states are also lagging in private coverage take-up through health insurance exchanges (Assistant Secretary for Planning and Evaluation (ASPE) 2014). As with Medicaid, private insurance coverage rates were much higher prior to ACA in expansion states. Expansion states are more likely than non-expansion states to provide navigators, which have been shown to be particularly helpful for non-English speaking groups and persons without internet connections (Blavin et al. 2014). Compared to other ethnic groups, Hispanics are the least reliant on the Internet and most reliant on navigators, friends and family in obtaining insurance through exchanges. In states like California, only 52 % of Hispanic households have broadband Internet access at home (Seipel 2014). As of this writing, Hispanics were lagging behind other minority groups in signing up for the health care plans (Alonso-Zaldivar 2014). Many states with high concentrations of Hispanics have too few Spanish-speaking enrollment counselors (Seipel 2014). In addition to the lack of bilingual navigators, some Hispanic families are concerned with privacy issues, whether they interact with navigators or government officials, or apply through the internet. The concern about privacy is due to the fear of deportation, especially among multi-citizenship households (Medrano 2014).

We focus on Hispanics who are middle age, 51–64 years of age. Although the effects of ACA for younger Hispanics are also of interest, the potential for reducing disparities is greater among those who are middle age and older. Approximately 12.6 million adult Hispanics in the United States were without coverage in 2011 (Kaiser 2013). The gap in coverage is particularly severe for Hispanics approaching retirement. Hispanic individuals in this age group are more susceptible to the negative effects of gaps in coverage due to lower health status and higher prevalence of chronic disease such as diabetes, which can be more easily addressed with coverage (Cantu et al. 2013). As noted earlier, insurance in exchanges are subject to community rating regulations which prevent charging penalties for chronic illnesses.

A large body of research points to a dramatic change in the face of aging in the United States. It is well established that while the older American population is still relatively homogenous, it is projected to substantially increase its racial and ethnic diversity in decades to come. The Hispanic population in the United States is still relatively young, but it is aging swiftly. The number of Hispanics aged 65 and older is expected to increase more than six times by 2050 to 17.5 million (Angel et al. 2012). Hispanics also have the highest life expectancy at birth, living approximately 3 years longer than non-Hispanics (Arias 2012).

In this paper, we examine and compare the health characteristics of aging Hispanics in expansion and non-expansion states to non-Hispanic whites. We investigatethe extent to which non-expansion, especially if it remains the status quo, hinders the disparity-reducing effects of ACA as it pertains to aging Hispanics. We examine whether aging uninsured Hispanics in non-expansion states are less healthy than those in expansion states. If so, disparities will remain due to the decision to not expand. On the other hand, if they are more healthy than non-Hispanic whites, adding them to the Medicaid and private exchange roles may reduce premiums by improving the average health of those insured.

Background

So far, enrollees in Medicaid and in private insurance through exchanges have been, on average, older (ASPE 2014). However, the early advantage of Medicaid for older persons has not been uniform due to individual states' decisions to expand, nor has enrollment through exchanges. Older persons in expanding states who now qualify for Medicaid have enrolled in Medicaid. In exchanges, older and middle aged enrollment has been strong despite the fact that older enrollees, who make more money on average than younger workers, will receive fewer subsidies on average than younger enrollees. The community rating and guaranteed issue provisions ensure that private coverage is affordable and attainable for older Americans. Yet, older *Hispanics* lag in enrollment (Blavin et al. 2014).

The Hispanic paradox may partially explain why Medicaid and private insurance lags among Hispanics. Despite having a low socio-economic profile, Hispanics as a group enjoy remarkably long lives, a phenomenon known as the Hispanic paradox. The life expectancy for Hispanics in the United States is 3 years higher than that of the total population. Hispanic men who live until 65 have a life expectancy of 84 years, and Hispanic women at 65 years of age have an expectancy of 87 years (Centers for Disease Control 2010). Research in economics suggests that healthier people have lower demand for insurance than those who are less healthy (Akerlof 1970). Thus, Hispanics, like younger people, may have lower demand for health insurance.

Although Hispanics live long lives, it is in spite of low access to health care. Hispanics, especially those of Mexican origin, are far less likely than Non-Hispanic whites to have health insurance, and often have more chronic illnesses, as noted earlier. Hispanics residing along the U.S.-Mexico border are more likely than other racial and ethnic groups to report no health insurance coverage of any kind, largely due to its high cost relative to income (Bastida et al. 2008). The need for coverage is likely to rise as the older Hispanic population continues to grow rapidly in the near term.

For citizens under 65 years of age, the United States does not have universal health care coverage (Eichner and Vladeck 2005b). As a result, Hispanics approaching retirement age are particularly at high risk. Approximately 14 % or 7.1 million people ages 50–64 years were without any coverage in 2005 (Smolka et al. 2007) This is a staggering figure in and of itself, but it masks an even more serious problem that arises from the fact that although the proportion of uninsured Americans is high across the board, the proportion of uninsured people is even higher among minority Americans. Among this minority, Mexican Americans face unique problems of inadequate coverage, especially in the cases of working-age adults near retirement age. While longer life expectancy is good news, there is a downside. Longer life often means that older adults are living longer with chronic diseases— diseases which might be prevented, delayed, or managed better with health care prior to Medicare. Hispanics typically spend a large portion of the years after age 50 in poor health (Cantu et al. 2013), which has clear negative implications for health care costs and family burden. In addition, older Hispanics suffer from high rates of diabetes, hypertension, and disability. Cognitive decline is also a major problem, perhaps related to extremely low levels of education. Furthermore, they have low rates of Medicaid coverage, low incomes, and little wealth.

Pre-retirement age Hispanics of Mexican-origin in particular face different barriers than majority Non-Hispanic white populations. They tend to work in low wage or service sector jobs, which rarely come with health insurance. Although employment disadvantages are one major source of this insurance disparity, other factors, including lower wealth and income, immigration history, citizenship status, and limited English language proficiency contribute to unequal options in quality coverage.

In the absence of any form of health insurance coverage, research shows that individuals have lower access to adequate health care including preventive care (Eichner and Vladeck 2005a). Many low-income Hispanics receive care, but they do so from federally sponsored community health centers that mostly provide primary care and only limited specialty medical care like mental health services (Vega and Gonzalez 2012). A recent Agency for Healthcare Research and Quality (AHRQ) report indicates that 70 % of Hispanics are receiving substandard care or have less access to care than non-Hispanic whites (Agency for Healthcare Research and Quality 2009). Such disparities in service use are magnified for residents along the U.S./Mexico border. Regardless of locale, uninsured Hispanics are far more likely than the insured to pay for health services out-of-pocket and to forego essential health services such as wellness care (Torres-Gil and Lam 2012). Hispanics are struggling with paying medical bills, and a catastrophic health care crisis often leads to medical bankruptcy (Mangan 2013).

Our previous research employing data from the Mexican Migration Project (MMP) shows that young Mexican males who migrate between the U.S. and

Mexico report good health (Brown et al. 2014). Consequently, they spend less on medical care while in the United States. The predicted costs are much lower for immigrants both initially and after time spent in the U.S. These findings are consistent with the literature, which shows that immigrants spend less on medical care than their native-born counterparts (Stimpson et al. 2013).

As of this research, at least one of the ACA provisions has been accepted by 34 states, and among these, 11 states have adopted all of the reforms (Keith and Lucia 2014).

Data and Methods

In the following analysis, we employ data from the 2011 Medical Expenditure Panel Survey (MEPS). MEPS is sponsored by the U.S. Department of Health and Human Services' Agency for Healthcare and Research Quality and includes detailed information on the cost and use of health care and health insurance coverage in all 50 states. We select a subset of the sample from a restricted state-level dataset that consists of 3,425 Hispanic and non-Hispanic white individuals ages 51–64, who were "in-scope" during 2010 and responded to the general health and insurance coverage items. Mean and chi-square tests of significance were used to compare racial/ethnic group mean scores on a variety of social, demographic, health, and health care variables separately for two insurance segments. A description of the operationalization of relevant variables follows:

Native born was coded as 1 if respondents replied that they were born in the U.S. and 0 if they replied that they were born elsewhere. *Years of education* were recorded as the number of years of education respondents reported having up to 12; 1–4 years of college are coded as 13–16, respectively; 5 and greater years of college are coded as 17. *Married* is coded as 1 if the respondent reports currently being married, and 0 if not. *Poverty level* is calculated as household income as a percent of the federal poverty level in the respondent's state of residence. *Employment status* was coded as a categorical variable with 0 being unemployed for all of 2011, 1 being employed for part of the year, and 2 for being employed for the entire year; we did not differentiate between part and full-time employment. *Number of years lived in the U.S.* is a direct operationalization of the reported number of years of residence in the U.S. *Lived in the U.S. less than 10 years* is coded as 1 if the respondent has resided in the U.S. for less than 10 years and 0 if for more than 10 years. *BMI* is a calculated field in MEPS.

BMI categories were created from this data with values less than 25 coded as 1 for Under or Normal Weight; between 25 and 30, coded as 2 for Overweight; and 30 or greater coded as 3 for Obese. *Physical* and *Mental Health* categories were created from the self-reported health questions in MEPS. Values of 1 and 2 were coded as Poor/Fair; 3 was coded as Good; and 4 and 5 were coded as Very Good/ Excellent. A *chronic disease index* was created by summing reported diagnoses of 11 chronic diseases including: asthma, cancer, chronic heart disease, high cholesterol, diabetes, high blood pressure, heart attack, arthritis, emphysema, other heart disease and stroke. The minimum value for this index was 0; the maximum possible was 11. *Disabled* was coded as 1 for respondents who reported receiving SSI due to disability, and 0 if not.

Cost barrier variables were created for medical care, dental care and prescriptions out of several MEPS items. The six items used to create the variables ascertain why the respondent did not get or delayed care (prescriptions) if they specified that they'd not gotten or had delayed that form of health care in the past year in a previous question. Response options used were: (1) could not afford care, (2) insurance would not approve/cover/pay, (3) doctor refused family insurance plan. The new variables were coded 1 if respondent did not get or delayed care because they specified one of the reasons above caused them to not get or delay care, and 0 if a different reason was given or they had no problems getting care when needed. A final cost barrier summary variable was created and coded as 1 if respondents reported delaying or not getting any form of care in the past year, and 0 if not.

We use non-Hispanic whites as the comparison group. Although we would like to focus on the uninsured exclusively, this would have reduced our sample size considerably, thus reducing the power needed to test for differences. Therefore, we focus on two groups split by whether they are in an expansion state. Over half (1951) of respondents were in expansion states. First, we consider those on Medicaid and those who are not but are under 133 % of poverty. Second, we consider privately insured persons and uninsured persons over 133 % of poverty, the threshold for Medicaid eligibility under ACA. In each case, the results can be interpreted as the final health characteristics for the Medicaid population and the private insurance population if expansion occurred and take-up is 100 %.

In each case, we conduct pairwise tests for differences between Hispanics and Non-Hispanic whites for each variable. If the variables are continuous, they are t-tests; if the variables are proportions, they are chi-square tests.

Results

Table 19.1 presents the results of the Medicaid analyses. In this table, we focus on those who are currently insured by Medicaid, are currently eligible but have elected not to receive coverage or are eligible to gain Medicaid under ACA. In both expanding and non-expanding states, Hispanics are less likely to be native-born and have less education and are more likely to be married compared to Non-Hispanic whites. However, in expanding states Hispanics in the Medicaid category have higher average income (assessed as a percentage of the federal poverty limit) than Non-Hispanic whites (107.6 vs. 86.7 %). This is somewhat surprising. One possible explanation is that expanding states have more generous welfare and Supplemental Security Income (SSI) benefits, leading to higher take-up rates for those programs, but lower incomes in comparison to working. Hispanics in expanding states are likely ineligible for welfare programs due to immigration restrictions and/or may be healthier, leading to higher employment rates and higher incomes relative to those

Table 19.1 Hispanic vs. non-Hispanic white Medicaid eligibility by state expansion

	Expanding states			Non-expanding states		
	Total $n = 371$	Hispanics $n = 195$	NH whites $n = 176$	Total $n = 269$	Hispanics $n = 125$	NH whites $n = 144$
Demographics						
Age (mean)	56.8	56.4	57.2	57.3	57.3	57.4
Male (%)	43.4	42.1	44.9	44.6	44.8	44.4
Native-born (%)[c]	56.0	*19.6*	96.0	61.9	*21.3*	96.5
Years of education (mean)[c]	10.4	*8.7*	12.2	10.5	*8.9*	11.9
Married (%)[c]	38.5	48.2	*27.8*	37.6	45.6	*30.6*
Poverty level (mean % of FPL)[b]	97.7	107.6	*86.7*	82.4	*74.2*	89.4
Employment status						
Employed for entire year	20.3	23.9	16.4	19.9	24.6	*16.1*
Intermittent employment	7.5	9.2	5.5	8.4	13.2	*4.4*
# Years lived in U.S. (mean)[c]	44.0	*32.6*	56.2	44.5	*29.9*	56.7
Lived in U.S. < 10 years (%)[b]	2.7	*5.1*	0	4.5	*9.6*	0
Health						
BMI categories[b]						
Overweight	34.8	*38.0*	31.3	33.5	37.6	29.9
Obese	41.0	*45.6*	35.8	45.0	45.6	44.4
Physical health categories						
Poor/fair	43.9	43.1	44.9	42.0	34.4	48.6
Good	32.1	33.9	30.1	33.1	36.0	30.6
Very good/excellent	24.0	23.1	25.0	24.9	29.6	20.8
Mental health categories[b]						
Poor/fair	25.1	18.5	*32.4*	21.9	9.6	*32.6*
Good	32.1	32.3	31.8	41.3	47.2	36.1
Very good/excellent	42.9	49.2	*35.8*	36.8	43.2	*31.3*
Diabetes diagnoses (%)[b]	27.2	*33.3*	20.5	24.5	24.0	25.0
Chronic disease index (mean)[c, d]	2.5	2.2	*2.9*	2.6	2.1	**3.2**
Disabled (on SSI due to)[c]	22.2	12.8	*32.6*	17.5	11.2	*22.9*
Cost barriers						
Cost barrier: medical care[c]	11.9	6.7	*17.6*	14.1	9.6	*18.1*
Cost barrier: dental care[b]	12.4	8.2	*17.1*	19.3	11.2	*26.4*
Cost barrier: Rx[a]	7.3	4.1	*10.8*	10.8	8.8	12.5
Cost barrier: care and/or Rx[c]	20.0	13.3	*27.3*	27.1	17.6	*35.4*

Table 19.1 (continued)

	Expanding states			Non-expanding states		
	Total *n = 371*	Hispanics *n = 195*	NH whites *n = 176*	Total *n = 269*	Hispanics *n = 125*	NH whites *n = 144*
Insurance status						
Currently covered by Medicaid[a]	57.1	63.1	*50.6*	27.5	*18.4*	35.4
Not currently on Medicaid but eligible under expansion[a]	42.9	36.9	*49.4*	72.5	*81.6*	64.6
Medicaid gap (100–133 % FPL)	20.5	21.5	19.3	32.3	33.6	31.3
Currently covered by private insurance	0	0	0	0	0	0
No current coverage and eligible for subsidy in ACA exchanges	0	0	0	0	0	0

[a]Ten percent level of significance
[b]Five percent level of significance
[c]One percent level of significance
[d]*Possible range 0–11*

on welfare programs. Among the employed, non-Hispanic whites are paid more. SSI program participation rates are higher for non-Hispanic whites than Hispanics in both expanding and non-expanding states. In non-expanding states, Hispanics report lower income than Non-Hispanic whites, as would be expected.

In terms of health variables, Hispanics in expanding states are more likely to report high BMI scores (overweight or obese) than non-Hispanic whites, although no statistically significant differences were revealed in non-expanding states. Non-Hispanic whites were more likely to report poor/fair mental health than Hispanics in both expanding and non-expanding states and less likely to report very good/excellent health. Hispanics were also more likely to have received a diabetes diagnosis than non-Hispanic whites in expanding states, but no significant difference (practically or statistically) was found in non-expanding states.

In both expanding and non-expanding states, non-Hispanic whites have more chronic disease diagnoses than Hispanics and have a far greater likelihood (2.5 times) of receiving Supplemental Security Income due to disability, as noted earlier.

In terms of cost barriers, non-Hispanic whites reported that they had a greater number of cost barriers that delay or keep them from receiving medical care, dental care and/or prescription drugs in the past year. This could be partly due to a health advantage for Hispanics, but it also could be due to greater interaction with the health care system on the part of non-Hispanic whites. Regular visits to the doctor lead to more visits to specialists and more prescriptions.

In expanding states, Hispanics are more likely to already be covered by Medicaid, while non-Hispanic whites are most likely to benefit from expansion. In non-expanding states, non-Hispanic whites are most likely to be covered, while Hispanics are most likely to benefit from expansion. There is a stark effect of current Medicaid coverage for both groups combined in expansion states compared to the groups in non-expansion states (57.1 vs. 27.5).

Private Insurance Analyses

Private insurance through exchanges is the other key plank of the ACA. We examined privately insured and uninsured Hispanics and non-Hispanic whites whose incomes are over 133 % of the poverty line in Table 19.2. As noted earlier, if sign-ups proceed as planned, this will be the privately insured population in a few years. Like in the previous section, Hispanics in both expanding and non-expanding states are less likely to be native born and are more likely to have lower levels of education than non-Hispanic whites. Hispanics also have much lower average household income than non-Hispanic whites.

Hispanics are also more likely than non-Hispanic whites to report a "poor/fair" self-assessment of health. On the other hand, Hispanics are more likely to report "good" health, but are less likely to report "very good or excellent" health. Hispanics in this category also have a greater number of chronic conditions and tend to be overweight or obese in both expanding and non-expanding states. Hispanics in this income category are much poorer than non-Hispanic whites.

Although Hispanics are more likely to report diabetes diagnoses in both expanding and non-expanding states than non-Hispanic whites, the rates are much lower than those reported in the lower income Medicaid analyses in expanding states. This may be the consequence of the true diabetes prevalence, or it may indicate that many were not receiving medical care in 2011 and therefore do not know that they have diabetes (Fisher-Hoch et al. 2010).

In expansion states, non-Hispanic whites are more likely to report cost barriers in obtaining prescription drugs than Hispanics. In terms of insurance variables in both expanding and non-expanding states, approximately 4–4.5 % who are currently covered by private insurance would be eligible for Medicaid under expansion due to income < 133 % FPL. Hispanics in both expanding and non-expanding states who fall above the income threshold for Medicaid expansion are more likely than non-Hispanic whites to qualify for subsidies for insurance purchased through the exchanges (31.2 vs. 8.7 % in expanding states; 31.5 vs. 12.4 % in non-expanding states).

Discussion

Our results provide a comparison of health characteristics and initial coverage status of pre-retirement age Hispanics and non-Hispanic whites in expansion and non-expansion states. First and foremost, Medicaid coverage is much lower for Hispanics in non-expanding states. Furthermore, 81.6 % of Hispanics in non-expansion states

Table 19.2 Hispanic vs. non-Hispanic white eligibility for premium subsidies by state expansion

	Expanding states			Non-expanding states		
	Total $n = 1,533$	Hispanics $n = 368$	NH whites $n = 1,165$	Total $n = 1,145$	Hispanics $n = 257$	NH whites $n = 888$
Demographics						
Age (mean)[b]	57.0	56.5	*57.1*	57.0	56.4	*57.2*
Male (%)	49.6	49.5	49.6	48.5	49.0	48.3
Native-born (%)[c]	79.7	*31.8*	94.8	82.9	*34.5*	96.8
Years of education (mean)[c]	13.4	*10.6*	14.3	13.3	*11.3*	13.9
Married (%)	70.8	71.5	70.6	72.4	74.3	71.9
Poverty level (mean % of FPL)[c]	516.2	*363.6*	564.4	465.0	*320.3*	506.9
Employment status						
Employed for entire year	73.7	73.5	73.8	75.7	81.2	74.1
Intermittent employment	6.8	8.2	6.4	6.7	5.7	6.9
# Years lived in U.S. (mean)[c]	51.7	*38.3*	55.8	52.3	*37.5*	56.6
Lived in U.S. < 10 years (%)[c]	1.4	*4.1*	0.6	1.1	*4.3*	0.1
Health						
BMI categories[c]						
Overweight	38.4	*45.7*	36.1	37.7	*43.6*	36.0
Obese	35.0	*36.1*	34.7	35.6	*37.4*	35.0
Physical health categories***						
Poor/fair	14.5	*18.2*	13.3	13.1	*16.0*	12.3
Good	30.9	37.5	*28.8*	31.8	40.1	*29.4*
Very good/excellent	54.7	*44.3*	57.9	55.1	*44.0*	58.3
Mental health categories						
Poor/fair	6.5	5.4	6.9	5.3	4.7	5.5
Good	28.8	28.3	29.0	29.0	33.5	27.7
Very good/excellent	64.6	66.3	64.1	65.7	61.9	66.8
Diabetes diagnoses (%)[c]	10.7	*16.0*	8.9	13.9	*19.8*	12.2
Chronic disease index (mean)[c,d]	1.7	1.4	*1.8*	1.8	1.5	*1.9*
Disabled (on SSI due to)	0.3	0.3	0.3	0.6	0.4	0.7
Cost barriers						
Cost barrier: medical care	2.4	2.2	2.4	3.9	3.9	3.9
Cost barrier: dental care	4.2	4.6	4.0	5.2	5.1	5.2
Cost barrier: Rx[a]	2.6	1.1	**3.1**	2.9	2.3	3.0

Table 19.2 (continued)

	Expanding states			Non-expanding states		
	Total $n=1,533$	Hispanics $n=368$	NH whites $n=1,165$	Total $n=1,145$	Hispanics $n=257$	NH whites $n=888$
Cost barrier: care and/ or Rx	7.4	6.8	7.6	8.7	9.3	8.6
Insurance status						
Currently covered by medicaid	0.3	0.5	0.2	0	0	0
Not currently on med- icaid but eligible under expansion	4.2	2.7	4.7	4.6	6.2	4.2
Medicaid gap (100– 133 % FPL)	2.1	1.6	2.2	2.3	*4.3*	1.7
Currently covered by private insurance[c]	85.5	*66.9*	91.3	83.3	*68.5*	87.6
No current coverage and eligible for subsidy in ACA exchanges[c]	14.6	*31.2*	8.7	16.7	*31.5*	12.4

[a]Ten percent level of significance
[b]Five percent level of significance
[c]One percent level of significance
[d]*Possible range 0–11*

are under 133 % of poverty would gain Medicaid if ACA were expanded in the future. This compares with 64.6 % of non-Hispanic whites. In expansion states, the potential for Hispanics to gain Medicaid is much lower—36.9 % could potentially gain Medicaid. This difference between expanding and non-expanding states is due to the higher poverty limit thresholds for qualifying for non-Hispanic whites prior to ACA. It is possible that ACA will be implemented in many more states in coming years, but until then, our results indicate that the potential for ACA to reduce dis- parities, especially as they pertain to Hispanics on the cusp of retirement, is severely diminished. Nevertheless, the potential gains in Medicaid expansion states are not trivial. Second, there are advantages for aging Hispanics in the private exchanges. The community rating regulations mean that they will not be penalized with higher premiums for having chronic diseases such as diabetes, which are common among older Hispanics. And provisions in the bronze and catastrophic plans ensure that out-of-pocket expenses for preventive care are low. However, the barriers erected in non-expansion states, such as regulations delaying access to navigators, low num- ber of bi-lingual navigators, lack of health insurance literacy, etc., mean that the benefits could be delayed for many years (Blavin et al. 2014).

Our results indicate that pre-retirement-age Hispanics are much more likely to be employed, have less education and are less likely to be on SSI across all states, regardless of expansion decision. Surprisingly, Hispanics have higher income than non-Hispanic whites in expansion states, albeit only 107 % of the poverty level on average. However, the reverse is true in non-expanding states, as well as for private insurance in non-expansion and expansion states. Whereas Hispanics with incomes over 133 % of poverty in expansion states average 363.6 % of poverty, Hispanics with incomes over 133 % of poverty in non-expansion states have incomes that average 320.3 % of poverty. Both groups of Hispanics would be able to receive subsidies in the health care exchanges.

In terms of health characteristics, Hispanics in the Medicaid group as shown in Table 19.1 are slightly healthier. Despite the fact that Hispanics are more likely to be obese and overweight in the expansion states and are more likely to have diabetes, they have significantly fewer chronic illnesses. They are also two-thirds less likely to be on SSI than non-Hispanic whites. Hispanics in the private market are also more likely to be overweight, obese and/or have diabetes than non-Hispanic whites. In terms of self-reported health, the results are ambiguous, with Hispanics being more likely to be in "good" health, but less likely to be in "very good/excellent" health. Hispanics are more likely to suggest "poor or fair," but are again less likely to have a chronic illness. These differences in assessments of their (respondents) overall health status may be attributed to cultural factors or the lack of linguistic equivalence in survey instruments (Angel 2013).

In the Medicaid market, non-Hispanic whites are more likely to report financial difficulties in receiving health care in the expanding and non-expanding states. This may be due to the fact that one has to use health care to be aware that he/she needs a prescription filled or be referred to a specialist. Hispanics may not have interacted with health care providers due to better health (Ruiz et al. 2012), barriers such as language or culture (Afable-Munsuz et al. 2013), or just cost.

The decision by the Supreme Court, and subsequent decisions by individual states to not expand the Medicaid component of ACA, has greatly limited the capacity of ACA to reduce disparities, especially as they pertain to Hispanics. Although fewer Hispanics live in states that are not expanding, these initially started with much less generous thresholds for obtaining Medicaid. The majority of Hispanics in expanding states already had Medicaid before ACA. Other than differences that result from their status as immigrant, the health characteristics do not greatly differ between Hispanic and non-Hispanics whites, although Hispanics in the Medicaid market are slightly healthier than non-Hispanic whites.

The ACA represents only a beginning in reducing health insurance disparities among aging Hispanics. Future research should assess the impact of changes on many parts of our health care system, including coverage in the exchanges, access to doctors and hospitals in the provider network, and the expansion of Medicaid on the health of aging Hispanics. As we move forward it will be necessary to include those who continue to lack the care that they need most in states not participating in Medicaid expansion, especially among late-life immigrants. Among Hispanics and other racial and ethnic groups, gender differences in insurance coverage are particularly important as one approaches Medicare eligibility. Women waiting to age into Medicare may delay or forgo care which could lead to more serious health problems. A large body of evidence compiled by the Institute of Medicine shows that insurance coverage increases medical care use and improves health (Institute of Medicine 2001). Without adequate health care coverage, older Latinos have

Table 19.3 Disability and medicaid coverage

Non-Hispanic whites			Hispanics		
	No SSI for disability (%)	SSI for disability (%)		No SSI for disability (%)	SSI for disability (%)
Medicaid eligible	98.8	1.2	Medicaid eligible	99.5	0.5
Medicaid recipient	48.2	51.8	Medicaid recipient	75	25.0

a greater likelihood of compromised health and financial hardship. Policymakers might consider more closely the affordability of private insurance purchased through the exchanges for Hispanic men and women. Even with the assistance of premium credits, the cost may remain prohibitive. Among Hispanics close to retirement age, the problem is potentially most serious for the near-poor who will not qualify for expanded Medicaid and who have limited financial resources. Though the Affordable Care Act holds the promise for increasing health care access for poor and minority group women, many barriers to universal coverage remain (Prickett and Angel 2011). Finally, given concerns regarding the sustainability of the insurance pool including individuals over age 50 versus younger enrollees, fiscal implications, including estimates of the costs and benefits, of increasing enrollments in Medicaid and state health insurance market places merit attention.

In conclusion, Hispanics approaching retirement will transform aging in the Americas. The current older Hispanic cohort has more favorable mortality rates than other racial and ethnic groups, despite generally lower levels of education and income. This means that adequate insurance prior to Medicare participation and access to health care is critical for healthful aging. In the near future, it will be crucial to understand the mechanisms to remove financial barriers in the health care system. This is critical for continuity of care and treatment of those close to retirement age. Health care reform in the U.S. and Mexico, particularly the expansion of insurance has the potential of preventing disease and promoting healthier lives in Hispanic communities (Table 19.3).

Acknowledgement This research was supported by a grant from the LBJ School Policy Research Institute.

References

Afable-Munsuz, A., Gregorich, S. E., Markides, K. S., & Pérez-Stable, E. J. 2. (2013). Diabetes risk in older Mexican Americans: Effects of language acculturation, generation and socioeconomic status. *Journal of Cross-Cultural Gerontology, 28,* 359–373.

Agency for Healthcare Research and Quality. (2009). *2008 National Healthcare Disparities Report*. United States Department of Health and Human Services. Rockville. http://www.ahrq.gov/qual/qrdr08.htm. Accessed 1 May 2014.

Akerlof, G. A. (1970). The market for lemons: Quality uncertainty and the market mechanism. *The Quarterly Journal of Economics, 84*(3), 488–500.

Alonso-Zaldivar, R. (24 March 2014). Hispanics on sidelines: Largest US minority group lags in signing up for health care plans. *The Tribune*.

American Factfinder. (2012). U.S. Census Bureau; 2012 Population Estimates, Table PEPSRRH; generated by H. Shelton Brown; using American FactFinder. http://factfinder2.census.gov. Accessed 7 Jan 2012.

Angel, J. L., Torres-Gil, F., & Markides, K. (2012). *Aging, health, and longevity in the Mexican-origin population*. New York: Springer.

Angel, R. J. (2013). After Babel: Language and the fundamental challenges of comparative aging research. *Journal of Cross-Cultural Gerontology, 28*, 223–238.

A. P. Press (17 March 2014a). Battling highest uninsured rate in nation, Texas makes deadline push for Obamacare enrollment. *Fox News Latino*.

A. P. Press (10 March 2014b). Poll: Uninsured rate drops, but Hispanics lag in sign-ups. *Tribune-Review*.

Arias, E. (2012). United States Life Tables, 2008 *National Vital Statistics Reports*, 61, 3. Hyattsville. National Center for Health Statistics. http://www.cdc.gov/nchs/data/nvsr/nvsr61/nvsr61-03.pdf. Accessed 1 May 2014.

ASPE. (2014). *Health insurance marketplace: March enrollment report for the period: October 1, 2013—March 1, 2014* (office of the assistant secretary for planning and evaluation (ASPE) issue briefs. Washington, DC: Department of Health and Human Services.

Bastida, E., Brown, H. S., & Pagan, J. A. (2008). Persistent disparities in the use of health care on the US/Mexico border: An ecological perspective. *American Journal of Public Health, 98*, 1987–1995.

Blavin, F., Zuckerman, S., Karpman, M., & Clemans-Cope, L. (18 March 2014). Why are Hispanics slow to enroll in ACA coverage? Insights from the health reform monitoring survey. *Health Affairs Blog.*

Brown, H. S., Angel, J. L., & Wilson, K. (2014). *Mexican immigrant health: The effects of insurance coverage*. Unpublished manuscript.

Cantu, P. A., Hayward, M. D., Hummer, R. A., & Chiu, C. T. (2013). New estimates of racial/ethnic differences in life expectancy with chronic morbidity and functional loss: Evidence from the national health interview survey. *Journal of Cross-Cultural Gerontology, 28*(3), 283–297.

Centers for Disease Control. (2010). *United States life tables by Hispanic origin* (*No. Series 2, Number 152*). Hyattsville: U.S. Department of Health and Human Services; Centers for Disease Control and Prevention; National Center for Health Statistics.

Chang, D. (03 Dec 2014). 8 million uninsured Hispanics nationwide eligible for health coverage. *The Miami Herald.*

Eichner, J., & Vladeck, B. C. (2005a). Medicare as A catalyst for reducing health disparities. *Health Affairs, 24*(2), 365–375.

Eichner, J., & Vladeck, B. C. (2005b). Medicare as a catalyst for reducing health disparities. *Health Affairs, 24*(2), 365–375.

Fisher-Hoch, S., Rentfro, A., Salinas, J., Perez, A., Shelton Brown, H., Reininger, B., et al. (2010). Socioeconomic status and prevalence of obesity and diabetes in a Mexican-American community, Cameron County, Texas. *Preventing Chronic Disease, 7*(3), 1–10.

Heberlein, M., Brooks, T., Guyer, J., Artiga, S., & Stephens, J. (2012). *Performing under pressure: Annual findings of a 50-state survey of eligibility, enrollment, renewal, and cost-sharing policies in medicaid and CHIP, 2011–2012*. Washington, DC: Kaiser Commission on Medicaid and the Uninsured.

Institute of Medicine. (2001). *Coverage matters: Insurance and health care*. Washington, DC: National Academies.

Kaiser. (2013). *Health coverage for the Hispanic population today and under the affordable care act* Kaiser Commission.

Keith, K., & Lucia, K. W. (2014). *Implementing the affordable care act: The state of the states.* New York: The Commonwealth Fund.

Mangan, D. (2013). Medical bills are the biggest cause of US bankruptcies: Study. *CNBC.*

Medrano, L. (21 Jan 2014). Obamacare and Latinos: Why a crucial constituency is wary of signing up. *The Christian Science Monitor.*

Prickett, K. C., & Angel, J. L. (2011). The new health care law: How will women near retirement fare? *Journal of Women's Health Issues, 22,* 331–337.

Ruiz, J. M., Steffen, P., & Smith, T. B. (2012). Hispanic mortality paradox: A systematic review and meta-analysis of the longitudinal literature. *American Journal of Public Health, 103,* e52–e60.

Seipel, T. (20 Dec. 2014). Few Latinos signing up for California health insurance exchange. *San Jose Mercury News.*

Smolka, G., Purvis, L., & Figueiredo, C. (2007). *FYI: Characteristics of uninsured 50- to 64-year-olds in California.* Washington, DC: AARP Public Policy Institute.

Stimpson, J. P., Wilson, F. A., & Su, D. (2013). Unauthorized immigrants spend less than other immigrants and US natives on health care. *Health Affairs (Project Hope), 32*(7), 1313–18.

Torres-Gil, F., & Lam, D. (2012). The evolving nexus of policy, longevity, and diversity: Agenda setting for Latino health and aging. In J. Angel, K. Markides, & F. Torres-Gil (Eds.), *Aging, health and longevity in the Mexican-origin population* (pp. 327–333). New York: Springer.

U.S. Census Bureau. (2012). Population estimates. Table PEPSRRH; generated by H. Shelton Brown; using American FactFinder. http://factfinder2.census.gov. Accessed 7 Jan 2012.

Vega, W., & Gonzalez, H. (2012). Aging in place: Issues and potential solutions. In J. Angel, K. Markides, & F. Torres-Gil (Eds.), *Aging, health and longevity in the Mexican-origin population* (pp. 193–205). New York: Springer.

Chapter 20
Racial and Ethnic Disparities in Willingness to Pay for Improved Health: Evidence from the Aging Population

Olufolake O. Odufuwa, Robert P. Berrens and R. Burciaga Valdez

Introduction

Considerable research effort has been focused on estimating individual willingness to pay (WTP) for mortality risk reductions in the economics literature (Alberini et al. 2004, 2006; Krupnick et al. 2002; Milligan et al. 2010). Relatedly, WTP for reduced morbidity or improved health is another important area of inquiry because of the large proportion of individuals affected and its possibility of progressing towards mortality. Morbidity is a case of being in less than 'perfect' or 'good' health and can either be chronic or acute (Freeman 2003). While acute morbidity only lasts for a couple of days, chronic morbidity is of a longer term or indefinite time period. Changes in actual or expected morbidity may affect an individual's risk-mitigation behavior. This may in turn bias estimates of the benefits for life-saving policies, such as estimates for the Value of a Statistical Life (VSL). The VSL is an aggregation of individuals' WTP for given changes in risk reduction. An economic valuation framework attempts to represent the preferences of individuals, as measured using either revealed preference approaches such as wage-risk observations or stated preference approaches such as contingent valuation and choice experiments. Contingent valuation (CV) is the most commonly used stated preference method for valuing changes in morbidity status, where WTP responses for a proposed scenario are elicited from individuals in a survey sample (Freeman 2003; Van Houtven et al. 2003).

O. O. Odufuwa (✉) · R. P. Berrens · R. B. Valdez
Department of Economics, MSC 05 3060, University of New Mexico, Albuquerque, NM 87131, USA
e-mail: oodufuwa@unm.edu

© Springer International Publishing Switzerland 2015
W. A. Vega et al. (eds.), *Challenges of Latino Aging in the Americas,*
DOI 10.1007/978-3-319-12598-5_20

Consideration of WTP estimates for improved health are relevant for policy interventions and are based on consumer preferences consistent with utility theory. WTP estimates, whether for mortality risk reduction or morbidity reduction, can be used to estimate the optimal scale of proposed health policy interventions relative to a fixed budget (Dickie and List 2006). In general, focusing on WTP estimates for different groups and settings requires taking into account the nature of the risk involved and other individual characteristics. The expected influence of such sources of disparities has also been proposed in some valuation studies (Sunstein 2004; EPA 2000, 2010; Viscusi 2010). Income, age and health status have been the most widely researched sources of disparities (Milligan et al. 2010; Krupnick et al. 2002; Alberini et al. 2004). Race and ethnicity are other possible sources of disparities in WTP estimates. Fully exploring how WTP estimates vary with race and ethnicity would require taking into account all possible economic and social mechanisms contributing to such disparities (Viscusi 2010). All these sources of disparities may also influence WTP for improved health.

The objective of this research is to examine racial and ethnic disparities in WTP valuations for improved health or reduced morbidity among a sample of the aging population. According to the U.S. Census Bureau, the U.S. population is aging rapidly. The population of older adults is also expected to more than double by 2050, with increasing racial and ethnic diversity (Vincent and Velkoff 2010). The Hispanic population which is the largest minority group has been projected to more than double their share of the population to 29 % by 2050. The older population also has unique medical needs relative to younger adults, and is more likely to suffer from chronic illnesses (U.S. Centers for Disease Control and Prevention, CDC 2007). More importantly, minority population among the aging is known to suffer more chronic illnesses than non-Hispanic Whites in part due to lifestyle factors and less access to medical services. As a result, one of the priorities of the CDC is addressing the health needs of the country's older and minority population. This analysis explores how this group values improved health. The impact of an individual's health status and risky health behaviors on WTP is also examined. Not considering prior morbidity or illnesses may lead to models of health-seeking behavior poorly predicting individuals' investment in their health (DeShazo and Cameron 2005).There are very few WTP for morbidity studies and there has been no research jointly analyzing WTP for morbidity valuations of racial minorities compared to non-Hispanic Whites, while also controlling for the impact of risky health behaviors.

To address this gap, the econometric analysis uses data from a valuation module of the Health and Retirement Study (HRS) 2000 (wave 5) data, which focuses on America's expanding aging population. The valuation module uses a non-standard CV format, which includes an initial selection question (for the absence or presence of positive WTP) and a structured sequence involving two different WTP response formats. Handling these two formats requires a pragmatic approach, retaining as many responses as possible (making full use of the sample) and accounting for possible selection bias using the Heckman two-step modeling technique.

After controlling for selection bias and other covariates, evidence across all econometric models does not point to the existence of racial and ethnic disparities in a minority grouping in relative WTP for improved health compared to non-Hispanic Whites. Hispanics, Blacks, American Indian and Asians groups, collectively, are more likely to have a positive WTP for improved health than non-Hispanic Whites. While their WTP is significantly lower than for non-Hispanic White groups in absolute terms, it is a higher percent of their annual household income. Results indicate that current morbidity does matter as health status affects whether or not an individual will have a positive WTP for improved health, but not the amount of money they would be willing to pay. Of note, individuals diagnosed with cancer and lung diseases are more likely to exhibit positive WTP for improved health than healthy individuals. However, risky health behaviors have no observable impact on WTP valuations.

Theoretical Framework of Utility for Health

WTP measures an individual's willingness to sacrifice a desired attribute (wealth) for future consumption in order to obtain another desired attribute—improved survival (Shepard and Zeckhauser 1984). This improved survival is synonymous with improved health or a reduction in morbidity. In a simple conceptual framework, each individual is assumed to have preferences described as follows:

$$U = U\left[H, X(H)\right]$$ (20.1)

The individual's utility is a function of current health status, H and a vector of all other goods, X that contribute to utility. H is determined exogenously. As shown in (20.1), it is assumed that the utility derived from all other goods is dependent on the individual's health status. Individuals are assumed to maximize (20.1) subject to a budget constraint:

$$Y = PX$$ (20.2)

where Y is an income level determined exogenously and P is a vector of prices. The indirect utility function corresponding to this utility maximization process can be written as follows:

$$V = V(Y, P, H)$$ (20.3)

This reflects the maximum amount the individual would be willing to give up for improved health:

$$V(Y, P, H) = V(Y - WTP, P, H^*)$$ (20.4)

In (20.4), the WTP for improved health is a Hicksian compensating welfare measure (Freeman 2003). It shows the change in current income (from Y to $[Y - WTP]$) the individual is willing to let go of for an improved health from his current health status (from H to H^*). WTP can also be defined explicitly as the difference between two expenditure functions.

$$WTP_c^+ = \left| e(\bar{P}, V, H) - e(\bar{P}, V, H^*) \right| \qquad (20.5)$$

where $\bar{P}$ is a vector of prices, with $H^* > H$ indicating that H^* is an improved health status.

The WTP framework in (20.5) above implicitly indicates that WTP for a health improvement should increase the more severe the current condition is. In addition to current health status, other behavioral factors (such as risky health behaviors) and socioeconomic factors are likely to influence WTP for health improvements. These will be included in the econometric estimations of (20.5). Large improvements in health could affect an individual's income level and their marginal utility of income (Reed Johnson et al. 1997). However, for modeling and estimation ease, it is assumed that the individual's marginal utility of income is constant. It is also expected that an individual's WTP for health improvement should increase at a decreasing rate as current health conditions move towards perfect health.

Hypotheses

This study investigates racial and ethnic disparities in WTP for improved health. Considerable prior research has revealed significant health disparities for minorities, as compared to the US population as a whole or with non-Hispanic whites. Such health disparities have been attributed to differences in social and economic determinants such as low socioeconomic status and lack of access to care (Koh et al. 2011). Various national initiatives aimed at reducing racial and ethnic health disparities underscore the importance of analyzing WTP valuations for health improvements among minorities. In addition, this analysis jointly tests for the impact of the individual's health status and risky health behaviors on WTP, expressed in the following natural log WTP function:

$$\ln WTP = f(\beta^{H^j} X^{H^j}, \beta^{H^k} X^{H^k}, \beta^{SE} X^{SE}) \qquad (20.6)$$

with: $j = \begin{cases} high\ blood\ pressure, diabetes, cancer, \\ lung\ disease, heart\ conditions, stroke \end{cases}$

$k = \{tobacco\ use, alcohol\ use\}$

In (20.6), X^{H^j} is a vector of the presence of chosen chronic illnesses in category j representing the individual's health status, X^{H^k} is a vector of chosen risky health behaviors in category k, X^{SE} is a vector of demographic and socioeconomic variables, including minority status, and the β's are conformable vectors of estimable coefficients.

Health Status and WTP Hypotheses

As implied from (20.5), an individual's WTP for a health improvement should depend on current health status. Health status refers to the range of manifestation of any disease in an individual, including symptoms and its functional limitation (Rumsfeld 2002). Health status may often be measured by the diagnosis of various chronic illnesses such as cancer, diabetics, heart conditions, lung diseases, respiratory disease, and stroke. An individual's current morbidity and knowledge of it could determine whether or not he will be willing to pay more for improved health.

As noted, most previous health and WTP analyses have focused on mortality risk reductions, and not on health improvement or morbidity reductions. For instance, using a CV survey from Canada, Krupnick et al. found that with the exception of cancer, WTP is not affected by health status (Krupnick et al. 2002). Alberini et al. also found that individuals diagnosed with cancer, chronic heart and lung conditions have higher WTP to reduce mortality risk (Alberini et al. 2004). Rather, this analysis focuses on WTP for health improvement and expands the list of chronic health illnesses to include not only cancer, lung and heart diseases but also diabetes, high blood pressure and stroke. Diabetes and high blood pressure can lead to heart problems and other chronic health problems. The chance of stroke and other medical conditions is also higher in individuals involved in risky health behaviors, such as smoking and heavy drinking (Markus 2012).

Individuals with poor health may face further worsening of their health either due to current or other illnesses. In this case, the difference between the two expenditure functions in (20.5) is assumed to be greater for an individual diagnosed with any form of chronic illness. Therefore, WTP for improved health is expected to be positive and higher than for healthy individuals. Against the null of no effect on the level of WTP, a set of testable hypotheses across various health statuses is presented in the top half of Table 20.1.

Table 20.1 Hypotheses

Category *j* hypotheses—health status		
Category *j*	Null hypothesis (H_o)	Alternative hypothesis (H_1)
High blood pressure	$\beta^{HBP}=0$	$\beta^{HBP}>0$
Diabetes	$\beta^{DIAB}=0$	$\beta^{DIAB}>0$
Cancer	$\beta^{CANCER}=0$	$\beta^{CANCER}>0$
Lung disease	$\beta^{LUNG}=0$	$\beta^{LUNG}>0$
Heart conditions	$\beta^{HEART}=0$	$\beta^{HEART}>0$
Stroke	$\beta^{STROKE}=0$	$\beta^{STROKE}>0$
Category *k* hypotheses—risky health behaviors		
Category *k*	Null hypothesis (H_o)	Alternative hypothesis (H_1)
Tobacco	$\beta^{TOBAC}=0$	$\beta^{TOBC}>0$
Alcohol	$\beta^{ALCH}=0$	$\beta^{ALCH}>0$

Risky Health Behaviors and WTP Hypotheses

The relationship between risky health behaviors and WTP is also analyzed. Risky health behaviors such as tobacco use, alcohol consumption and sedentary lifestyles are major causes of death. The CDC reports that adverse health effects from cigarette smoking accounts for nearly one of every five deaths each year (CDC 2004). In their research identifying the leading cause of mortality in the US, Mokdad et al. found that 18.1 % of total deaths were caused by tobacco consumption, 3.5 % from alcohol use and 15.2 % from poor diet and sedentary lifestyles (Mokdad et al. 2004). Khwaja et al. also analyzed whether health valuations varied between current smokers and former smokers (Khwaja et al. 2006). However, their findings did not show any significant difference in WTP values between the two groups of individuals. The HRS wave used for this analysis includes a question on vigorous physical activity, such as sports, heavy housework or a job that involves physical labor. It is assumed that the aging population is more likely to be involved in light physical activity such as walking, gardening or golfing. Such light physical activities are not included in the HRS wave 5 data, or in this analysis.

Based on these health reports, an individual's probability of dying is assumed to be higher if they engage in risky health behaviors. As a result, the difference between the two expenditure functions in (20.5) is expected to be greater for an individual that engages in risky health behaviors. Against the null hypothesis of no effect on WTP values for a health improvement, a second set of testable hypotheses across both risky health behaviors is presented in the bottom half of Table 20.1.

Data, Variable Descriptions and Summary Statistics

Data

This paper uses the RAND HRS data. The HRS is a national longitudinal study funded by the National Institute on Aging (NIA) and conducted by the University of Michigan's Institute for Social Research (ISR). The HRS study follows age eligible individuals and their spouses every two years since initial survey administration. To make the data more accessible to researchers, the RAND Center for the Study of Aging created the RAND HRS data files containing cleaned and processed variables, with consistent and intuitive naming conventions (St.Clair et al. 2011).

The RAND HRS Wave 5 data is used, which is synonymous with the HRS 2000 data, because it includes a health valuation module. In the module, respondents were asked to compare their current state of health to 'perfect' health in a series of CV WTP questions. Individual respondent-level survey results are used for this analysis.

Dependent Variable—WTP for Improved Health

The HRS 2000 survey includes a set of modules applied to a subset of respondents. At the end of the main survey, respondents were randomly assigned a number which was used to determine if the respondent would proceed to the health valuation module. Based on this randomization procedure, 914 respondents were asked CV questions related to their WTP for perfect health given their current health. A brief introduction explained that the financial resources government and universities allocate to medical research requires knowledge about how people with different health conditions feel about their health problems. Responses help identify the most important health problems for medical research (HRS 2000). Then, respondents were asked an initial 'yes' or 'no' health valuation selection question as follows:

> Imagine that you will live for 10 more years in your current state of health. Assuming that your current medical expenses and insurance premiums stay the same as they are now, would you be willing to pay more every month for additional medical treatment if it allowed you to live those ten years in perfect health?

Out of the 914 survey respondents, 547 answered yes, 332 answered no, and 35 declined. The 547 respondents that expressed positive WTP ($WTP_{4HEALTH} > 0$) continued with the CV questions. However as presented in Fig. 20.1, the HRS valuation module uses a very non-standard format which consists of the initial selection question and a structured series or sequence involving two different WTP response formats. These are open ended (OE) and double-bounded dichotomous-choice (DB-DC) formats. Analyzing this requires a pragmatic approach for handling the two formats, retaining as many valuation responses as possible and accounting for possible selection bias in the initial screening question for positive WTP.

Open-Ended (OE) Questions

As shown in Fig. 20.1, the OE question asked:

> What is the greatest amount you would be willing to pay each month to live in perfect health?

This provides continuous positive WTP data. Responses represent the individual's maximum WTP monthly for improved health, which is synonymous with the Hicksian compensating welfare measure in (20.5). A total of 280 gave actual values of WTP. The other 267 either refused to answer or could not ascertain their maximum WTP. OE questions are relatively straightforward and perhaps the simplest to interpret (Freeman 2003). Compared to other elicitation formats, they do not have any starting point bias problems, where the suggestions of an initial starting point can influence final WTP. In addition, standard statistical techniques can be used to analyze continuous positive responses (Pearce et al. 2006). However, they tend to yield protest zeroes, invalid large responses and sometimes unreliable responses (Freeman 2003; Mitchell and Carson 1989; Carson et al. 2001; Boyle 2003). Of

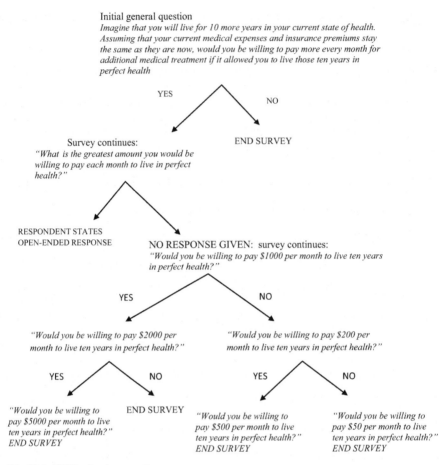

Fig. 20.1 WTP valuation questions

concern, mean OE WTP values are usually sensitive to the presence of large individual bids. Respondents can influence the outcome of a proposed change by stating a value that exceeds their true WTP (Boyle 2003). This problem of unreliable large responses can be potentially addressed by either using a rule of thumb about the relationship between the stated WTP value and the respondent's income, or using robust statistical estimators such as an α-trimmed mean where α is determined by the analyst (Mitchell and Carson 1989; Freeman 2003). In this sample, a small number of respondents reported extremely high OE WTP values. Carson suggests that WTP values should not exceed 5 % of household income (Alberini and Cooper 2000). While this suggestion was made for environmental goods, this study follows this rule of thumb in select models. Responses above 5 % of income are not dropped but trimmed to the upper limit.

Double-Bounded Dichotomous Choice (DB-DC) Questions

Following Fig. 20.1, the interviewer's instruction was to continue with DB-DC questions with respondents who could not provide OE responses. Thus, 267 respondents continued with the survey. Various types of discrete choice formats are commonly used for eliciting WTP responses. The DB-DC format does not elicit WTP values directly, but provides a bounded interval within which the respondent's WTP value lies. Respondents were asked their willingness to pay a specific dollar price, or payment amount, for a given change. Follow-up questions were asked, in which the amounts presented depends on the response to the previous question. If the response is positive, the respondent is asked a second discrete choice question of a specific higher amount but if the response is negative, a second question of a specific lower amount is asked. The upper and lower bounds are found when respondents provide a positive response to one of the questions and a negative response to the other (Hanemann and Kanninen 2001).

The initial dichotomous question (DC) asked

Would you be willing to pay $ 1000 per month to live ten years in perfect health?

Based on the response, follow up questions were asked with higher or lower payment amounts, which allow WTP classification into seven categories: less than $ 50, between $ 50 and $ 200, $ 200 and $ 500, $ 500 and $ 1,000, $ 1,000 and $ 2,000, $ 2,000 and $ 5,000, and over $ 5,000.

To begin, DC questions could minimize non-response and protest zero situations because the format mimics a familiar market context for the respondent of simply accepting or rejecting a good with a posted payment amount. However, there has been mixed performance of the DB-DC elicitation format with some response anomalies, which appear to be due to possible strategic changes in a respondent's answers on the follow-up questions, especially after the first DC question is asked. There is a tendency for respondents who answered 'yes' to one of the follow-up questions to answer 'no' to the next follow up question. Possible explanations include the possibility that respondents may assume a lower follow-up amount corresponds to a lower quality of the good or that a higher follow-up amount is a form of exploitation (Haab and McConnell 2002). As a result, mean WTP may be biased depending on the presence of such strategic behaviors. Another important factor to be taken into consideration with DB-DC questions is the selection of the payment amounts. With a lower range of payment amounts, the estimated mean WTP could be biased downwards (Freeman 2003). In addition, DB-DC data require more complex statistical techniques, with results very sensitive to statistical assumptions made (Pearce et al. 2006).

This study takes into account possible selection effects from the initial question, and further avoids having to ignore large subsamples by having multiple types of WTP data (both continuous OE and bounded categorical DB-DC). For example, studies using the HRS valuation module may ignore possible selection effects and available OE data, to simply estimate double-bounded maximum likelihood models on the DB-DC data. This throws away the majority of the sample and would greatly limit the ability to explore sub-sample effects (e.g., minority status). Alternatively, this analysis converts all respondents in the DB-DC subsample to a single OE value. The highest payment amount with a 'yes' response is conservatively assumed as their maximum WTP. Thus, responses are coded such that respondents no longer have double-bounded WTP but only one maximum WTP. This provides a conservative estimator and allows us to combine all WTP responses in the data, and retain the maximum possible sample.

Further, Heckman selection models are used, where the selection equation uses the data from the initial binary question on positive WTP, and then the continuous WTP data is used for the outcome equation. To control for possible large individual bids in these calculated open-ended WTP data, values in select models are also trimmed to an upper limit of 5 % of reported annual household income. All estimations are done with and without this upper-limit trimming.

Explanatory Variables

Ethnic minorities have higher rates of chronic illnesses, such as cancer and obesity, and higher incidence of risky health behaviors (Mead et al. 2008). Here, race and ethnicity (MINRACE) is coded in terms of racial minority (1 if respondent is Hispanic, Black, American Indian and Asian; 0 otherwise). Health status is measured by the diagnosis of chronic health conditions. Six chronic health problems are analyzed. These are high blood pressure (HBP), diabetes (DIAB), cancer (CANCER), lung disease (LUNG), heart problem (HEART), and stroke (STROKE). The specific health questions asked are:

> "Has a doctor ever told you that you have high blood pressure or hypertension?", "Has a doctor ever told you that you have diabetes or high blood sugar?", "Has a doctor ever told you that you have cancer or a malignant tumor, excluding minor skin cancers?", "Has a doctor ever told you that you have chronic lung disease such as chronic bronchitis or emphysema?", "Has a doctor ever told you that you had a heart attack, coronary heart disease, angina, congestive heart failure, or other heart problems?" and "Has a doctor ever told you that you had a stroke?"[1]

In addition to these health variables, respondents were also asked to rank their health on a 1–5 scale (SRHEALTH), with 1 representing excellent health and 5 poor health. Respondent's current smoking (TOBAC) and alcohol consumption status (ALCH) are the two risky health behaviors analyzed.

> "Do you smoke cigarettes now?"; and "Do you ever drink any alcoholic beverages such as beer, wine or liquor?"

---

[1] The lifetime diagnosis question of *"have you ever…?"* of these chronic illnesses is used. While not available in the HRS data, the current diagnosis *"do you currently…"* of the chronic illnesses would be preferable given that health status of some respondents previously diagnosed with illnesses may have improved with consistent usage of prescribed treatments and drugs.

Table 20.2 Summary statistics—dependent variables (Absolute WTP). (Source: RAND HRS data (Wave 5). n = sample size)

Variable	Description	Total sample mean (St. dev)	Minorities mean (St. dev)	Whites mean (St. dev)
$WTP_{4HEALTH} > 0$	Whether or not a respondent's WTP is positive (selection question: 1 = Yes, 0 otherwise)	0.62 (0.49) ($n=877$)	0.67 (0.47) ($n=189$)	0.61 (0.49) ($n=688$)
WTP_{OE}	Non-trimmed open-ended (OE) monthly WTP values ($)	7,673.74 (84,334.16) ($n=280$)	15,093.40 (122,149.20) ($n=67$)	5,339.86 (68,498.75) ($n=213$)
$WTP_{TRIM-OE}$	Trimmed monthly OE WTP values ($)	424.51 (1,170.46) ($n=280$)	131.90 (170.74) ($n=67$)	516.56 (1,326.01) ($n=213$)
WTP_{DCOE}	Calculated monthly WTP from both OE and dichotomous choice responses ($)	4,772.93 (62,778.69) ($n=506$)	9,366.47 (92,790.99) ($n=116$)	3,406.65 (50,624.46) ($n=390$)
$WTP_{TRIM-DCOE}$	Trimmed calculated WTP from both OE and dichotomous choice responses ($)	546.94 (1,137.97) ($n=506$)	340.27 (761.37) ($n=116$)	608.41 (1,221.98) ($n=390$)

Following similar prior CV studies, other socioeconomic covariates are included such as age (AGE), marital status (MARITAL), gender (FEMALE), total annual household income (INCM) and highest education degree attained (EDUC).[2]

Descriptive Statistics and Models

Descriptive Statistics

Four different individual monthly WTP measures are provided. The first two are the actual OE WTP responses while the last two are calculated WTP derived from a combination of both OE and DB-DC responses. These are untrimmed WTP from open-ended questions (WTP_{OE}), trimmed WTP from OE questions ($WTP_{TRIM-OE}$), calculated WTP from both OE and DB-DC responses (WTP_{DCOE}), and trimmed calculated WTP from both OE and DB-DC responses ($WTP_{TRIM-DCOE}$).

Tables 20.2 and 20.3 present descriptive statistics for the total sample, minorities and non-Hispanic Whites. Both tables present conventional statistics based on

---

[2] Income includes before-tax income from earnings, unemployment, Social Security and public benefits, retirement income, interests and dividends, child support and income from other sources, with exception of non-cash benefits (e.g., food stamps).

Table 20.3 Summary statistics—explanatory variables. (Source: RAND HRS data Wave 5)

Variable	Description	Total sample mean (St.dev)	Minorities mean (St.dev)	Whites mean (St.dev)
Health Status (1 = Yes; 0 otherwise)				
HBP	Whether the respondent was ever diagnosed with high blood pressure	0.45 (0.50)	0.54 (0.50)	0.43*(0.50)
DIAB	Whether the respondent was ever diagnosed with diabetes	0.13 (0.33)	0.23 (0.42)	0.10*(0.30)
CANCER	Whether the respondent was ever diagnosed with cancer	0.12 (0.32)	0.10 (0.29)	0.12 (0.33)
LUNG	Whether the respondent was ever diagnosed with lung diseases	0.07 (0.25)	0.04 (0.20)	0.08*(0.27)
HEART	Whether the respondent was ever diagnosed with any heart condition	0.19 (0.39)	0.17 (0.38)	0.19 (0.40)
STROKE	Whether the respondent was ever diagnosed with stroke	0.04 (0.21)	0.03 (0.18)	0.05 (0.21)
Risky health behaviors(1 = Yes; 0 otherwise)				
ALCH	Whether the respondent ever drinks any alcohol	0.50 (0.50)	0.36 (0.48)	0.54*(0.50)
TOBAC	Whether the respondent currently smokes	0.15 (0.36)	0.20 (0.40)	0.14 (0.35)
Other covariates				
MINRACE	Proportion of minority races (Hispanics, African-Americans, American Indian, Asian; %)	0.22 (0.41)	–	–
SRHEALTH	Respondent's self-rated health rank(1 = excellent, … , 5 = poor)	2.70 (1.10)	3.20 (1.10)	2.57*(1.06)
INCM	Annual household income (/$ 1000)	62.08 (194.42)	38.76 (50.36)	68.49*(217.52)
EDUC	Respondent's highest education degree level attained 1 = no degree, 2 = GED and high school, and 3 = associates, bachelors and graduate degrees).	2.47 (1.88)	1.70 (1.84)	2.68*(1.84)
AGE	Respondent's age (in years)	65.55 (10.61)	64.05 (10.32)	65.96*(10.66)
FEMALE	Proportion of females (%)	0.64 (0.48)	0.61 (0.49)	0.65 (0.48)
MARITAL	Proportion of married respondents (%)	0.69 (0.46)	0.60 (0.49)	0.71 (0.45)

N = 877. Minorities n = 189. Non-Hispanic Whites n = 688. T-test for Minorities mean vs. Whites conducted where * represents significance at 10 percent level. Two observations dropped due to missing values.

a normal distribution. The number of observations, conventional mean responses and standard deviations are reported. Comparing minorities to their non-Hispanic White counterparts, mean WTP_{OE} is \$ 15,093 and \$ 5,340 for minorities and non-Hispanic Whites respectively. However, after trimming WTP_{OE} to an upper limit of 5% of annual household income, mean $WTP_{TRIM-OE}$ is \$ 132 for minorities and \$ 517 for non-Hispanic Whites. On the other hand, mean WTP_{DCOE} is \$ 9,366 and \$ 3,407 for minorities and non-Hispanic Whites, respectively but \$ 340 and \$ 608 for $WTP_{TRIM-DCOE}$. The wide difference between the trimmed and untrimmed responses for minorities indicates that the large individual bids or outliers are more likely to be from the minorities in the sample. Significant racial and ethnic differences in means exist for self-rated health rank and income. Table 20.3 shows that minorities' self-rated health rank, absolute income level and highest education degree attained are significantly lower than their White counterparts. With the exception of high blood pressure and diabetes, a higher percentage of the non-Hispanic White sample have been diagnosed with cancer, lung disease, heart disease and stroke compared to the minority grouping. 54% of the non-Hispanic White sample consume alcohol compared to just 36% of minorities. These observed and significant differences in both the health status and behavioral variables between the two racial groups may influence their stated WTP. Average income for minorities is \$ 38,760 while average income for Whites is \$ 68,489.

Critically, as shown in Table 20.4, monthly WTP measures are converted to annual measures and estimated as a proportion of annual household income. Results across all possible WTP measures in Table 20.4 reveal that minorities' annual WTP as a percentage of annual income is an higher percentage of their household income, compared to non-Hispanic Whites. Minorities' annual WTP_{OE} as a function of annual household income is 24.8% versus 1.1% for non-Hispanics Whites. This difference is found to be highly significant. On the other hand, annual WTP_{DCOE} as a function of annual household income is 15.07% for minorities compared to 0.99% for non-Hispanic whites, a difference significant at the 10% level. This suggests that racial and ethnic minorities are likely to have a higher valuation for health improvement if they had higher household incomes.

Econometric Models

An initial visual inspection of maximum WTP values was conducted by plotting a histogram. Since highly skewed values tend to produce heteroskedastic effects, natural log of WTP is used to achieve a more uniform spread. The Breusch-Pagan test was used to test for additional heteroskedaticity effects. Comparing against the critical χ-squared value showed that taking the natural log of WTP did not completely eliminate the heteroskedastic effect. The natural log of WTP and heteroskedastic-consistent robust standard errors were therefore used for all estimations.

WTP values are only observed if the respondent provided a positive response to the initial selection question. As a result, unobserved characteristics or qualities may exist and cause individuals to self-select into either the positive or zero WTP

Table 20.4 Comparisons of annual WTP as a percentage of annual household income (Relative WTP). (Source: RAND HRS data Wave 5)

Variable	Description	Total sample mean (St. dev) (n)	Minorities mean (St. dev) (n)	Whites mean St. dev) (n)
Annual WTP_{OE}/ Income	Non-trimmed annual WTP values as a percentage of annual income	6.80 (99.19) ($n=279$)	24.76* (200.91) ($n=67$)	1.13* (14.21) ($n=212$)
Annual $WTP_{TRIM-OE}$/ Income	Trimmed annual WTP values as a percentage of annual income	0.10 (0.15) ($n=279$)	0.09 (0.14) ($n=67$)	0.09 (0.15) ($n=212$)
Annual WTP_{DCOE}/ Income	Calculated annual WTP as a percentage of annual income	4.22 (73.78) (n=505)	15.07* (152.66) (n=116)	0.99* (10.87) (n=389)
Annual $WTP_{TRIM-DCOE}$/ Income	Trimmed calculated annual WTP as a percentage of annual income	0.15 (0.20) (n=505)	0.17 (0.22) (n=116)	0.15 (0.19) (n=389)

* represents significant difference at 10% level

groups. With this, possible sample selection bias may arise (Heckman 1979). When this happens, statistical analysis based on this non-randomly selected sample may lead to wrong conclusions. The Heckman technique therefore helps to correct for possible non-randomization in the sampling process. The Heckman two-step estimation consists of a selection (or participation equation) and an outcome equation. The selection equation uses the data from the initial binary question on positive WTP while the continuous WTP data is used for the outcome equation.

$$\textbf{\textit{Selection equation}}: y_t^* = f\left(\alpha^{H^J} X^{H^J}, \alpha^{H^k} X^{H^k}, \alpha^{SE} X^{SE}\right) + \epsilon_t \qquad (20.7)$$

$$y_t = 1 \; if \; y_t^* > 0; y_t = 0 \; if \; y_t^* \le 0$$

$$\textbf{\textit{Outcome equation}}: \ln WTP = f\left(\beta^{H^J} X^{H^J}, \beta^{H^k} X^{H^k}, \beta^{SE} X^{SE}\right) + u_t \qquad (20.8)$$

y_t^* in (20.7) represents whether or not the respondent provided a positive response to the first WTP question. $y_t = 1$ with positive WTP; and 0 otherwise. WTP is a continuous variable in the outcome equation. Estimating (20.8) gives a set of β coefficients, corresponding to health status, risky health behaviors and other socioeconomic variables. u_t is assumed to be normally distributed with mean 0 and standard deviation σ. The outcome equation estimates the WTP equation, conditional on being observed. For values of $y_t = 1$, estimating (20.7) gives a set of α coefficients. ϵ_t is assumed to be normally distributed with mean 0 and standard deviation σ. The error terms, ϵ_t and u_i are assumed to have a correlation of ρ. The selection bias results if $\rho \neq 0$, implying that applying standard regression techniques to the outcome equation would yield biased results.

Results

χ-squared values for Wald test results on the data in Table 20.5 provide evidence of selection bias due to sample selection on the initial valuation question (i.e., whether $WTP_{4HEALTH} > 0$). This implies that applying standard regression techniques would not be appropriate. Thus, four specifications of the Heckman two-step model (each with both selection and outcome equation results) are presented in Table 20.5 (Models 1–4). All four specifications have the same selection equation but differ in outcome equations, in terms of the WTP dependent variable used. In addition to the independent variables in the outcome equation, a well-identified Heckman selection model should contain at least one independent variable not in the outcome equation (StataCorp 2009). This identifying variable usually significantly affects selection but not the outcome (Heckman 1979). In a variety of separate preliminary models, the estimated coefficient on the respondent's self-rated health rank (SRHEALTH) was significant and positive on selection (i.e., $WTP_{4HEALTH} > 0$), but never significant in the outcome models. This variable is used as the identifying variable in the Heckman selection models, and its estimated coefficient is shown to be positive and significant across all models. In the selection equations, the probability of having a positive WTP amount is expressed as a function of self-rated health rank, racial and ethnic grouping, health status, risky health behavioral status and other socioeconomic covariates. The outcome equation in Model 1 examines the effect of health status and risky health behaviors on WTP_{OE}. Model 2 includes all the same explanatory terms used in Model 1 but with $WTP_{TRIM-OE}$ as the dependent variable. WTP_{DCOE} and $WTP_{TRIM-DCOE}$ are used in Models 3 and 4 respectively.

Across all specifications of the Heckman model, evidence reveals that Hispanics, Blacks, American Indian and Asian groups, collectively, are more likely to have a positive WTP than their non-Hispanic White counterparts. However they are not willing to pay increasing dollar amounts for improved health. Specifically, all four outcome equations show that WTP responses are significantly lower for minorities. In Model 1, WTP_{OE} for the specified minority groups, collectively, is 77% lower than that of their non-Hispanic White counterparts, 97% lower using $WTP_{TRIM-OE}$ in Model 2, 70% lower using WTP_{DCOE} in Model 3, and 89% lower using $WTP_{TRIM-DCOE}$ in Model 4. This could be as a result of the higher percentage of non-Hispanic Whites diagnosed with most of the chronic illnesses. While clearly lower in absolute terms, we caution that this may not signify minorities' lower valuation of improved health because their relative mean WTP is a higher percent of their household income when compared to non-Hispanic Whites, as earlier reflected in Table 20.4.

The results from Models 1–4 in Table 20.5 allow evaluation of the hypotheses on the influence of health status and risky health behaviors on WTP, as presented in Table 20.1. Starting with the set of hypotheses on j health status categories as stated in Table 20.1, evidence across all four outcome equations in Models 1–4 support the null hypotheses of no effect of health status on WTP outcomes. In particular, a diagnosis of any of the chronic health illnesses does not significantly affect the amount

Table 20.5 Heckman selection and WTP estimations. (Source: RAND HRS data (Wave 5)

Outcome Equation	(1) WTP$_{OE}$	SE	(2) WTP$_{TRIM-OE}$	SE	(3) WTP$_{DCOE}$	SE	(4) WTP$_{TRIM-DCOE}$	SE
MINRACE	−0.772**	0.323	−0.967***	0.289	−0.703**	0.292	−0.885***	0.262
HBP	0.033	0.229	−0.031	0.221	0.307	0.204	0.234	0.191
DIAB	−0.174	0.312	−0.023	0.292	−0.117	0.291	−0.021	0.274
CANCER	−0.063	0.331	−0.136	0.312	−0.107	0.301	−0.087	0.283
LUNG	1.073	0.660	−0.023	0.445	0.030	0.469	−0.634*	0.351
HEART	0.170	0.282	−0.132	0.268	0.303	0.255	0.154	0.231
STROKE	−0.613	0.435	−0.932*	0.561	−0.500	0.517	−0.734	0.486
TOBAC	−0.259	0.293	−0.064	0.291	−0.067	0.278	−0.022	0.263
ALCH	0.326	0.218	0.327	0.214	0.089	0.199	0.152	0.190
INCM	0.001***	0.000	0.001**	0.000	0.001**	0.000	0.001**	0.000
EDUC	0.284*	0.170	0.229	0.163	0.364**	0.150	0.356**	0.143
Selection equation								
Minority	0.278*	0.146	0.297**	0.139	0.255**	0.122	0.284**	0.117
SRHEALTH	0.137**	0.055	0.108**	0.050	0.090**	0.041	0.052	0.034
HBP	−0.138	0.110	−0.123	0.108	−0.029	0.091	−0.025	0.088
DIABETES	−0.069	0.156	−0.080	0.151	−0.172	0.136	−0.154	0.130
CANCER	0.350**	0.169	0.347**	0.164	0.349**	0.143	0.369***	0.140
LUNG	0.662**	0.330	0.584**	0.247	0.727***	0.255	0.559***	0.180
HEART	0.035	0.144	0.026	0.137	0.006	0.122	−0.011	0.113
STROKE	0.085	0.267	0.166	0.271	0.324	0.240	0.348	0.225
TOBAC	0.076	0.157	0.080	0.155	0.174	0.132	0.185	0.128

Table 20.5 (continued)

	(1)		(2)		(3)		(4)	
ALCH	0.032	0.110	0.021	0.108	−0.037	0.092	−0.051	0.090
INCM	0.0010.06	0.001	0.002**	0.001	0.001	0.001	0.001**	0.001
EDUC	0.066	0.086	0.057	0.084	0.026	0.071	0.021	0.068
Wald test (ρ = 0)	18.37		71.70		79.76		109.75	
Mean WTP	$ 679.97		$ 455.87		$ 1,394.16		$ 812.41	
95%CI for Mean WTP	$ 485.03–$ 953.26		$ 341.71–$ 608.17		$ 1,033.41–$ 1,880.84		$ 630.20–$ 1,047.29	
Median WTP	$ 137.00		$ 122.73		$ 192.48		$ 160.77	
N	612		612		838		838	

AGE, MARITAL, FEMALE are included in the Heckman estimations for all models but not reported in this table.
Significance level: * $p < 0.10$; ** $p < 0.05$; *** $p < 0.01$

of dollars the aging population is willing to pay for improved health. However, health status is found to be significant across all selection equations, especially for cancer and lung diseases. Individuals diagnosed with cancer are 35 % more likely to have a positive WTP in the selection equations in Models 1–3, and 37 % more likely in Model 4. These estimated coefficients are highly significant at the 5 % level. Similarly, individuals diagnosed with lung diseases are 66 % more likely to have a positive WTP in the selection equation in Model 1, 58 % more likely in the selection equation in Model 2, 73 % more likely in the selection equation in Model 3 and 56 % more likely in the selection equation in Model 4. Turning to the set of hypotheses on risky health behaviors as stated in Table 20.1, evidence across all outcome equations also support the null hypotheses of no effect of risky health behaviors on WTP outcomes. However, both risky health behaviors are also not statistically significant across all selection equations. Tobacco use or alcohol consumption does not affect whether the aging population would have a positive WTP (from the selection equations), neither does it affect the amount of dollars they are willing to pay (from the outcome equations) for improved health.

As expected, the estimated coefficients of annual household income exhibit a positive and significant relationship with WTP for improved health. This positive relationship holds in both selection and outcome equations across all models. Individual model specifications for separate sub-samples of minorities and non-Hispanic whites were also explored. However the Heckman specification did not converge for the minorities' sub-sample. This could likely be due to the relatively small sub-sample of minorities compared to non-Hispanics whites. Minorities make up 22 % of the sample. The higher the level of education attained, the higher the amount of money the aging would be willing to pay for improved health, as reflected in the outcome models.

Further, to test robustness of the results in Table 20.5, both trimmed model specifications are re-estimated with monthly WTP responses trimmed to 2 % of reported annual household income. From these additional estimations, previous magnitude and significance for estimates remain largely unchanged. Trimmed models are preferable because they control for outliers in WTP responses.

Table 20.5 also presents mean and median individual monthly WTP estimates for all four models. Note that conventional mean WTP observations and calculations reported in Tables 20.2 and 20.3 could be biased because they are based on a normal distribution assumption. Separate analysis of the initial WTP responses from the sample respondents reveal a log-normal distribution hence the mean, median and confidence intervals should not be calculated conventionally. Mean and median estimates of WTP for the overall sample are therefore calculated using the maximum likelihood method

$$Mean = \exp(\mu + \sigma^2 / 2) \qquad (20.9)$$

$$Median = \exp(\mu) \qquad (20.10)$$

where μ and σ^2 are the mean and variance of the distribution of logarithms of the lognormal WTP distribution respectively (Land 1972). Mean and median WTPs for the overall sample calculated using the maximum likelihood method are reported in Table 20.5 for all four models. Mean individual monthly WTP for WTP_{OE} and $WTP_{TRIM-OE}$ is \$ 680 and \$ 456 respectively while the mean individual monthly WTP for WTP_{DCOE} and $WTP_{TRIM-DCOE}$ is \$ 1,394 and \$ 812, respectively. These payment amounts reflects the average monthly amount the aging population as a whole is willing to pay for improved health, with reasonable confidence interval ranges. The median WTPs are also reported in Table 20.5 where median WTP for WTP_{OE} and $WTP_{TRIM-OE}$ is \$ 137 and \$ 123 and median WTP for WTP_{DCOE} and $WTP_{TRIM-DCOE}$ is \$ 192 and \$ 161 respectively. Similarly, confidence intervals for mean WTP estimates should not be calculated using standard estimation techniques due to its log-normality. Thus, the widely used Cox method for estimating confidence intervals for log-normal means is used (Parkin et al. 1990; Land 1972). The Cox method is based on estimating a confidence interval about the mean of the log-normal distribution using the first two moments of the sample mean and variance of the log transformed WTP distribution and then exponentiating the results (Land 1972). 95% confidence intervals for the respective mean WTP estimates are reported in Table 20.5.

Conclusions

This analysis uses CV data from the RAND HRS (wave 5), a survey focused on older Americans, to analyze racial and ethnic disparities in WTP responses for improved health. The effect of health status and risky health behaviors on WTP for improved health is also investigated. Responses to an initial question in the valuation module of the survey show the presence of selection bias, therefore the Heckman econometric estimation approach is used to control for such bias. Researchers using the HRS survey may ignore possible selection effects and estimate DB-DC models on only the much smaller DB-DC portion of the data. However given the unique layered elicitation format design in the HRS survey, such an approach leaves out much of the available data, ignores selection bias and runs into possible concerns or limitations of the DB-DC approach. This study contributes a pragmatic alternative approach to account for the selection bias and more fully utilize available HRS data. Within that approach, evidence from comparing the mean and median WTP estimates from trimmed and untrimmed models also confirms that OE WTP responses are influenced by large individual bids. Therefore trimming WTP to a maximum of 5 % of household income in select models helped control for such effects.

Evidence across all Heckman specifications indicates that although minorities are more likely to have a positive WTP than non-Hispanic Whites, the amount of dollars they are willing to pay in absolute terms for improved health is lower. However their relative mean WTP is a higher percentage of their annual household income compared to non-Hispanic Whites. In addition, risky health behaviors such as alcohol and tobacco consumption do not affect WTP valuations. Health status (a previous or current diagnosis of cancer and lung diseases) does affect whether or not an individual will have a positive WTP but not the specific dollar amount. This suggests that perhaps although people place a higher value on cancer and lung diseases than other chronic health conditions, some other constraints may prevent them from being willing to pay higher amount of dollars for improved health.

Finally, while advocating for exploring the heterogeneity in WTP, Viscusi correctly notes that there is substantial reluctance in analyzing the existence of racial and ethnic disparities in WTP valuations as it would also require fully exploring possible economic and social mechanisms contributing to such disparities, if any (Viscusi 2010). For example, if historical or current discrimination has contributed (with possible intra-generational transmission effects) to lower education and income then one might expect lower absolute WTP. As evident in this analysis, the lower observed absolute WTP estimates for minorities compared to non-Hispanic Whites should not portray that minorities value improved health less than their non-Hispanic Whites counterparts.

References

Alberini, A., & Cooper, J. (2000). *Applications of the contingent valuation method in developing countries: A survey. FAO economic and social development paper 146*. Rome: FAO

Alberini, A., Cropper, M., Krupnick, A., & Simon, N. B. (2004). Does the value of a statistical life vary with age and health status? Evidence from the US and Canada. *Journal of Environmental Economics and Management, 48,* 769–792.

Alberini, A., Cropper, M., Krupnick, A., & Simon, N. B. (2006). Willingness to pay for mortality risk reductions: Does latency matter? *Journal of Risk & Uncertainty, 32,* 231–245.

Boyle, K. J. (2003). *Contingent valuation in practice. Primer on nonmarket valuation* (pp. 111–169). Boston: Kluwer Academic.

Carson, R. T., Flores, N. E., & Meade, N. F. (2001). Contingent valuation: Controversies and evidence. *Environmental and Resource Economics, 19,* 173–210.

Deshazo, J., & Cameron, T. A. (2005). The effect of health status on willingness to pay for morbidity and mortality risk reductions. California center for population research, On-line working paper series. CCPR-050-05.

Dickie, M., & List, J. (2006). Economic valuation of health for environmental policy: Comparing alternative approaches. Introduction and overview. *Environmental and Resource Economics, 34,* 339–346.

Freeman, A. M., III. (2003). *The measurement of environmental and resource values: Theory and methods* (2nd ed.). Resources for the future, Washington, DC.

Haab, T. C., & Mcconnell, K. E. (2002). *Valuing environmental and natural resources: The econometrics of non-market valuation*. Cheltenham: E. Elgar Pub.

Hanemann, M., & Kanninen, B. (2001). *The statistical analysis of discrete-response CV data*. Oxford: Oxford University Press.

Health And Retirement Study (HRS). (2000). Core final version 1.0. Data description and usage.

Heckman, J. (1979). Sample selection bias as a specification error. *Econometrica, 47,* 153–161.

Khwaja, A., Sloan, F., & Salm, M. (2006). Evidence on preferences and subjective beliefs of risk takers: The case of smokers. *International Journal of Industrial Organization, 24,* 667–682.

Koh, H. K., Graham, G., & Glied, S. A. (2011). Reducing racial and ethnic disparities: The action plan from the department of health and human services. *Health Affairs (Millwood), 30,* 1822–1829.

Krupnick, A., Cropper, M., Alberini, A., Heintzelman, M., et al. (2002). Age, health and the willingness to pay for mortality risk reductions: A contingent valuation survey of Ontario residents. *Journal of Risk and Uncertainty, 24,* 161–186.

Land, C. E. (1972). An evaluation of approximate confidence interval estimation methods for lognormal means. *Technometrics, 14,* 145.

Markus, H. (2012). Stroke: Causes and clinical features. *Medicine, 40,* 484–489.

Mead, H., Cartwright-Smith, L., Jones, K., Ramos, C., et al. (2008). *Racial and ethnic disparities in U.S. Health care: A Chartbook. The commonwealth fund.*

Milligan, M. A., Bohara, A. K., & Pagan, J. A. (2010). Assessing willingness to pay for cancer prevention. *International Journal of Health Care Finance and Economics, 10,* 301–314.

Mitchell, R. C., & Carson, R. T. (1989). *Using surveys to value public goods: The contingent valuation method.* Resources for the future, Washington, DC.

Mokdad, A. H., Marks, J. S., Stroup, D. F., & Gerberding, J. L. (2004). Actual causes of death in the United States, 2000. *Journal of American Medical Association, 291,* 1238–1245.

Parkin, T. B., Chester, S. T., & Robinson, J. A. (1990). Calculating confidence-intervals for the mean of a lognormally distributed variable. *Soil Science Society of America Journal, 54,* 321–326.

Pearce, D., Atkinson, G., & Mourato, S. (2006). *Cost-benefit analysis and the environment: Recent developments.* Paris: Organisation for Economic Co-operation and Development.

Reed Johnson, F., Fries, E. E., & Spencer Banzhaf, H. (1997). Valuing morbidity: An integration of the willingness to pay and health status index literatures. *Journal of Health Economics, 16,* 641–665.

Rumsfeld, J. S. (2002). Health status and clinical practice—when will they meet? *Circulation, 106,* 5–7.

Shepard, D. S., & Zeckhauser, R. J. (1984). Survival versus consumption. *Management Science, 30,* 423–439.

St.Clair, P. et al. (2011). RAND HRS Data Documentation, Version L. Labor & Population Program. *RAND Center for the Study of Aging.*

Statacorp. (2009). *Stata 11 base reference manual.* College Station: Stata Press.

Sunstein, C. R. (2004). Valuing life: A plea for disaggregation. *Duke Law Journal, 54,* 385–445.

U.S. Centers for Disease Control and Prevention. (2004). The health consequences of smoking [electronic resource]. A Report of the Surgeon General, Atlanta, GA.

U.S. Centers for Disease Control and Prevention. (2007). *The state of aging and health in America* Whitehouse Station, New Jersey.

U.S. Environmental Protection Agency. (2000). Guidelines for preparing economic analyses. EPA 240-R-00-003. Washington, DC.

U.S. Environmental Protection Agency. (2010). Valuing mortality risk reductions for environmental policy: A White Paper. (SAB Review Draft). Washington, DC.

Van Houtven, G., et al. (2003). Valuation of morbidity losses: Meta-analysis of willingness to pay and health status measures. Final report, prepared for the food and drug administration by research Triangle Institute.

Vincent, G. K., & Velkoff, V. A. (2010). The next four secades [electronic resource]: The older population in the United States: 2010 to 2050. Washington, DC.

Viscusi, W. K. (2010). The heterogeneity of the value of statistical life: introduction and overview. *Journal of Risk and Uncertainty, 40,* 1–13.

Chapter 21
The Economic Security of Latino Baby Boomers: Implications for Future Retirees and for Healthcare Funding in the U.S.

Zachary D. Gassoumis, Kathleen H. Wilber and Fernando M. Torres-Gil

The United States is currently facing dramatically increasing healthcare costs, driven in part by population aging and the retiring baby boom population. At the same time, racial/ethnic diversity is growing across the U.S., with particular rises seen among the Latino population. This chapter ties these two trends together, linking healthcare spending to shifting racial/ethnic demographics through an analysis of Latino economic security.

Healthcare spending is growing internationally, but the U.S. is seeing far higher increases than other developed nations. The per capita spending as of 2008 was over $ 7,500, equivalent to 16.0 % of GDP, well over the OECD median of $ 2,995 per capita, equivalent to 8.7 % of GDP (Squires 2011). Federal healthcare spending in the U.S. has been growing both in real dollars and as a percentage of the federal budget, totaling over $ 909 billion or 26.3 % of the federal budget in 2013 (Office of Management and Budget 2014). While this growth cannot continue unfettered, implementing cuts with 78 million baby boomers entering retirement is a challenging prospect.

On the funding side, federal healthcare spending in the U.S. comes largely from general taxation revenues, which are in turn dependent on economic productivity. An important component of this is income from individuals, primarily the U.S. workforce. According to projections from the U.S. Census Bureau, working-age Latinos will grow from 24 million in 2010 (15 % of the working-age population) to 39 million in 2030 (22 % of the working-age population). By contrast, the working-age non-Latino white population will shrink from 107 to 95 million. This is part of the continuing progression of the U.S. to a majority-minority nation; based on

Z. D. Gassoumis (✉) · K. H. Wilber
Davis School of Gerontology, University of Southern California, Los Angeles, USA
e-mail: gassoumi@usc.edu

F. M. Torres-Gil
Luskin School of Public Affairs, University of California, Los Angeles, USA

© Springer International Publishing Switzerland 2015
W. A. Vega et al. (eds.), *Challenges of Latino Aging in the Americas,*
DOI 10.1007/978-3-319-12598-5_21

current Census Bureau projections, non-Latino whites will drop below 50% of the population in 2042.

Despite growing numbers of racial/ethnic minorities across the U.S. population, the number of minority elders remains relatively modest. As of the 2010 census, only 7% (2.8 million people) of the U.S. population aged 65 and older was Latino. This means that a very ethnically diverse workforce is providing the taxation revenue to the federal government to support healthcare and social services for a vastly non-Latino white aging population. With an increasingly diverse workforce on the horizon, it will be important for the viability of these services to eliminate—or at least reduce—income disparities across racial/ethnic groups. This in turn will increase the earnings power and tax base of the U.S. workforce and improve taxation revenues for the federal government.

Economic disparities between racial/ethnic groups in the U.S. are still pronounced (Gassoumis et al. 2011; Smith 1997; Wu 2013). Among several possible sources of income disparities between racial/ethnic groups, the most widely recognized is education. Educational attainment, a metric that has strong links with both income and wealth, varies considerably by race/ethnicity. Asians have much higher rates of college graduation than the non-Latino white population, whereas black/African Americans and Latinos fare worse than non-Latino whites (U.S. Census Bureau 2012a). Immigrant and citizenship status is the second most widely recognized driver of income disparity that is particularly relevant to racial/ethnic minority groups (e.g., Chiswick 1978; Gassoumis et al. 2010). Although these sociodemographic factors are strongly associated with economic disparities within the U.S., racial/ethnic differences are also linked to inherent structural inequalities between minority and majority groups (Sandefur and Pahari 1989). The degree to which economic disparities are due to structural forces in contrast to individual-level sociodemographic characteristics, however, has received little attention from scholars studying the economics of aging.

To better understand economic security in retirement, this chapter examines income in the years immediately preceding retirement. Wealth is also analyzed to provide a broader look at economic disparities between racial/ethnic groups. We focus on Latinos, the largest racial/ethnic minority group in the U.S. The Latino population also offers a diverse demographic group to study that includes a broad range of educational attainment—typically well below the U.S. population average—and large variations based on immigration and citizenship status (Gassoumis et al. 2010).

The purpose of this chapter is twofold: to assess the degree to which racial/ethnic structural economic disparities are due to the societal factor of ingrained race-ethnic disadvantages versus individual-level factors, and to examine whether the interplay between those two factors has changed across generational cohorts. These disparities are tested immediately before retirement age—when they are expected to be largest due to the effects of cumulative inequality (Ferraro and Shippee 2009)—among two generational cohorts: the Baby Boom Generation (born 1946–1964) and the Silent Generation (born 1925–1945). These analyses and interpretations focus primarily on Latinos, the largest and fastest growing racial/ethnic group in the U.S.

and one that is subject to particularly high heterogeneity in sociodemographic factors due to high levels of immigration and relatively low naturalization rates.

The Structural Inequality of Race/Ethnicity

Individual economic success, as with most individual-level achievement, is attained within the context of a broader society. As such, it is subject to the social constraints that exist within that society (Alba and Nee 2003). These social constraints on economic status can emerge from various factors, including split labor market dynamics and their resultant race/ethnicity-based occupational pigeonholing and discrimination (Bonacich 1972; Restifo et al. 2013), citizenship-based employment constraints for immigrant groups (e.g., Bratsberg et al. 2002), and broad-scale discriminatory workplace dynamics (Green 2003).

Social scientists have long observed these structural inequalities in economic variables after adjusting for sociodemographic characteristics, both in income (e.g., Waters and Eschbach 1995) and housing/housing wealth (e.g., Krivo and Kaufman 2004). Based on the work of classic social theorists, structural inequalities would be expected to lessen over time, as society's valuation of performance supersedes the importance of racial/ethnic differences; indeed, Sandefur and Pahari (1989) found evidence that the second half of the twentieth century saw a closing of the gap in the earnings of racial/ethnic minority groups. Although disparities due to structural inequalities still exist, they appear to have been decreasing in the recent past.

Cumulative Inequality

Under the theory of cumulative inequality (Ferraro and Shippee 2009; Ferraro et al. 2009), sociodemographic, economic, and other factors experienced early in life are expected to have ongoing and cumulative effects throughout the lifecourse. Cumulative inequality and cumulative advantage/disadvantage (Dannefer 1987)—the original theory on which cumulative inequality was based—is most commonly applied to issues of biomedical and health disparities (e.g., Shuey and Willson 2008; Wickrama et al. 2013). However, it has great relevance to income, wealth, and other economic factors that are similarly shaped by sociodemographic characteristics and exhibit cumulative properties over the lifecourse, and is frequently applied to studies of these factors (e.g., Crystal and Shea 1990; Gregoire et al. 2002).

Core to the cumulative inequality framework is that inequalities between demographic groups are manifested and compounded over the lifecourse, with their trajectories shaped by the accumulation of risk and resources. However, an individual's trajectory can crucially be modified by human agency (Ferraro and Shippee 2009), through means such as education and behavioral changes. The provision for human agency within the cumulative inequality framework is crucial in studies involving immigrant populations. The act of immigrating is, in and of itself, an example of that agency, and the attainment of citizenship may induce a further alteration of one's trajectory.

Mechanisms of Human Agency: Education and Naturalization

Two primary mechanisms of human agency will be addressed in this chapter: education and, for immigrants, naturalization. Education is largely accumulated early in the lifecourse, but the degree of educational attainment is determined in part by human agency. Although additional formal education can be sought later in life, this can be more difficult for racial/ethnic minority immigrant populations, many of whom enter the U.S. without any secondary education. When attained by immigrants, however, it comes with higher earnings (Bratsberg and Ragan 2002).

Almost all immigrants enter the U.S. as non-citizens; those who wish to attain citizenship must do so through the process of naturalization. Naturalization comes with an economic cost and is not open to everyone. There is currently no broadscale mechanism in place for individuals who entered the U.S. undocumented to obtain documentation, let alone citizenship. But to those eligible who have the necessary means, naturalization has the potential to make a downward trajectory of cumulative inequality curve upward by removing barriers to occupational mobility.

This chapter first identifies the magnitude of racial/ethnic structural disadvantage in the years preceding retirement for the Baby Boom Generation, the most recent generational cohort to reach retirement age. Next, it compares their structural disadvantage during this age range with that of members of the Silent Generation cohort when they were the same age. These analyses are performed both for income and wealth.

Methods

This chapter assesses the levels of income and wealth of two cohorts in the years before they reached retirement age: a subset of the Baby Boom Generation (born 1946–1953) and a subset of the Silent Generation (born 1938–1945). To capture these two groups at similar points in their lives, 2008 data were used for the baby boomers, when they were aged 54–62, and 2000 data were used for the Silent Generation cohort, when they were at the same ages.

Data sources provided by the U.S. Census Bureau were used for income analyses, to ensure national generalizability. Baby boomer income data from 2008 were taken from the American Community Survey (ACS), the Census Bureau's annual snapshot that samples roughly 1 % of the of the U.S. population. Although the ACS does not provide birth year, it provides quarter of birth and age at the time of the survey. Since data for the ACS are collected throughout the year, people born from January through June were more likely to have had their birthday at the time of their data collection, whereas people born from July through December were more likely not yet to have had their birthday. To achieve a sample that best resembled those born 1946–1953, the analysis sample included those aged 54 who were born in July through December, those aged 55–61, and those aged 62 who were born in January through June.

Silent Generation income data were taken from the 2000 Decennial Census' 5 % public-use microdata sample, which is a compilation of the roughly 6 % of the U.S. population that completed the long-form version of the 2000 Census. The decennial census in the U.S. collects a snapshot as of April 1st of the calendar year in which the data are collected. Year of birth information is not released, but age is provided. So that the sample best captured those born 1938–1945, it included everyone aged 54–61 as of April 1, 2000.

Wealth data for both cohorts were drawn from the Health and Retirement Study (HRS), a large-scale, longitudinal, nationally representative sample of the U.S. population over the age of 50. Wealth and demographic data were taken from the 2008 HRS wave for the baby boom cohort and from the 2000 HRS wave for the Silent Generation cohort.

ACS and Census data were extracted from the Integrated Public Use Microdata Series (IPUMS) out of the University of Minnesota (Ruggles et al. 2010), and HRS data were extracted from the RAND HRS Data File (RAND HRS Data 2011) and the RAND Enhanced Fat Files. For all data sources, individuals living in group quarters were excluded and nationally representative sample weights were applied. There were incomplete (missing) data for 104 cases (2.6 % of total) from the 2000 HRS sample and 23 cases (0.7 % of total) from the 2008 HRS sample; since less than 5 % of the cases in each sample had missing data, case-wise deletion was employed to deal with the missingness (Allison 2002). The final sample sizes were: 1,063,588 for the 2000 Census sample, 326,157 for the 2008 ACS sample, 3,868 for the 2000 HRS sample, and 3,094 for the 2008 HRS sample.

Measures

The outcome variables were individual income and household wealth. When constructing the individual income measure for the HRS samples, the two components measured at the household level (income from assets, and other household income) were divided by two for individuals whose households contained a spousal partner. All 2000 income and wealth data are presented in 2008 dollars, using inflationary weights from the Consumer Price Index for All Urban Consumers (CPI-U).

Both income and wealth are constructs known to contain considerable skew, but since zero and negative values are possible, logarithmic transformation is unwieldy. Therefore, an alternative transformation was used—the inverse hyperbolic sine— that transforms data in a manner similar to the logarithmic transformation for values above 1 and provides comparable transformations of all values less than 1 (Burbidge et al. 1988; Pence 2006). Following the work of Burbidge and colleagues, the value of 0.0001 was used for θ (Gale and Pence 2006).

Key predictor variables were age, gender (female), race/ethnicity (non-Latino white, non-Latino black, Latino, non-Latino Asian/Pacific Islander, or non-Latino other), citizenship status (born as a U.S. citizen, naturalized U.S. citizen, or not a U.S. citizen), education (less than high school, high school degree, some college, or college degree or higher), marital status (married, divorced/separated, widowed, or single/never married), and labor force participation (employed, unemployed, or not in the labor force). The HRS samples did not include sufficiently large numbers of non-Latino Asian/Pacific Islanders to warrant separating them out, so they were included in the non-Latino other category for the wealth analyses. In the HRS, citizenship status and year of naturalization were asked in 2006, 2008 and 2010; data from all three of these time points were used to construct the citizenship status variables for both of the HRS samples.

Analyses

Descriptive statistics are presented for all variables across the four samples; individual income and household wealth are further broken down by race/ethnicity. Two nested OLS regression models are presented for each sample. The first model regresses age, gender, and race/ethnicity on income (for the Census and ACS samples) or wealth (for the HRS samples). The second model adds four additional predictor variables: citizenship status, educational attainment, marital status, and employment status. Unstandardized and standardized parameter estimates are presented with their corresponding t statistics and significance levels, and fit statistics (R^2) are presented for each model. All analyses were run in SAS version 9.2 (SAS Institute Inc., Cary, NC).

Individual Income and Household Wealth, by Race/ Ethnicity

Table 21.1 presents mean and median levels of individual income (from the 2000 Census to 2008 ACS) and household wealth (from the 2000 to 2008 waves of the HRS) for each cohort, broken down by race/ethnicity. Median income for Latinos in the Silent Generation sample was 49.3 % ($ 16,617 vs. $ 33,728) that of non-Latino whites, and 54.3 % ($ 19,000 vs. $ 35,000) for the baby boom sample. Non-Latino blacks had moderately higher median incomes: compared to non-Latino whites, 67.7 % ($ 22,832) for the Silent Generation and 65.7 % ($ 23,000) for the Baby Boom Generation. Those in the non-Latino other category were in a similar range (63.9 % [$ 21,564] compared to non-Latino whites for the Silent Generation, 70.3 %

Table 21.1 Income and wealth, by cohort and race/ethnicity

	Silent generation: 2000[a]		Baby boomers: 2008[a]	
	Mean	Median	Mean	Median
Individual income, in 2008 dollars				
Race/ethnicity				
Non-Latino White	$ 50,715	$ 33,728	$ 50,526	$ 35,000
Non-Latino Black	$ 33,232	$ 22,832	$ 31,142	$ 23,000
Non-Latino Asian/P.I.	$ 45,674	$ 25,369	$ 42,081	$ 25,000
Non-Latino Other	$ 34,788	$ 21,564	$ 37,829	$ 24,600
Latino	$ 27,619	$ 16,617	$ 28,469	$ 19,000
Household wealth, in 2008 dollars				
Race/ethnicity				
Non-Latino White	$ 570,849	$ 231,807	$ 546,159	$ 239,000
Non-Latino Black	$ 128,029	$ 43,761	$ 159,297	$ 31,600
Non-Latino Other[b]	$ 342,603	$ 107,902	$ 873,314	$ 97,025
Latino	$ 136,685	$ 58,764	$ 205,734	$ 63,000

[a] Income values are calculated from U.S. Census Bureau data sources (2000 Census [n=1,063,558] and 2008 American Community Survey [n=326,157]); wealth values are calculated from two waves of the Health and Retirement Study (2000 [n=3868] and 2008 [n=3094])
[b] Too few Asian/Pacific Islanders exist in the Health and Retirement Study samples to justify reporting them separately; they have been included in the "Other" category. Note: P.I. = Pacific Islander

[$ 24,600] for the baby boomers), as were non-Latino Asians (75.2 % [$ 25,369] for the Silent Generation, 71.4 % [$ 25,000] for the baby boomers).

In contrast to individual income, household wealth data showed lower median wealth for non-Latino blacks than for Latinos. Non-Latino blacks had median wealth that was 18.9 % ($ 43,761 vs. $ 231,807) that of non-Latino whites in the Silent Generation cohort and 13.2 % ($ 31,600 vs. $ 239,000) for those in the boomer cohort, whereas Latinos in the Silent Generation had 25.4 % ($ 58,764) of the levels of non-Latino white wealth and Latino boomers had 26.4 % ($ 63,000) the level of non-Latino white wealth. The non-Latino other category, which for the HRS-based wealth analysis included Asians/Pacific Islanders, had considerably higher wealth levels than Latinos and non-Latino blacks but were still below the levels of non-Latino whites: 46.5 % ($ 107,902) for the Silent Generation and 40.6 % ($ 97,025) for the Baby Boom Generation.

Across all of the racial/ethnic groups, members of the baby boom cohort had median levels of individual income that were higher or approximately equivalent to that of the Silent Generation cohort at the same ages. This was most notable among Latinos and members of the non-Latino other category, both of which had median incomes 14 % higher than those of the same age 8 years earlier. Different results were seen for levels of household wealth. While median wealth for Latinos and non-

Latino whites were mildly higher among the baby boom cohort than they had been for the Silent Generation cohort (7.2 and 3.1% higher, respectively), the level was considerably lower for both non-Latino blacks and others (27.8 and 10.1% lower, respectively).

Sociodemographic Characteristics

Full characteristics for the four samples are presented in Table 21.2. Due to the application of population weights to all samples, there is considerable similarity across most sociodemographic variables. Apart from the Silent Generation sample's HRS data (which had a lower proportion female, at 48.7%), the samples ranged from 50.6 to 51.9% female. The proportion of Latinos in the samples ranged from 7.1 to 8.6%, and the proportion of non-Latino blacks ranged from 9.5 to 11.1%. A larger degree of variation existed for citizenship status, with the HRS samples less likely than the Census and ACS samples to have naturalized citizens (4.8 vs. 7.1% for the Silent Generation; 5.7 vs. 8.1% for the boomers) and non-citizens (1.9 vs. 4.5% for the Silent Generation; 3.4 vs. 4.7% for the boomers).

Educational attainment was similar between the Census Bureau and HRS samples, with the boomers showing slightly higher levels of academic achievement than their Silent Generation counterparts (e.g., roughly 30% of boomers held a college degree or higher, compared with roughly 24% of the Silent Generation cohort). Marital status varied slightly between samples, but all four followed the general pattern of roughly one-fifth divorced or separated and two-thirds married. Finally, roughly two-thirds (64–68%) of all four samples were employed at the time of the survey.

Regression Results: Income

The first regression model for the Silent Generation cohort's individual income expands on the bivariate statistics reported in Table 21.2. Model 1 reveals that being older (standardized β [st. β] = −0.068) and being female (st. β = −0.382) are associated with lower levels of income, as is being a race ethnicity other than non-Latino white (see Table 21.3). Model 2 demonstrates that when additional sociodemographic characteristics are considered, the direct effect of age, gender, and race/ethnicity is dramatically mitigated. The effect of age became negligibly positive, and the effect of being female dropped by 19% (st. β = −0.308). Profoundly, the effect of being Latino fell by 77% (from −0.124 to −0.029, compared to non-Latino whites), non-Latino black by 76% (from −0.075 to −0.018, compared to non-Latino whites), and Asian by 26% (from −0.031 to −0.023, compared to non-Latino whites). Of the variables added into Model 2, being a non-citizen had a negative effect (st. β = −0.047), whereas there was a negligible effect of having naturalized,

Table 21.2 Demographic and economic characteristics, by analysis dataset/cohort

	Income samples	
	Silent generation (2000 census)	Baby boomers (2008 ACS)
	$n=1,063,558$	$n=326,157$
Age	57.3 (2.26)	57.8 (2.34)
Female	51.99	51.93
Race/ethnicity		
Non-Latino White	78.22	75.61
Non-Latino Black	9.47	10.09
Non-Latino Asian/Pacific Islander[a]	3.31	4.14
Non-Latino Other[a]	1.88	1.65
Latino	7.12	8.50
Citizenship status		
Born U.S. citizens	88.38	87.17
Naturalized citizens	7.08	8.09
Non-citizens	4.54	4.74
Education		
< High school	19.20	11.69
High school degree	30.62	27.54
Some college	25.77	30.16
College degree and beyond	24.40	30.61
Marital status		
Married	71.04	67.12
Divorced/separated	17.57	19.90
Widowed	5.99	4.87
Never married	5.40	8.10
Labor force participation		
Employed	63.75	67.60
Unemployed	2.19	2.92
Not in labor force	34.06	29.49
Individual income, in 2008 dollars	$ 46,949 ($ 65,662)	$ 46,135 ($ 61,523)
Transformed income, in 2008 dollars[b]	17,231 (10,604)	17,552 (10,089)
Household wealth, in 2008 dollars	N/A	N/A
Transformed wealth, in 2008 dollars[b]	N/A	N/A

	Wealth samples	
	Silent generation	Baby boomers
	(2000 HRS)	(2008 HRS)
	n = 3,868	n = 3,094
Age	57.9 (2.30)	57.9 (2.40)
Female	48.72	50.62

Table 21.2 (continued)

	Wealth samples	
	Silent generation (2000 HRS) n=3,868	Baby boomers (2008 HRS) n=3,094
Race/ethnicity		
Non-Latino White	79.11	76.85
Non-Latino Black	10.89	11.06
Non-Latino Asian/Pacific Islander[a]	N/A	N/A
Non-Latino Other[a]	2.29	3.51
Latino	7.70	8.57
Citizenship status		
Born U.S. citizens	92.64	91.68
Naturalized citizens	4.84	5.70
Non-citizens	1.93	3.42
Education		
< High school	18.41	10.84
High school degree	34.81	30.30
Some college	23.29	29.26
College degree and beyond	23.49	29.59
Marital status		
Married	66.68	66.22
Divorced/separated	20.56	21.02
Widowed	7.89	6.55
Never married	4.88	6.21
Labor force participation		
Employed	66.32	65.65
Unemployed	2.46	4.45
Not in labor force	31.21	29.90
Individual income, in 2008 dollars	$ 56,331 ($ 145,741)	$ 51,273 ($ 79,160)
Transformed income, in 2008 dollars[b]	19,054 (10,209)	18,550 (10,041)
Household wealth, in 2008 dollars	$ 483,933 ($ 1,694,548)	$ 485,655 ($ 1,275,721)
Transformed wealth, in 2008 dollars[b]	33,221 (17,762)	31,609 (20,770)

Note: Units are percentages, by column, or mean (standard deviation) for continuous variables
ACS American community survey; *HRS* health and retirement study
[a] Too few Asian/Pacific Islanders exist in the HRS samples to justify reporting them separately; they have been included in the "other" category
[b] Transformed using the inverse hyperbolic sine

Table 21.3 Linear regression on income (IHS-transformed), by cohort

	Silent generation (2000 census; $n = 1,063,558$)							
	Model 1			Model 2				
	B	St.β	t	b	St.β	t		
Age	−318	−0.068	−76.75***	20	0.004	5.96***		
Female	−8,111	−0.382	−433.33***	−6,533	−0.308	−412.89***		
Race/ethnicity								
Non-Latino White (ref.)	–	–	–	–	–	–		
Non-Latino Black	−2,706	−0.075	−84.10***	−651	−0.018	−24.13***		
Non-Latino Asian	−1,813	−0.031	−34.58***	−1,366	−0.023	−27.63***		
Non-Latino Other	−3,315	−0.042	−48.02***	−1,582	−0.020	−27.83***		
Latino	−5,095	−0.124	−139.21***	−1,207	−0.029	−35.20***		
Citizenship status								
Born U.S. citizens (ref.)				–	–	–		
Naturalized citizens				154	0.004	4.46***		
Non-citizens				−2,419	−0.047	−58.15***		
Education								
< High school (ref.)				–	–	–		
High school degree				1,995	0.087	85.77***		
Some college				4,172	0.172	172.30***		
College degree and beyond				8,439	0.342	338.95***		
Marital status								
Married (ref.)				–	–	–		
Divorced/separated				712	0.026	34.44***		

Table 21.3 (continued)

	Silent generation (2000 census; n = 1,063,558)					
	Model 1			Model 2		
				b	St.β	t
Widowed				2,319	0.052	69.66***
Never married				−317	−0.007	−9.19***
Labor force participation						
Employed (ref.)				–	–	–
Unemployed				−6,040	−0.083	−114.38***
Not in labor force				−9,001	−0.402	−528.52***
Adjusted R²	0.174				0.446	

	Baby boomers (2008 ACS; n = 326,157)					
	Model 1			Model 2		
	B	St.β	t	b	St.β	t
Age	−196	−0.045	−27.56***	84	0.019	14.80***
Female	−6,059	−0.300	−182.42***	−4799	−0.238	−179.96***
Race/ethnicity						
Non-Latino White (ref.)	–	–	–	–	–	–
Non-Latino Black	−3,083	−0.092	−55.43***	−1,018	−0.030	−22.70***
Non-Latino Asian	−2,409	−0.048	−28.78***	−1,436	−0.028	−18.41***
Non-Latino Other	−2,485	−0.031	−19.05***	−848	−0.011	−8.24***
Latino	−4,789	−0.132	−79.79***	−1,337	−0.037	−24.25***
Citizenship status						
Born U.S. citizens (ref.)				–	–	–
Naturalized citizens				−481	−0.013	−8.37***

Table 21.3 (continued)

	Baby boomers (2008 ACS; $n = 326{,}157$)			
	Model 1		Model 2	
Non-citizens			−3,157	−0.066	−44.68***
Education					
< High school (ref.)			−	−	−
High school degree			1,444	0.064	30.12***
Some college			3,464	0.158	72.26***
College degree and beyond			7,639	0.349	157.59***
Marital status					
Married (ref.)			−	−	−
Divorced/separated			99	0.004	2.94**
Widowed			1,333	0.028	21.40***
Never married			−945	−0.026	−19.22***
Labor force participation					
Employed (ref.)			−	−	−
Unemployed			−9,024	−0.151	−115.21***
Not in labor force			−10,274	−0.464	−342.97***
Adjusted R^2	0.119		0.454		

Note: *IHS* inverse hyperbolic sine, *ACS* American Community Survey, *b* = *parameter estimate*; St. *β* = standardized parameter estimate. The analyses were run using a transformed version of the dependent variable (using the IHS transformation)
*$p<0.05$; **$p<0.01$; ***$p<0.001$

both when compared to those who were born as U.S. citizens. As is typical, higher educational attainment had a strongly positive relationship with income, and having left the labor force had the strongest negative effect on income (st. $\beta=-0.402$), compared to those who were employed. For the Silent Generation cohort, the addition of sociodemographic characteristics beyond age, gender, and race/ethnicity increased the model's explanatory power from 17.4% of variance to 44.6% of variance.

Income results for the Baby Boom Generation cohort were similar to those for the Silent Generation cohort. In Model 1, the boomer cohort also exhibited a negative effect of age, being female, and falling into any racial/ethnic group other than non-Latino white. There was less of an effect of gender (st. $\beta=-0.300$) and age (st. $\beta=-0.045$) than there was for the Silent Generation, but with the exception of the non-Latino other group, the racial/ethnic differences were more pronounced. Upon the addition of sociodemographic variables in Model 2, the effect of being female dropped again by 21%, Latino by 72%, non-Latino black by 67%, and Asian by 42%. Educational attainment again had a considerably positive effect on income; the effect of being a non-citizen was even greater (st. $\beta=-0.066$), and being a naturalized citizen was linked to lower individual income (st. $\beta=-0.013$), both compared to those who were born as U.S. citizens. The negative effects of being out of the labor force and being unemployed were also stronger: -0.464 and -0.151, respectively, compared to those who were employed. The percentage of variance explained by Model 2 for the baby boomer cohort was comparable to that for the Silent Generation cohort (45.4%), but the core demographics of Model 1 explained considerably less of the variance in individual income among the boomers (11.9%) than it had among the Silent Generation.

Regression Results: Wealth

In contrast to income, household wealth in the Silent Generation cohort's Model 1 was positively impacted by age (st. $\beta=0.030$), as expected (see Table 21.4). Being female had a negative association with wealth (st. $\beta=-0.093$), as did being of a race/ethnicity other than non-Latino white. Adding the additional sociodemographic variables in Model 2 increased the positive impact of age on wealth (st. $\beta=0.044$) and dropped the impact of being female to 0. The effect of being Latino on household wealth decreased by 44% (from -0.201 to -0.112, compared to non-Latino whites) and being non-Latino black decreased by 31% (from -0.286 to -0.197, compared to non-Latino whites). Non-citizens did not have significantly less household wealth than those born in the U.S., and naturalized citizens had higher levels than those born in the U.S. (st. $\beta=-0.038$). The standardized effect of educational attainment, compared to those who had not completed high school, ranged from 0.198 for a high school degree to 0.402 for a college degree or beyond. In sharp contrast to the modest effect of marital status on income, it was strongly related to household wealth among the Silent Generation cohort. Never being married

Table 21.4 Linear regression on wealth (IHS-transformed), by cohort

	Silent generation (2000 HRS; n=3,868)					
	Model 1			Model 2		
	b	St.β	t	B	St.β	t
Age	232	0.030	1.99*	339	0.044	3.16**
Female	-3,293	-0.093	-6.12***	-39	-0.001	-0.08
Race/ethnicity						
Non-Latino White (ref.)	–		–	–		–
Non-Latino Black	-16,337	-0.286	-18.85***	-11,249	-0.197	-14.01***
Non-Latino Other	-4,134	-0.035	-2.30*	-4,827	-0.041	-2.95**
Latino	-13,399	-0.201	-13.23***	-7,473	-0.112	-7.23***
Citizenship status						
Born U.S. citizens (ref.)				–		–
Naturalized citizens				3,159	0.038	2.66**
Non-citizens				-3,060	-0.024	-1.63
Education						
<High school (ref.)				–		–
High school degree				7,381	0.198	10.05***
Some college				10,377	0.247	13.10***
College degree and beyond				16,859	0.402	20.72***
Marital status						
Married (ref.)				–		–
Divorced/separated				-11,085	-0.252	-17.76***
Widowed				-8,046	-0.122	-8.57***

Table 21.4 (continued)

Silent generation (2000 HRS; n=3,868)

	Model 1			Model 2		
	b	St.β	t	b	St.β	t
Never married				−8,782	−0.106	−7.70***
Labor force participation						
Employed (ref.)				−	−	−
Unemployed				−4,498	−0.039	−2.87**
Not in labor force				−1,359	−0.035	−2.47*
Adjusted R^2		0.122			0.290	

Baby boomers (2008 HRS; n=3094)

	Model 1			Model 2		
	b	St.β	t	b	St.β	t
Age	480	0.055	3.20**	502	0.058	3.65***
Female	−1,029	−0.025	−1.43	1,639	0.039	2.48*
Race/ethnicity						
Non-Latino White (ref.)	−	−	−	−	−	−
Non-Latino Black	−16,467	−0.249	−14.31***	−9,419	−0.142	−8.79***
Non-Latino Other	−7,929	−0.070	−4.06***	−6,214	−0.055	−3.36***
Latino	−12,184	−0.164	−9.47***	−5,647	−0.076	−4.03***
Citizenship status						
Born U.S. citizens (ref.)				−	−	−
Naturalized citizens				2,824	0.032	1.81
Non-citizens				−3,516	−0.031	−1.80

Table 21.4 (continued)

	Baby boomers (2008 HRS; $n = 3,094$)		
	Model 1	Model 2	
Education			
< High school (ref.)		–	–
High school degree		6,186	0.137	5.06***
Some college		11,459	0.251	9.25***
College degree and beyond		19,510	0.429	15.40***
Marital status			
Married (ref.)		–	–	–
Divorced/separated		−12,941	−0.254	−15.68***
Widowed		−9,862	−0.117	−7.29***
Never married		−11,311	−0.131	−8.32***
Labor force participation			
Employed (ref.)		–	–	–
Unemployed		−4,864	−0.048	−3.08**
Not in labor force		−2,245	−0.049	−3.03**
Adjusted R²	0.085	0.263	

Note: *IHS* inverse hyperbolic sine; *HRS* Health and Retirement Study; b = *parameter* estimate; St. tudy; imate. The analyses were run using a transformed version of the dependent variabl(using the IHS transformation)

*$p<0.05$; **$p<0.01$; ***$p<0.001$

(st. $\beta = -0.106$) and being widowed (st. $\beta = -0.122$) had modestly negative effects on wealth, compared to those who were married, and being divorced or separated had the largest negative effect in the model (st. $\beta = -0.252$). The amount of variance in household wealth explained by the model more than doubled, from 12.2% in Model 1 to 29.0% after the additional sociodemographic variables were added in Model 2.

For the baby boomer cohort, age was again positively related to household wealth in Model 1 (st. $\beta = 0.055$), but gender was not a significant predictor (see Table 21.4). Although being Latino or non-Latino black had less of an effect on wealth for the boomers than it had for the Silent Generation cohort, they were still significant negative predictors (st. $\beta = -0.164$ and -0.249, respectively, compared with non-Latino whites). The additional sociodemographic characteristics in Model 2 reduced the effect of being Latino by 54% (to -0.076) and the effect of being non-Latino black by 43% (to -0.142). There was no significant effect of citizenship, but the effects of educational attainment and marital status were similar to those seen for the Silent Generation: education ranged from st. $\beta = 0.137$ for those with a high school degree to 0.429 for those with a college education or beyond, compared to those with less than a high school degree; being widowed or never married had standardized effects of -0.117 and -0.131, respectively, and being divorced/separated had roughly double the effect (st. $\beta = -0.254$). The variance explained increased from 11.9% in Model 1 to 45.4% in Model 2.

Discussion

Healthcare costs in the U.S. continue to rise dramatically, and population aging makes a reversal of this trend unlikely in the near future. Within the current policy landscape, a large proportion of these costs are borne by governmental payers (e.g., Medicare, Medicaid, the Veterans Health Administration, the Children's Health Insurance Program). The future high costs of health care will be funded by the taxes of a workforce that is increasingly comprised of Latinos and other racial/ethnic minority groups. To maximize the earnings of the U.S. population and, by extension, the U.S. tax base, economic disparities between racial/ethnic groups must be eliminated, or at least decreased.

To study racial/ethnic disparities, this chapter used data for two generational cohorts—the Silent Generation and Baby Boom Generation—to evaluate the degree to which racial/ethnic economic disparities in the pre-retirement years are due to structural disadvantages. After adjusting for the effects of sociodemographic variables—age, gender, citizenship status, education, marital status, and labor force participation—the structural effects of race/ethnicity on income and wealth were considerably reduced. When comparing the two cohorts, however, the expected reduction in structural effects from the Silent Generation to the Baby Boom Generation was seen for wealth but not for income, confirming hypothesis 4 but failing to confirm hypothesis 2.

Although it pales in comparison to the unadjusted racial/ethnic disparities, primarily for income, the presence of a structural disadvantage is still very real for the broad racial/ethnic groups identified in this chapter. The reduction of structural disparities in wealth from the Silent Generation to the Baby Boom Generation follows the expectation articulated by Sandefur and Pahari (1989) that these disparities would be reduced over time. As such, a continued trajectory in the direction of reduced structural disparities can be expected, which signals good news for the younger members of the Baby Boom Generation, Generation X, and future generational cohorts. There currently exist large gaps in wealth levels between racial/ethnic groups, even after sociodemographic adjustment, and a future reduction in structural inequalities can help decrease those gaps. This will be increasingly important as the proportion of the U.S. population that is not non-Latino white continues to grow.

The lack of findings that support the expected reduction of structural income disparities between the two generational cohorts may reflect a true absence of such a reduction. If this is the case, it would signal that occupational discrimination of some sort is still in place, but that similar discrimination has less bearing on the ability of racial/ethnic minorities to convert their income into wealth. It is possible, however, that the absence of a reduction in structural income discrimination is due to the years in which this study's data were collected. The data for baby boomers were collected in the heart of the "Great Recession," and may reflect unique income dynamics that were in play at the time due to heightened levels of unemployment and income insufficiency (Myers et al. 2012). Further research that spans different economic climates will be necessary to discern how structural disparities are shifting in the twenty-first century.

It must be noted that, despite the relatively modest structural disparities that exist, real differences in income and wealth based on sociodemographic disparities are stark. The Asian population serves as a good example of the compensatory power of human agency-driven factors such as education and naturalization. Despite being subject to structural racial/ethnic disparities in income that are more pronounced than those identified for Latinos and non-Latino blacks, the attainment of high educational levels and naturalization among most Asian immigrants elevates the overall income of the Asian population to a level that is above that of the other minority groups. Attainment of similar educational levels and naturalization rates can be mechanisms for other racial/ethnic minorities to counteract the economic pressures of structural disparity. Additionally, improvements in financial literacy and savings behaviors are crucial to balancing the especially disparate wealth levels, and early interventions are important to mitigate the forces of cumulative inequality.

Limitations

Although these analyses used large, nationally representative datasets, limitations still exist. The 2008 economic climate could have led to results that are not representative for income, as discussed above, or for wealth due to disparate effects of the recession by race/ethnicity (Kochhar et al. 2011) as well as age (Gassoumis 2012). This distinction could not be made in the analyses, since they were unable to distinguish between period and cohort effects. The relatively short timeframe between the cohorts may also have led to an inability to detect effects. Additionally, the analysis approach did not integrate nonparametric approaches and therefore could not take into account growing levels of inequality between the upper and lower ends of the income and wealth distribution.

For the sake of parsimony, this chapter considers the sociodemographic variables—age, gender, marital status, education, and citizenship status—to have a homogenous effect across different racial/ethnic groups. This assumption has been shown to be invalid in other samples for education (Sandefur and Pahari 1989); future studies on larger samples should allow for heterogenous effects across racial/ethnic groups wherever possible. This is underscored by the likelihood that educational attainment, a key component of income and income disparity, can have different ramifications for immigrants who may have received their education abroad versus U.S.-born populations who likely received a domestic education; the effect of 1 year of education may vary based on where and when that education was attained.

HRS serves as a rich dataset for studying microeconomic factors, but the participants tend to have above average levels of economic security; therefore, the results for wealth may reflect patterns that do not generalize to the U.S. population. Also, none of the datasets used have information on documentation status of immigrants, which is a key component in income and wealth building as well as the ability to naturalize. This as well as other measures not included in the model—including, but not limited to, generation status (e.g., 1, 1.5, 2, 3 +) and time since immigration for the foreign-born—may have impacts on income and wealth that could mitigate the findings of the above analyses.

Implications for Healthcare Funding and for Future Retirees

Findings reveal that the lion's share of the racial/ethnic disparity in income is due to individual factors, including education and citizenship status. This shows great promise for the prospect of decreasing future disparities in the U.S., which will bolster the tax base of an increasingly racially/ethnically diverse nation. Equalizing racial/ethnic minority disparities in educational attainment and citizenship status, all else being equal, would have had the impact of increasing the income base of pre-retirement boomers by tens of billions of dollars, based on the effects reported above. Back-of-the envelope calculations suggest that this could have added roughly $ 5 billion per year in income tax revenue for the U.S. federal government, and roughly an additional $ 4 billion in FICA tax revenues for Social Security and Medicare programs. While this is far from a solution to funding the escalating cost of U.S. healthcare, it would go a ways to closing the funding gap. Furthermore, these figures represent only the income effects for one generational cohort; the effects on the entirety of the U.S. workforce would be twice this high.

These findings show promise for the future U.S. tax base and on an individual level for the next generation of Latinos, black/African Americans, and Asians approaching retirement. However, it also points to the necessity for equalizing the sociodemographic profiles of these groups through education and, among immigrants, naturalization.

Education is a particularly important sociodemographic characteristic in terms of earnings potential. Yet considerable educational disparities persist among racial/ethnic minority youth. An often cited barrier to educational attainment is the cost, especially for many low-income, minority, and immigrant parents who may have an aversion to borrowing money for education or simply be inexperienced with credit markets (Burdman 2005; McDonough and Calderone 2006). Increasing access to grant funding (Gross et al. 2013) and financial aid (Kim et al. 2009) is crucial for improving the accessibility of postsecondary education—especially in a climate of soaring higher education costs—and providing information about education financing may make these options more attractive to those who are wary (McDonough and Calderone 2006). But for families who are unwilling to incur debt, approaches to saving for educational expenses in advance may be the best option. Mechanisms such as the Individual Development Account have been shown to be effective routes to encouraging savings for postsecondary education (Sherraden et al. 2003; Zhan and Schreiner 2005) and may increase the feasibility of attending college for low-income minorities. Documentation status may provide an additional barrier for postsecondary financing to many immigrants who would otherwise take advantage of credit markets as a route to educational attainment (Gonzales 2011), primarily among "dreamers" or the 1.5 generation (Rumbaut 2004).

Even net of educational attainment, however, disparities in wealth are particularly severe along racial/ethnic lines. While structural disparities seem to be decreasing, there is still a long way to go before the gap between racial/ethnic minorities and non-Latino whites is eradicated. Differences exist in savings patterns (Fisher and Hsu 2012) and investment decisions (Plath and Stevenson 2005) that exacerbate this disparity. However, culturally sensitive and culturally targeted financial literacy programs have proven promising for improving the level of comfort with financial concepts and products (Forte 2013; Spader et al. 2009), and offer one approach to reducing the sharp racial/ethnic wealth divides.

Acknowledgments This research was undertaken largely under a Ford Foundation-funded project, Latinos and Economic Security, based at the UCLA Center for Policy Research on Aging. Further support was provided from the USC Edward R. Roybal Institute on Aging and through a dissertation fellowship from the Center for Retirement Research at Boston College.

References

Alba, R., & Nee, V. (2003). *Remaking the American mainstream: Assimilation and contemporary immigration*. Cambridge: Harvard University Press.

Allison, P. D. (2002) *Missing data*. Thousand Oaks: Sage.

Bonacich, E. (1972). A theory of elite antagonism: The split labor market. *American Sociological Review, 37*(5), 547–559.

Bratsberg, B., & Ragan, J. F., Jr. (2002). The impact of host-country schooling on earnings: A study of male immigrants in the United States. *The Journal of Human Resources, 37*(1), 63–105.

Bratsberg, B., Ragan, J. F., Jr., & Nasir, Z. M. (2002). The effect of naturalization on wage growth: A panel study of young male immigrants. *Journal of Labor Economics, 20*(3), 568–597. doi:10.1086/339616.

Burbidge, J. B., Magee, L., & Robb, A. L. (1988). Alternative transformations to handle extreme values of the dependent variable. *Journal of the American Statistical Association, 83*(401), 123–127.

Burdman, P. (2005). *The student debt dilemma: Debt aversion as a barrier to college access (Report No. CSHE.13.05)*. Berkeley: Center for Studies in Higher Education..

Chiswick, B. R. (1978). The effect of Americanization on the earnings of foreign-born men. *The Journal of Political Economy, 86*(5), 897–921.

Crystal, S., & Shea, D. (1990). Cumulative advantage, cumulative disadvantage, and inequality among elderly people. *The Gerontologist, 30*(4), 437–443. doi:10.1093/geront/30.4.437.

Dannefer, D. (1987). Aging as intracohort differentiation: Accentuation, the Matthew effect, and the life course. *Sociological Forum, 2*(2), 211–236.

Ferraro, K. F., & Shippee, T. P. (2009). Aging and cumulative inequality: How does inequality get under the skin? *The Gerontologist, 49*(3), 333–343. doi:10.1093/geront/gnp034.

Ferraro, K. F., Shippee, T. P., & Schafer, M. H. (2009). Cumulative inequality theory for research on aging and the lifecourse. In V. Bengtson, M. Silverstein, N. Putney, & D. Gans (Eds.), *Handbook of theories of aging* (2nd ed., pp. 413–433). New York: Springer.

Fisher, P. J., & Hsu, C. (2012). Differences in household saving between non-Hispanic white and Hispanic households. *Hispanic Journal of Behavioral Sciences, 34*(1), 137–159. doi:10.1177/0739986311428891.

Forte, K. S. (2013). Educating for financial literacy: A case study with a sociocultural lens. *Adult Education Quarterly, 63*(3), 215–235. doi:10.1177/0741713612460267.

Gale, W. G., & Pence, K. M. (2006). Are successive generations getting wealthier, and if so, why? Evidence from the 1990s. *Brookings papers on economic activity, 1*, 155–234.

Gassoumis, Z. D. (2012). *The recession's impact on racial and ethnic minority elders: Wealth loss differences by age, race and ethnicity*. Los Angeles: USC Edward R. Roybal Institute on Aging.

Gassoumis, Z. D., Wilber, K. H., Baker, L. A., & Torres-Gil, F. M. (2010). Who are the Latino baby boomers? Demographic and economic characteristics of a hidden population. *Journal of Aging & Social Policy, 22*(1), 53–68.

Gassoumis, Z. D., Lincoln, K. D., & Vega, W. A. (2011). *How low-income minorities get by in retirement: Poverty levels and income sources*. Los Angeles: USC Edward R. Roybal Institute on Aging.

Gonzales, R. G. (2011). Learning to be illegal: Undocumented youth and shifting legal contexts in the transition to adulthood. *American Sociological Review, 76*(4), 602–619.

Green, T. K. (2003). Discrimination in workplace dynamics: Toward a structural account of disparate treatment theory. *Harvard Civil Rights-Civil Liberties Law Review, 38*(1), 91–157.

Gregoire, T. K., Kilty, K., & Richardson, V. (2002). Gender and racial inequities in retirement resources. *Journal of Women & Aging, 14*(3/4), 25–39.

Gross, J. P., Torres, V., & Zerquera, D. (2013). Financial aid and attainment among students in a state with changing demographics. *Research in Higher Education, 54*(4), 383–406. doi:10.1007/s11162-012-9276-1

Kim, J., DesJardins, S. L., & McCall, B. P. (2009). Exploring the effects of student expectations about financial aid on postsecondary choice: A focus on income and racial/ethnic differences. *Research in Higher Education, 50*(8), 741–774. doi:10.1007/s11162-009-9143-x.

Kochhar, R., Fry, R., & Taylor, P. (2011). *Wealth gaps rise to record highs between whites, blacks and Hispanics*. Washington, DC: Pew Social & Demographic Trends.

Krivo, L. J., & Kaufman, R. L. (2004). Housing and wealth inequality: Racial-ethnic differences in home equity in the United States. *Demography, 41*(3), 585–605.

McDonough, P. M., & Calderone, S. (2006). The meaning of money: Perceptual differences between college counselors and low-income families about college costs and financial aid. *American Behavioral Scientist, 49*(12), 1703–1718. doi: 10.1177/0002764206289140.

Myers, D., Calnan, R., Jacobsen, A., & Wheeler, J. (2012). *California roller coaster: Income and housing in the boom and bust, 1990–2010*. Los Angeles: Population Dynamics Research Group, USC Sol Price School of Public Policy.

Office of Management and Budget. (2014). *Budget of the U.S. Government: Fiscal year 2015 historical tables*. Washington, DC: U.S. Government Printing Office.

Pence, K. M. (2006). The role of wealth transformations: An application to estimating the effect of tax incentives on saving. *Contributions to Economic Analysis & Policy, 5*(1), 1–24.

Plath, D. A., & Stevenson, T. H. (2005). Financial services consumption behavior across Hispanic American consumers. *Journal of Business Research, 58*(8), 1089–1099. doi:10.1016/j.jbusres.2004.03.003.

RAND HRS Data, Version L (2011, December). Produced by the RAND Center for the Study of Aging, with funding from the National Institute on Aging and the Social Security Administration. Santa Monica, CA.

Restifo, S. J., Roscigno, V. J., & Qian, Z. (2013). Segmented assimilation, split labor markets, and racial/ethnic inequality: The case of early-twentieth-century New York. *American Sociological Review, 78*(5), 897–924. doi:10.1177/0003122413501071.

Ruggles, S., Alexander, J. T., Genadek, K., Goeken, R., Schroeder, M. B., & Sobek, M. (2010). *Integrated public use microdata series: Version 5.0* [*Machine-readable database*]. Minneapolis: University of Minnesota.

Rumbaut, R. G. (2004). Ages, life stages, and generational cohorts: Decomposing the immigrant first and second generations in the United States. *International Migration Review, 38*(3), 1160–1205.

Sandefur, G. D., & Pahari, A. (1989). Racial and ethnic inequality in earnings and educational attainment. *Social Service Review, 63*(2), 199–221.

Sherraden, M., Schreiner, M., & Beverly, S. (2003). Income, institutions, and saving performance in individual development accounts. *Economic Development Quarterly, 17*(1), 95–112. doi:10.1177/0891242402239200.

Shuey, K. M., & Willson, A. E. (2008). Cumulative disadvantage and black-white disparities in life-course health trajectories. *Research on Aging, 30*(2), 200–225.

Smith, J. P. (1997). Wealth inequality among older Americans. *Journals of Gerontology Series B: Psychological Sciences and Social Sciences, 52B*(Special Issue), 74–81.

Spader, J., Ratcliffe, J., Montoya, J., & Skillern, P. (2009). The bold and the bankable: How the Nuestro Barrio telenovela reaches Latino immigrants with financial education. *The Journal of Consumer Affairs, 43*(1), 56–79. doi:10.1111/j.1745-6606.2008.01127.x.

Squires, D. A. (2011). *The U.S. health system in perspective: A comparison of twelve industrialized nations* (*Publication No. 1532*). Washington, DC: The Commonwealth Fund.

U.S. Census Bureau. (2012). *Educational attainment of the population 18 years and over, by age, sex, race, and Hispanic origin: 2012* (*Tables No. 1-03 through 1-06, current population survey, annual social and economic supplement, 2012*). Washington, DC: U.S. Census Bureau.

Waters, M. C., & Eschbach, K. (1995). Immigration and ethnic and racial inequality in the United States. *Annual Review of Sociology, 21*, 419–446. doi: 10.1146/annurev.so.21.080195.002223.

Wickrama, K. A. S., Mancini, J. A., Kwag, K., & Kwon, J. (2013). Heterogeneity in multidimensional health trajectories of late old years and socioeconomic stratification: A latent trajectory class analysis. *Journals of Gerontology Series B: Psychological Sciences and Social Sciences*, *68*(2), 290–297. doi:10.1093/geronb/gbs111.

Wu, K. B. (2013). *Income and poverty of older Americans, 2011 (Fact Sheet No. 287)*. Washington, DC: AARP Public Policy Institute.

Zhan, M., & Schreiner, M. (2005). Saving for post-secondary education in individual development accounts. *Journal of Sociology and Social Welfare, 32*(3), 139–163.

Chapter 22
Present State of Elder Care in Mexico

Luis Miguel Gutiérrez Robledo, Raúl Hernán Medina Campos
and Mariana López Ortega

Introduction

Mexico is currently halfway through its demographic bonus, a so-called period of time during the demographic transition of a population when the group under age 15 is reduced to less than 30 % but the proportion of elderly is still not large enough to become burdensome (by definition, 15 %) (Hakker 2007), thus allowing for a window of opportunity to prepare economically and socially for the moment when the elderly become a large proportion of the population. However, for Mexico to take advantage of its demographic bonus, it still has a lot of work ahead.

Life expectancy in Mexico in 1895 was 24.4 years in average, and barely different between men and women (24.3 and 24.5, respectively) (Zavala 1992). It remained practically unchanged until 1930, when the average increased to 33.9 years (Instituto de Geriatría 2010). This means that Mexico began its transition at least 80 years later than Europe. However, by 1950 life expectancy was already 47.6 years and by 1980, 66.3 years. According to estimations by the National Population Council (*Consejo Nacional de Población*, CONAPO), life expectancy at birth in 2014 is 74.3 years on average, 77.5 for women and 72 for men (Consejo Nacional de Población 2013). The last estimation of life expectancy at age 65 by the same organization, dating from 2010, was 18.3 years for women and 16.8 years for men (Consejo Nacional de Población 2011). Although these numbers are still behind the leading developed countries, it is the pace of increase that must be underscored: in the last 80 years, Mexicans gained nearly 40 years in life expectancy at birth, whereas it took Europe 160 years to accomplish a similar gain from about 45 to 85 years (Oeppen 2002). In other words, Mexico started later and from behind but has gained life expectancy twice as fast as Europe. Life expectancy in Mexico is predicted to reach nearly 80 years by 2050 (Consejo Nacional de Población 2013).

L. M. Gutiérrez Robledo (✉) · R. H. Medina Campos · M. López Ortega
National Institute of Geriatrics, Mexico City, Mexico
e-mail: gutierrezrobledoluismiguel@gmail.com

© Springer International Publishing Switzerland 2015
W. A. Vega et al. (eds.), *Challenges of Latino Aging in the Americas,*
DOI 10.1007/978-3-319-12598-5_22

379

The population 60+ has grown in the past decade at a rate close to 4% and currently represents about 11% of the whole population. In two decades or less, the elderly population will reach 20 million people and by 2040, 1 out of every 4 Mexicans will be 60 years or older. Today, the aged live predominantly in urban areas (74%), are married or have a partner (60%), and have low levels of educational attainment, with 50% having only completed elementary school and 27.2% having never attended formal education (Consejo Nacional de Población 2011).

As most Latin American countries, Mexico is experiencing a "mixed" epidemiological transition with a sharp increase in prevalence of chronic diseases and a marked decrease in communicable diseases in some regions, while still suffering from moderate or high incidence of the latter in the least developed regions of the country. In addition, the country's health care system is currently not capable of responding adequately to the changing social and health care demands of an aged population. Within the country, differences among states and regions and between urban and rural areas are sharp and mainly obey to lags in economic growth which determine marked inequalities in socioeconomic development.

Data from the Organisation for Economic Cooperation and Development (OECD) reveals that, in 2010, Mexico's health expenditure as a percentage of GDP was 6.2%; of this, 47.3% came from public sources. In terms of health workforce, while the number of physicians per capita increased substantially over the past two decades reaching two practicing physicians per 1000 population in 2010, this figure still lags behind the OECD average of 3.1. As for nurses, in the same year there were 2.5/1000 population. The number of hospital beds for acute care in Mexico was 1.6/1000 population (Organisation for Economic Cooperation and Development 2013).

Morbidity and Mortality Profile in Elderly Mexican People

As in the rest of the world, women outnumber men at old age in Mexico. There are 112 females for every 100 males in the 60–74 age group, and among the oldest old (85+) there are 135 females for every 100 males. Furthermore, the group of older women is expected to continue to grow, since females aged 40–59 are now a large group of the population, at 13.1 million (11% of the total) (Consejo Nacional de Población 2013). Considering that functional dependence affects predominantly women, this portends that if sanitary programs and policies do not focus on chronic care and avoidance of functional dependence, the care for this group of elders-to-be will be neither attainable nor sustainable, and acute services will be overburdened with them. On the other hand, aging males face their own particular challenges, such as lower survival, higher migration rates and greater rates of violent deaths resulting from accidents, suicides and addictions (Torres 2010).

According to official Health Ministry records, the leading five causes of hospital-bound morbidity in persons 65+ in Mexico are heart diseases (including ischemic heart disease, hypertension and others), accidents (mainly fractures), diabetes mellitus, cancer and kidney disease (Dirección General de Información en Salud 2014). Multimorbidity is also an issue, since 68.6% of the Mexican population 60+ has at least one chronic disease, and 34.1% has two or more chronic diseases, according to the seven-country survey on Health, Well-being and Aging in Latin America (*Salud, Bienestar y Envejecimiento*, SABE) (Menéndez 2005).

Regarding mental health, the Mexican Health and Aging Survey (MHAS), a population-based, longitudinal study of community-dwelling elderly in Mexico, revealed the prevalence of depressive symptoms to be 37.9% (Ávila 2007), of cognitive decline 28.7% and of Alzheimer's disease 6.1% among persons 60+; the latter's incidence was estimated at 27.3/1000 persons year for the same age group (Mejía 2011).

The 2012 National Survey on Health and Nutrition (*Encuesta Nacional de Salud y Nutrición*, ENSANUT), showed that 34.9% of the people 60+ had had at least one fall in the previous year. Meanwhile, 26.9% of the elderly reported having at least one disability in activities of daily living, with those over the age of 80 having the greatest prevalence (47.5%). Accordingly, the elderly are the group with the highest rate of health care use. In particular, among the oldest old (80+) the rate of health care use rises up to 15.7/100 population, compared to 14.7/100 population between 0 and 4 years—the second largest group of health care users. The main causes for medical consultation in people 70 years and over are diabetes, cardiovascular disease and obesity, which pooled represent 33% of all consultations, followed by acute respiratory diseases (12.9%) (Instituto Nacional de Salud Pública 2013). In spite of this heavy use of health care services by the elderly, providers are in general lacking geriatric expertise, which results in poor quality of care for the main users. In turn, this leads to poor control of chronic diseases and functional decline.

The mortality profile has also changed substantially during the last century: in the 1920s the Mexican population was decimated by war, flu epidemics and migration, so that only 5% of the population reached age 75. In contrast, it is estimated that nowadays 75% of the population will reach age 75 (Camposortega 2014).

A total of 587,826 deaths were registered as occurring in 2012, of which 364,704 (62%) were people aged 60+, the main causes being ischemic heart disease, diabetes and cancer; only 5.7% of the total mortality occurred in children under age 5. As for the causes, 3% of the deaths were attributed to communicable causes (codes A and B from the International Classification of Diseases, tenth edition [ICD-10]), while 85% were caused by non-communicable diseases (ICD-10 codes C through R) and the remaining 12% were originated by external injuries (ICD-10 codes S through Z) (Dirección General de Información en Salud 2014).

An analysis of mortality in elderly people accounting for the Social Inequalities Index (SII) found no difference in causes of death for age or sex. However, there were important socioeconomic differences: elderly people living in small villages and in underprivileged conditions (in terms of the SII) were found to suffer from a higher number of deaths from malnutrition and unknown causes. Of note, "senility" may have been a cause of death in up to 6% of all deaths in people 85+ (Velasco, Personal communication).

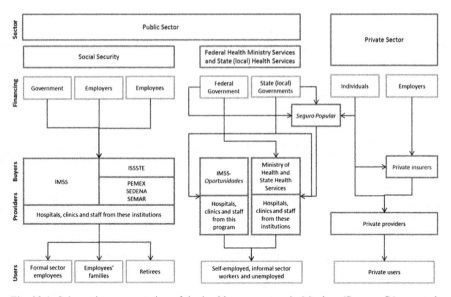

Fig. 22.1 Schematic representation of the health care system in Mexico. (Source: Gómez et al. 2011)

Health Care System

Health protection is granted as a right to all Mexicans by the fourth article of the Constitution. In theory, the older population has the same access to health care provision and services as the rest of the population through the National Health System. However, several issues make access unequal and hinder the achievement of universal coverage that the system strives for.

The National Health System has been highly fragmented since its creation. Health services and users are divided according to the health institution that provides the service (Fig. 22.1). There are three main providers: (a) social security institutions, (b) public services offered by the Ministry of Health, and (c) the private sector. They offer different service packages, work independently and parallel to each other and are financed through different funding sources (Gómez et al. 2011).

Social security is further divided into a number of institutions that provide services to workers from different sectors. The Mexican Institute of Social Security (*Instituto Mexicano del Seguro Social*, IMSS) covers nongovernment employees working in the formal sector of the economy. Government employees are covered by the Institute of Social Security and Services for State Employees (*Instituto de Seguridad y Servicios Sociales para los Trabajadores del Estado*, ISSSTE) and other institutions for specific sectors such as oil company workers (*Petróleos Mexicanos*, PEMEX), military (*Secretaría de la Defensa Nacional*, SEDENA) and naval (*Secretaría de Marina*, SEMAR) officers and employees of the state (local) governments. Services offered by these institutions extend to the worker's spouse, children and parents (Gómez et al. 2011).

The recently introduced Popular Health Insurance (*Seguro Popular*) is an income-based health care insurance administered by the Ministry of Health which provides coverage to those not insured by any of the social security institutions, including people working in the informal sector and unemployed persons (e.g. homemakers). *Seguro Popular* provides financial protection for more than 50 million Mexicans and is improving access to health services and reducing the prevalence of catastrophic and impoverishing health expenditures, especially for the poor.

According to the 2010 census, 64.55% of the whole population declared having health insurance of any kind, while 33.8% remain uninsured. *Seguro Popular* accounts for 36.1% of the insured population. The two main social security institutions (IMSS and ISSSTE) account for 48.8 and 9.9% respectively, while private insurance companies account for 2.3% of the insured population (Instituto Nacional de Estadística y Geografía 2011).

Access to health care in Mexico is still linked for the most part to affiliation to any of the social security institutions, and is thus linked to having a job in the formal sector. This is particularly troublesome in face of Mexico's aging population, because 34.3% of the elderly continue to work, and more than a third of them (37.4%) do so in the informal sector, with no access to social security. In total, 27.4% of the population 60+ remains uninsured (Instituto Nacional de Estadística y Geografía 2011).

Long-Term and Social Care

A formal, long-term care system aimed at the elderly is lacking in Mexico. Some federal and state programs cater for older adults, most of which are responsibility of the Ministry of Social Development and other state agencies. Additionally, there are several private providers (both for-profit and not-for-profit) offering services such as daycare and institutionalization.

A few social services are available through the National Institute for Older Persons (*Instituto Nacional para las Personas Adultas* Mayores, INAPAM) and the National System for Integral Family Development (*Sistema Nacional para el Desarrollo Integral de la Familia*, DIF), a public family-welfare institution with federal and state agencies. INAPAM offers career placement fairs, training, daycare and cultural centers, shelters, socio-cultural activities, health education, psychological and educational services, as well as training for people 65+ to become "certified elderly caretakers" (Secretaría de Desarrollo Social 2014). DIF has a daycare program (*Programa de Atención de Día*) which provides services for a small number of individuals. The supply at state level varies according to the local needs, the available resources or the commitment of the local governments with this age group (Sistema Nacional para el Desarrollo Integral de la Familia 2014).

Most of the government-provided services are an extension of poverty reduction programs or other social care strategies led by the Ministry of social development.

Three specific poverty reduction strategies are on place that have focused or extended their services towards the older population. First, *Programa 65 y más* (65+ Program, formerly 70+), by the Ministry of Social Development, provides a non-taxed fee roughly equivalent to US$ 40 to people 65+ who are not claimant of *Oportunidades* (see below). Its objectives are to provide economic support and social participation activities organized by local authorities (Secretaría de Desarrollo Social 2013).

The second strategy is also supported by the Ministry of Social Development through the extension of the *Oportunidades* (Opportunities) program in order to support individuals 70 years and older who are members of affiliated families. *Oportunidades* is a wide-scoped, comprehensive, human development program aimed at reducing poverty, malnutrition and poor health in rural areas of the country, including money transfers, alimentary aid, education, and medical consultations. Families receive a monthly non-taxed allowance which varies depending on a number of features, such as number of children and the school grade they are in, up to a maximum of US$ 205 per family. The extension for families with older adults consists of an additional US$ 25 per older adult. Participants are required to attend a bimonthly medical consultation in order to get the support (Secretaría de Hacienda y Crédito Público 2013).The third strategy refers to the System of Social Protection in Health, provided through *Seguro Popular*. Its main objective is to ensure effective access to timely, quality health care without discrimination and free at the point of care (Ley General de Salud 2014). Out-of-pocket payments at the point of care are reduced by establishing an annual pre-payment fee for each insured family. The two lowest income brackets of the population are exempt from this fee. The benefits package for men and women 60+ covers diverse areas such as vaccination (anti-influenza and anti-pneumococcal), mental health, oral health, osteoporosis, prevention and control of tuberculosis, and diagnosis of diabetes, hypertension and obesity (Comisión Nacional de Protección Social en Salud/Seguro Popular 2012).

Some initiatives have also been developed locally by the state governments. The first and oldest of these initiatives began in Mexico City in 2001 as a comprehensive program of food support, health care and free medications, but has been modified in the course of the years. In its current form, the Alimentary Pension Program for Older Adults (*Programa de Pensión Alimentaria Adultos Mayores de 68 años*) provides a non-taxed, non-contributory, US$ 66 pension estimated to have reached 100% coverage of adults 68 years and older living in Mexico City in 2012 (Instituto para la Atención de los Adultos Mayores en el Distrito Federal 2013). The pension is provided through an electronic card that can be used in most commercial establishments in the city for purchasing food, medications and other goods.

On March 18, 2014, a bill in the Congress was voted that will assign a monthly pension of roughly US$ 43 to all those persons 65+ who are not covered by any of the aforementioned programs (Nieto 2014).

Within the private insurance sector, long-term care insurance schemes have begun to emerge but little is known about premiums, coverage and population insured. Given the fact that these schemes are expensive, they are only affordable to a small fraction of the population.

As for institutionalization options, there are currently few public institutions for older adults who are dependent because of medical or mental health problems. A larger number of private institutions exist, offering day and institutional care, as well as a wide range of community services and activities. Unfortunately, a national registry that gathers information and monitors all public and private institutions is lacking.

The current legal framework related to health and social care is extensive but does not enhance or regulate long-term care in itself. Most of the regulatory documents are Official Mexican Standards (*Normas Oficiales Mexicanas*, NOMs), which are a type of regulatory document with a lower hierarchy than a law. Some NOMs regulate medical ambulatory or hospital care for the ill and disabled, but only one of them is aimed at regulating social care for older adults at risk or in vulnerable situations, as defined by lack of family, family rejection, physical or psychological abuse, or lack of economic resources (Secretaría de Salud 2012). Regulatory laws are urgently needed in order to standardize care, and the services provided by public and private institutions.

Few studies have evaluated the quality of the institutions and/or the impact they have on older adults' wellbeing. One study concluded that quality of services provided is unacceptable both in private (profit and nonprofit) and public institutions, due to many factors including unqualified personnel, inadequate facilities and unsafe conditions, to name a few. The authors point out the lack of adequate normative framework and the lack of supervision as the cause and consequence of the increase in the number of institutions providing inadequate or substandard care (Gutiérrez et al. 1996).

Informal Care

Most of the elderly live at home with their spouse or partner, children, grandchildren or other close relatives. The family provides all or most of the care. No special benefits for family caregivers are available, neither fiscal nor social security-wise. Within the families, the responsibility of care often falls upon female members, which in turn limits women's opportunities at professional or personal development. Nevertheless, reduced fertility rates, national and international migration, women's increasing participation in the workforce and activities outside of the household, among other factors, have changed family size and composition and pose future challenges to the availability of household care and support.

A country-wide study on care and its associated burden was lacking until 2012, when the Survey on Work and Shared Responsibility (*Encuesta Laboral y de Corresponsabilidad Social*, ELCOS) was undertaken by INEGI and the National Institute for Women. While not specifically focused on care of the elderly, ELCOS's main objective was to produce indicators on care provided at home, both by household members and external persons, in order to determine whether women are overburdened by caregiving and whether this overburden represents an obstacle to their insertion in the workforce. Noticeably, this survey is only representative of 32 urban areas (population > 100,000), including the three major metropolitan areas: Mexico City, Guadalajara and Monterrey.

An official report from ELCOS has yet to be published, but preliminary results posted on the survey's official website already reveal some interesting insights into the care needs and caregiving phenomenon occurring in Mexican households. The total number of persons needing care and/or support in urban areas is 12,061,361 (25.1% of the surveyed population). Of these, 86.6% are under age 15, 9.9% have temporary health conditions and 6.1% have permanent limitations. Out of those who reported having permanent limitations, 468,926 (63%) were older adults, and of these, 63% were women and 88.1% needed permanent care or support from another person (Instituto Nacional de Estadística y Geografía 2014).

In spite of the fact that most of the care in Mexico is provided at home, evaluations of the health system did not take it into account. From 2010 onwards, the National Health Accounts includes information regarding informal care in the Health Sector Satellite Account. In 2011, the total health expenditure (THE) was equivalent to 5.6% of the national gross domestic product. Unpaid health care provision accounted for 20.5% of the THE, while the public sector accounted for 37.9% and the private sector for 41.6%. Unpaid health care provisions is almost equivalent to the expenses in hospital services (24.4% of THE) and more than double the expenses in medications and other care goods (9.3% of THE) (Instituto Nacional de Estadística y Geografía 2013). Although the unpaid health care is not disaggregated by age of the care recipient or type of care provided, it can be assumed that a significant proportion of it is destined to the elderly.

Geriatric Health Care Workforce and Facilities

There is a serious shortage of human resources in health care, spanning not only actual care provision but also teaching and research. At present, the workforce dedicated to caring for the elderly is not enough and the geriatric skills of general practitioners are still suboptimal, even with most medical schools now including undergraduate training programs on geriatrics and/or gerontology as a part of their undergraduate medical curricula, including the National Free University (*Universidad Nacional Autónoma de México*, UNAM). More physicians and other health professionals need to be trained in geriatrics, and there is a need for more social scientists, policy experts, biologists and health practitioners to be trained in gerontology.

The first geriatric medicine residency program in Mexico was established in 1994. Currently, there are approximately 500 certified geriatricians. As of August 2010, seven Mexican universities offered specialization in geriatrics, a 2-year training program requiring 4 years of previous training in internal medicine (Ávila 2012). However, the slot occupation in these geriatric residency programs was only 70%. In response to this situation, a new, direct entry, 4-year geriatric medicine program was established in 2013 and is set to begin its first academic cycle in March, 2014.

For this program, 58 positions were created and funded by the Ministry of Health, with an occupation rate of 100 % (Secretaría de Salud 2014).

High-specialty programs are also available for geriatricians and other specialists to further specialize in geriatric cardiology, dementia, geriatric neurology and geriatric rehabilitation. Furthermore, a Master of Public Health program with focus on Aging was established in cooperation between the National Institute of Public Health (*Instituto Nacional de Salud Pública*, INSP) and the National Institute of Geriatrics (Instituto Nacional de Geriatría, *INGER*).

In spite of recent efforts for training geriatricians, considering the most conservative recommendations, Mexico would need to produce more than 2,000 geriatricians in the following 10 years to meet the needs of its population (Ávila 2012). Clearly, the current rate of production of human resources is insufficient.

As of 2010, Mexico had 21 geriatrics departments in general hospitals, most of them located in Mexico City. There are also some specialized geriatric services such as memory clinics, a geriatric cardiology ward and a geriatric rehabilitation ward (Ávila 2012).

Instituto Nacional de Geriatría

Mexico's National Institute of Geriatrics (*Instituto Nacional de Geriatría*, INGER) was created in 2008, with the purposes of promoting research on aging, training specialized human resources in elderly care, developing models of care for the elderly and innovating in the field of public policies for the aged population.

The Institute's Research division works on a number of lines, including biological mechanisms of aging, socioeconomic aspects of aging, clinical geriatrics, epidemiology and gerontechnology. There is also a line of health care systems research and development that covers different levels of care (emergency, acute elderly care, primary care, telemedicine). Additionally, in 2011 the Institute created the Aging, Health and Social Development Research Network (*Red Temática de Envejecimiento, Salud y Desarrollo Social*, ESDS). To date, the ESDS has accrued 150 researchers focused on different aspects of aging throughout the country.

In line with its purpose of innovating in public policies for the aging population, the INGER has developed and proposed an Action Plan on Aging and Health together with the UNAM and the Mexican National Academy of Medicine (Gutiérrez and Kershenobich 2012). Published in 2012, this position statement together with an epidemiological report published in 2013 with data from the 2012 ENSANUT survey, led to the introduction of a specific strategy for healthy and active aging in the government's Health Sectorial Program, which will direct the public health sector for the 2013–2018 period (Presidencia de los Estados 2013). Similarly, in collaboration with other institutions, the INGER has started a working group on Dependence and Aging, which is about to publish its first position statement with the main purposes of bringing dependence to the public policy arena and promoting the development of a long-term care system in Mexico. At the same time, the INGER is on its way to develop a new model of care for the elderly based on primary care. Finally, a new model of age-friendly primary care centers has been jointly developed by the Ministry of Health and the INGER and is soon to be released.

Another major division of the INGER is Education and Knowledge Translation, now working in the development of a large human resources network dedicated to elderly care. This project has been named FORHUM3, because it is a triple alliance program. In addition, the Institute is building its Virtual Library with a specific focus on age and aging.

Last, but not the least, the INGER is committed to promoting positive attitudes towards aging and the aged. It is also interested in promoting intergenerational relationships and social engagement in later life. From this point of view, it strives for the elimination of cumulative disadvantage across the lifespan in order to remove barriers to participation in late life. In Mexico, aging is almost always associated with negative notions of loss, decline and decay. The INGER contributes to convey the idea that many people are still able to live healthy, active and productive lives well into old age.

Preventive Health Approaches

Recent evidence indicates that the extension of healthy life expectancy is achievable, even at old age. With the knowledge that interventions already exist to delay the onset and progression of major fatal and disabling diseases, a preventative strategy with a life course perspective has been introduced in the Mexican public health policies (Presidencia de los Estados 2013).

Across the lifespan, mental health problems such as depression and dementia compromise physical health and quality of life alike. They often go unnoticed or misdiagnosed and may interfere with the effective treatment of other conditions, for example, by compromising therapeutic compliance or hampering help-seeking behaviors. They are also associated with suicide, which is increasingly becoming a public health issue. Structured interventions for depression and dementia have proven to be even more effective than psychoactive drugs, especially in the long term. Accordingly, the INGER is pursuing research and development of such interventions, aiming to implement and Action Plan on Alzheimer's disease and related disorders.

A Look Ahead: The Elderly as Drivers of Policy Making

Policy making is inevitably influenced by political strategy, which is in turn strongly driven by trends in public opinion. Historically, women's health and children's health have dominated the health policy global scenario, mostly as a long-standing consequence of the baby boom and the inequities in the social determinants of health beginning from an early age, but also very likely due to the fact that most of the voting population was made up of young people of child-bearing and child-rearing ages. As the population grows old, however, the group of older people will become a major source of voters, which will undoubtedly attract the attention of decision makers and politicians. Thus, the so-called grey vote is likely to become decisive in the democracies of the near future around the world.

In Mexico, according to the National Institute for Elections (*Instituto Nacional Electoral*), as of May 9, 2014 the nominal list of voters (the list which is used to identify voters and register their votes by electoral district) contained 78.84 million people, of which 12.12 million (15.3%) were 60+ years old (Instituto Nacional Electoral 2014). This means that the elderly currently represent 15.3% of the voting population in the country.

An example of the power of the grey vote has already occurred in the capital of the country. In 1997, a left-wing party won the elections for Mexico City Major. Under this government, the approach to the elderly population shifted from regarding it as a vulnerable group to regarding it as a prioritary one, and a law for the rights of the elderly was passed in 2000 which paved the way towards the introduction of the previously discussed non-contributory pension for the elderly in 2001 (López 2003). With this and several other progressive social policies that have been widely welcomed by the general population, it is not surprising that the same leftist party has maintained the rule over the city for 17 years (that is, three elections for city major in a row after the introduction of the pension). In Mexico City, the elderly currently represent 18.2% of the voting population.

It is reasonable to anticipate that the elderly will become major drivers of policy making not only as a sheer source of votes, but also as major players in society. As the population ages, more and more people are reaching old age in a better health, which means they are able to participate more in society. It is only a matter of time before they begin to organize into civil society organizations and other associations of the like with the purpose of bringing forward their interests and pushing for their rights.

Conclusions

Mexico's demographic transition is nearly complete, and the aging of its population is beginning to pose stress on the health care and social development systems, as well as on the families and the society in general. The country is largely unprepared to respond to the needs of its aging population. The absence of formal care and support services for the elderly has left informal caregivers, mostly women, overburdened with the responsibility of providing household care throughout the life course. Although the lack of strategies for the older population has not yet originated a major crisis in terms of service provision and financing, presumably due to the support capacity of the families and social networks, the expected growth of the elderly population will further strain households and institutions alike.

Public care offer for the elderly in Mexico is limited, and little is known about the extant private care offer. In any case, the latter is small and informal in nature and there is no real way to assess the quality of the care provided. Elderly care is still a gray area lacking regulation, information and evaluation. Even public institutions which provide some form of care do not conduct comprehensive monitoring and evaluation of their activities.

If Mexico wants to ensure quality care for its present and future generations of elderly people, long-term care, social care and improvement of health care are long overdue priorities that should be addressed immediately. Unfortunately, very little effort is being done in this direction at present. There is still some reluctance in Mexico to include an aging perspective into decision making and policy design, and a tendency to prioritize the perspectives of younger stages of life. However, as the population grows older the grey vote may become a major electoral force, which will hopefully compel politicians and decision makers to turn their looks towards the needs and rights of the elderly. An aging perspective needs to be included not only in health and social issues, but in general all across public policies aimed at well-being and development. Similarly, more spaces for the elderly to effectively participate and engage in civil society need to be made available nation-wide.

There is much that can and must be done today to improve elderly care in Mexico. Whatever the country's answer to this challenge, it will have to include a new interpretation of the elderly as a priority and not a vulnerable group. Furthermore, a health and social agenda is in order that regards care as a personal, civil right and no longer as a mere feature of welfare.

Acknowledgement The authors would like to express their gratitude to Elizabeth Caro López for her valuable advice and guidance in assessing the role of the elderly population as voters and drivers of decision making in public policy.

References

Ávila, F. J. A., Melano, C. E, Payette, H., & Amieva, H. (2007). Síntomas depresivos como factor de riesgo de dependencia en adultos mayores. *Salud Pública Mex, 49*, 367–375.

Ávila-Fematt, F., & Montaña-Álvarez, M. (2012). Enseñanza de la Geriatría en México. In D. Kershenobich-Stalnikowitz & L. M. Gutiérrez-Robledo. *Envejecimiento y salud: una propuesta para un plan de acción* (pp. 287–319). Ciudad de México: Universidad Nacional Autónoma de México.

Camposortega, S. (2014). Cien años de mortalidad en México. http://www.ejournal.unam.mx/dms/no10/DMS01005.pdf. Accessed 14 March 2014.

Comisión Nacional de Protección Social en Salud/Seguro Popular. (2012). *Catálogo Universal de Servicios de Salud 2012*. México: Secretaría de Salud.

Consejo Nacional de Población. (2011). *La situación demográfica de México 2011*. México: CONAPO.

Consejo Nacional de Población. (2013). Proyecciones de la Población Nacional 2010–2050. http://www.conapo.gob.mx/es/CONAPO/Proyecciones_de_la_Poblacion_2010–2050. Accessed 14 March 2014.

Dirección General de Información en Salud. (2014). Base de datos de egresos hospitalarios por morbilidad en Instituciones Públicas, 2004–2012. Sistema Nacional de Información en Salud (SINAIS), Secretaría de Salud, México. http://www.sinais.salud.gob.mx. Accessed 14 March 2014.

Gómez, D. O., Sesma, S., Becerril, V. M., Knaul, F. M., Arreola, H., & Frenk, J. (2011). Sistema de Salud de México. *Salud Pública Mex, 53*(suppl 2), S220–S232.

Gutiérrez, R. L. M., & Kershenobich, S. D. (2012). Envejecimiento y Salud: una Propuesta para un Plan de Acción. México: Academia Nacional de Medicina de México/Academia Mexicana de Cirugía/Instituto de Geriatría/Universidad Nacional Autónoma de México.

Gutiérrez, R. L. M., Reyes, O. G., Rocabado, Q. F., & López, F. J. (1996). Evaluation of long-term care institutions for the aged in the Federal District. A critical viewpoint. *Salud Publica Mex, 38*(6), 487–500.

Hakker, R. (2007). *The demographic bonus and population in active ages*. Brasilia: Brazilian Institute for Applied Economics/UNFPA.

Instituto de Geriatría. (2010). *Envejecimiento humano: Una visión transdisciplinaria*. México: Secretaría de Salud.

Instituto Nacional de Estadística y Geografía. (2011). *Principales resultados del Censo de Población y Vivienda 2010*. México: INEGI.

Instituto Nacional de Estadística y Geografía. (2013). *Sistema de Cuentas Nacionales de México: Cuenta Satélite del Sector Salud de México 2008–2011*. México: INEGI.

Instituto Nacional de Estadística y Geografía. (2014). Encuesta Laboral y de Corresponsabilidad Social (ELCOS). http://www.inegi.org.mx/est/contenidos/Proyectos/Encuestas/Hogares/especiales/ELCOS/. Accessed 19 March 2014.

Instituto Nacional de Salud Pública. (2013). *Encuesta Nacional de Salud y Nutrición 2012. Resultados Nacionales*. Cuernavaca: Instituto Nacional de Salud Pública.

Instituto Nacional Electoral. (2014). Instituto Nacional Electoral. http://www.ife.org.mx/portal/site/ifev2/Estadisticas_Lista_Nominal_y_Padron_Electoral/. Accessed 16 May 2014.

Instituto para la Atención de los Adultos Mayores en el Distrito Federal. (2013). *Evaluación interna del Programa de Pensión Alimentaria para Adultos Mayores de 68 años, Gestión 2012*. México: Gobierno del Distrito Federal.

Ley General de Salud. (2014). Cámara de Diputados. http://www.diputados.gob.mx/LeyesBiblio/pdf/142.pdf. Accessed 19 March 2014.

López, C. E. (2003). Nuevas políticas para adultos mayores: el caso del Distrito Federal—de la asistencia a la participación social. *Revista de Administración Pública, 109, 109*(38), 75–86.

Mejía, A. S., & Gutiérrez, R. L. M. (2011). Prevalence and incidence rates of dementia and cognitive impairment no dementia in the Mexican population: Data from the Mexican health and aging study. *J Aging Health, 23*(7), 1050–1074.

Menéndez, J., Guevara, A., Arcia, N., León, D. E. M., Marín, C., & Alfonso, J. C. (2005). Enfermedades crónicas y limitación funcional en adultos mayores: estudio comparativo en siete ciudades de América Latina y el Caribe. *Rev Panam Salud Publica, 17*(5/6), 353–361.

Nieto, F., & Jiménez, H. (2014). Avanzan pensión y seguro laboral. *El Universal*. http://www.eluniversal.com.mx/nacion-mexico/2014/impreso/avanzan-pension-y-seguro-laboral-214067.html. Accessed 19 March 2014.

Oeppen, J., & Vaupel, J. W. (2002). Broken limits to life expectancy. *Science, 296*, 1029–1031.

Organization for Economic Cooperation and Development. (2013). *Health at a glance 2013: OECD indicators*. Paris: OECD.

Presidencia de los Estados Unidos Mexicanos. (2013). Decreto por el que se aprueba el Programa Sectorial de Salud 2013–2018. *Diario Oficial de la Federación*. Morning Ed., Third Section, Dec 12.

Secretaría de Desarrollo Social. (2013). Acuerdo por el que se emiten las Reglas de Operación del Programa de Pensión para Adultos Mayores, para el ejercicio fiscal 2014. *Diario Oficial de la Federación*. Morning Ed., Second Section, Dec 29.

Secretaría de Desarrollo Social. (2014). Instituto Nacional de las Personas Adultas Mayores. http://www.inapam.gob.mx/es/INAPAM/Servicios. Accessed 19 March 2014.

Secretaría de Hacienda y Crédito Público, Secretaría de Desarrollo Social, Secretaría de Educación Pública, Secretaría de Salud, Instituto Mexicano del Seguro Social, Coordinación Nacional del Programa de Desarrollo Humano Oportunidades. (2013). Acuerdo por el que se emiten las Reglas de Operación del Programa de Desarrollo Humano Oportunidades, para el ejercicio fiscal 2014. *Diario Oficial de la Federación*. Morning Ed., Sixth Section, Dec 30.

Secretaría de Salud. (2012). Norma Oficial Mexicana NOM-031-SSA3–2012, Asistencia Social. Prestación de servicios de asistencia sociala a adultos y adultos mayores en situación de riesgo y vulnerabilidad. *Diario Oficial de la Federación*. Morning Ed., Fourth Section, Sept 13.

Secretaría de Salud, Secretaría de Educación Pública. (2014). Comisión Interinstitucional para la Formación de Recursos Humanos en Salud. *XXXVII Examen Nacional para Aspirantes a Residencias Médicas*. http://cifrhs.salud.gob.mx/. Accessed 19 March 2014.

Sistema Nacional para el Desarrollo Integral de la Familia. (2014). DIF Nacional, Servicios. http://sn.dif.gob.mx/servicios/. Accessed 19 March 2014.

Torres, A. LdelP., & Villa, B. J. P. (2010). Consideraciones sobre el envejecimiento, género y salud. [book auth.] Instituto de Geriatría. *Envejecimiento Humano: una Visión Transdisciplinaria*. s.l.: Secretaría de Salud.

Zavala de Cosío, M. E. (1992). Los antecedentes de la transición demográfica en México. *Historia Mexicana*. México: El Colegio de México, 42(1):103–128.

Chapter 23
Pension Reform, Civil Society, and Old Age Security in Latin America

Ronald J. Angel and Javier Pereira

Pension Reform, Civil Society, and Old Age Security in Latin America

Although the number of individuals who survived to old age in previous centuries was larger than one might imagine, it is only in modern times that individuals in the developed world routinely live to be 70, 80, or even older (Bharmal et al. 2012; Easterlin 2000; Elo 2009; Morbidity and Mortality Weekly Report 1999; Preston 1996; Thane 2005; Wilmoth 1998). During the twentieth century dramatic increases in life expectancy occurred in both developed and developing nations as the result of interacting economic, political, and social changes that reduced mortality at all ages (Elo 2009; Link and Phelan 1995; Phelan et al. 2010). This demographic transition, which has occurred at a far faster rate in developing nations than was the case in developed nations, has profound implications for individuals, families, and for societies at large. In this paper we examine some of the consequences of longer lives on the tie between generations in the context of radical pension reforms in Latin America, focusing specifically on Argentina, Chile, Mexico and Uruguay, nations with different pension systems, but that all introduced fully or partially privatized pension systems. In what follows we discuss the concept of retirement, and reflect on the potential impact of various options in the support of the elderly for intergenerational solidarity and for national economies. We end with a discussion of the potential role of civil society organizations, including non-governmental organizations (NGOs) and faith-based organizations (FBOs), in complementing state agencies in providing the support to the elderly.

R. J. Angel (✉)
The University of Texas at Austin, Austin, TX, USA
e-mail: rangel@austin.utexas.edu

J. Pereira
Universidad Católica del Uruguay, Montevideo, Uruguay

© Springer International Publishing Switzerland 2015
W. A. Vega et al. (eds.), *Challenges of Latino Aging in the Americas,*
DOI 10.1007/978-3-319-12598-5_23

Who Can Afford to Retire?

Retirement is a relatively new social phenomenon and stage in the life course. For most of human history older individuals have worked until they became too frail to do so, and then found themselves dependent on family or charity. As children migrated from rural areas to the city, or from Europe to the New World, the economic situation of many older individuals who were left behind became precarious. When they became infirm and were no longer able to support themselves, their fate was often the almshouse (Goose and Looijesteijn 2012; Thane 2005). Given low levels of productivity, few opportunities for individual or collective saving existed. In previous centuries in the developed world, as in much of the developing world today, a period at the end of the life course during which a physically capable adult could stop working and live off accumulated wealth or the efforts of others was a privilege confined to the upper classes.

That reality began to change slowly during the nineteenth century. In 1889 Otto Von Bismarck, Chancellor of the newly unified Germany, introduced pensions for all unemployed citizen 65 or older. Bismarck may have been motivated less by a social conscience than by a desire to blunt the appeal of Marxism among the seriously exploited working classes, but whether keeping Socialism at bay was the basic motivation for the old age welfare state or whether it emerged from the basic logic of capitalism, retirement increasingly became a new phase of the life course. During the twentieth century other nations, including most Latin American nations, introduced retirement schemes for workers in their formal sectors, the military, and workers in specific occupations (Kritzer et al. 2011; Mesa-Lago 2008; Müller 2000; Rofman and Oliveri 2011). These pension schemes provided what has been termed a "retirement wage", referring to a secure, if minimal, income that allows a worker to exit the labor force (Myles 1983, 2002). In developed nations retirement schemes have become an integral part of the welfare state. Even as fiscal crises force nations to tinker with various aspects of those systems to insure solvency and long-term sustainability, the basic commitment to old-age security remains firm, and for the most part coverage is universal.

In middle-income countries, on the other hand, although retirement systems are common, their coverage has remained more limited (Kritzer et al. 2011a; Mesa-Lago 2008). High rates of labor force informality, along with other problems with the structure and administration of plans reduce participation rates and the level of contributions and leave large segments of those populations without access to a retirement wage. The situation of these individuals is not much different than that of low-wage workers before Bismarck's time. Even with these limitations in coverage, though, and perhaps because of them, retirement systems and governments in Latin America are confronted by serious fiscal crises and problems with long-term sustainability that are exacerbated by the rapid population aging (CEPAL 2007; Cotlear 2011).

The phrase commonly used to characterized traditional pension systems that are paid for through payroll or general taxes is pay-as-you-go, a phrase which refers to the fact that payments to retirees come directly from the contributions of those currently employed. Such a system can function as long as the number of retirees relative to the number of workers remains small. When the number of retirees relative to those still working increases, the system becomes strained. One clear option is the adoption of individual retirement accounts in which one's contribution is invested in a personal savings or investment account rather than going to current retirees (Bertranou et al. 2009; Orenstein 2008, 2011). Such a system ties one's retirement security directly to one's own effort and essentially severs the solidary tie between generations. Nations can obviously shift completely from a publicly funded scheme to individual fully-funded accounts, or adopt some combination of the two. In most cases, some public provision for individuals with very low income or those in the informal sector who do not contribute to such plans must be made.

The Rise of Private Retirement Accounts

The shift to individual retirement accounts began in Chile in 1981 when that nation replaced its public pension system with a system based on individual retirement accounts. The new system was the brainchild of José Piñera, a Harvard educated economist who served as Chile's Secretary of Labor and Social Security from 1978 to 1980. In an article published in the *Cato Journal* in the mid-1990s Piñera summarized the essence of the new philosophy by noting ironically that,

> [a] specter is haunting the world. It is the specter of bankrupt state-run pension systems. The pay-as-you-go pension system that has reigned supreme through most of this century has a fundamental flaw, one rooted in a false conception of how human beings behave: it destroys, at the individual level, the essential link between effort and reward-in other words, between personal responsibilities and personal rights. Whenever that happens on a massive scale and for a long period of time, the result is disaster…. Two exogenous factors aggravate the results of that flaw: (1) the global demographic trend toward decreasing fertility rates; and, (2) medical advances that are lengthening life. As a result, fewer and fewer workers are supporting more and more retirees. Since the raising of both the retirement age and payroll taxes has an upper limit, sooner or later the system has to reduce the promised benefits, a telltale sign of a bankrupt system (Piñera 1995, p. 155).

Rather than fostering solidarity, Piñera believes that traditional pay-as-you-go systems undermine intergenerational relations as fewer and fewer workers are forced to support a growing number of retirees. He also argued that traditional social security systems do not redistribute income to the poor, but actually benefit privileged workers. Ultimately, in Piñera's view, traditional social security systems create insecurity rather than reducing it. Clearly, this view is very different than that held by the supporters of traditional pensions, but it is a widely-held perspective that is based on very real demographic and fiscal considerations. What seems clear is that this new dominant philosophy has potentially profound social consequences that one can view as either positive or negative depending on one's personal philosophy.

Piñera's arguments and reasoning were forceful and influenced many others all over the world (Orenstein 2008). During the 1990s, ten other Latin American countries either replaced their public pension systems entirely, e.g., Mexico, or introduced a mixed system of public and private plans, e.g., Uruguay (Bertranou et al. 2009; Mesa-Lago 2008). In addition to severing the bond of dependence between generations, the shift to private retirement accounts potentially undermines the redistributive component of traditional pensions in which public funding transfers income from those with higher earnings to those with lower earnings, a large fraction of whom are women (Barrientos 2006).

The speed with which nations adopted individual accounts reflected both the realization of the fiscal crisis that pay-as-you-go schemes faced, but also aggressive promotion by the World Bank, the International Monetary Fund, and a number of other actors that make up what Mitchell Orenstein has characterized as a transnational activist network (Orenstein 2008, 2011). One of the objectives of these reforms was the extension of coverage and the reduction in informality in the labor force (World Bank 1994). Yet in most nations coverage dropped after reforms were introduced (Kritzer et al. 2011; Mesa-Lago 2008), and the level of informality did not decrease (Tornarolli et al. 2012). The clear result of the reforms, though, is a vastly changed retirement environment with implications for intergenerational solidarity, equity, and national accounts. As we explain below, serious shortcomings with the original privatization plans have led to a second round of reforms designed to address some of these shortcomings.

Individualizing Risk

Before proceeding to our case studies we review the basic philosophies, objectives, and funding of different retirement income schemes and discuss their social and intergenerational implications. We discuss how funding and intergenerational solidarity are related and what the new reforms mean for the future of both. Traditional pension arrangements basically promise workers a guaranteed income calculated on the basis of some formula that requires that they have contributed for a required period (Orenstein 2008). These pensions are paid from general tax revenues or a pension fund into which workers and employers contribute. The pension fund is usually invested in government bonds. The essence of such an arrangement is, as we have noted, that the pensions received by retirees come directly from current workers or tax payers. Such a system creates an immediate bond of solidarity between generations since each generation pays into the system with the expectation that they will be supported in return.

Such plans are also referred to as "defined benefit" plans since a worker is entitled to a specific payment for life after he or she has contributed to the system for the required number of years. Individuals who do not contribute or who do not do so for the required number of years may be entitled to a minimal non-contributory pension. Such a system is based upon a collectivization, or socialization, of the risk of poverty in old age with solidarity between generations a core objective (United National Economic Conference for Europe 2008; United Nations 2002; Zaidi et al.

2012). As we noted earlier, this arrangement works well, or at least can remain solvent, as long as the population is young, with a large number of workers contributing to the support of relatively few retired individuals. It faces serious fiscal problems, though, as the number of retirees relative to the number of workers increases.

In contrast to "defined benefit" plans that assure a lifelong income, the new individualized investment plans are referred to as "defined contribution" plans to emphasize that their long-term value depends on the workers personal contribution. If one fails to save enough, or if the investment in which his or her contributions are invested experience a low return or even a loss, one's financial situation in old age can be dire. Clearly, such an approach shifts the burden for the support of retirees from the population of workers or the general taxpayer to the individual him or herself. This arrangement weakens or completely severs the tie between generations on which the solidarity principle of traditional pension systems is based. Despite its widespread adoption, the shift from public to private pension plans has been opposed by defenders of traditional public pension plans, including groups such as the International Labor Organization (ILO) and the International Social Security Association (ISSA). These organization, though, have not been as effective as the proponents and have not succeeded in slowing the trend toward privatization (Orenstein 2008, 2011).

Although the proponents of privatization expected these new arrangements to not only address the long-term sustainability of national pension systems and the problem of high rates of labor force informality, serious problems related to coverage, contributions, and gender equity persisted (Bertranou et al. 2009; Mesa-Lago 2008). The result has been a second round of reforms that reintroduce many of the public aspects of traditional public pension systems, including allowing individuals to remain in or return to traditional public plans that operate in conjunction with the new privatized systems. In 2008, for example, Uruguay began allowing some individuals with defined-contribution individual retirement plans to switch back to a defined-benefit plan (Bertranou et al. 2009). In the most extreme case, in 2008 Argentina completely abandoned its defined-contribution plan and moved all workers back to a public defined-benefit system (Ferro and Castagnolo 2010).

The Old Age Problem in Latin America

With significant variation, most Latin American countries today are experiencing a demographic transition that is resulting in rapidly aging populations and an increase in the proportion of individuals in the older age ranges relative to those in the younger age ranges, resulting in rapidly increasing old-age dependency ratios (CEPAL 2007; Cotlear 2011). In Europe this demographic transition began in the eighteenth century and took a century or more to complete, allowing social security and health systems, as well as other institutions time to adapt. In the nations of Latin America the demographic transition started later, but as a result of improved living standards, improved contraception, and better medical care, has proceeded at a far more rapid pace. In many low and middle income nations in the world, the doubling of the proportion of the population in the older age ranges that took a century or more in most in developed nations is taking slightly more than 20 years and in certain cases even fewer (CEPAL 2007; Cotlear 2011).

In combination with other social, economic, and political developments, rapidly aging populations present governments with major fiscal, political, and practical challenges (WHO and Milbank Memorial Fund 2000; World Health Organization 2003). Given this new demographic and social reality, new models for institutional and community-based eldercare that go beyond a reliance on formal state initiatives and employ the family and other informal resources within the context of specific national contexts are essential (Angel 2011; Qingwen and Chow 2011; World Health Organization 2000, 2002). Even developed nations face serious policy dilemmas as a result of rapid population aging that is occurring during a period that is likely to be characterized by protracted fiscal austerity (Pierson 2001). For all nations, financial resources are finite and the needs of the elderly inevitably compete with the needs of younger age groups (Angel and Angel 1997).

The shift from high infant and child dependency ratios to high old-age dependency ratios that characterizes the demographic transition has profound institutional, as well as political implications. Decreasing fertility rates may reduce educational and child care expenditures, but a growing older population increases the need for expensive acute and chronic medical care, as well as long-term care (Cotlear 2011; Lloyd-Sherlock 2000). For numerous reasons, then, rapid population aging presents the nations of Latin America with serious challenges related to the support and care of older individuals (MIDES 2012; Papadópulos and Falkin 2011). The speed with which Latin American populations are aging creates an even more serious problem since developing new aging policies and developing, deploying, and financing new eldercare models takes time.

The aging of populations accompanies other important demographic and social changes that affect families' and the states' capacities to provide care and social services to the elderly. Dynamic economies are inevitably characterized by high rates of rural to urban, interurban, and international migration, and the entry of women into the labor force. At the beginning of the century Latin American nations, as most other parts of the world, were predominantly rural. Today most Latin American countries are highly urban as rural residents move to metropolitan areas in search of greater opportunities. Migration, lowered fertility, smaller families, and the entry of women into the labor force means that the family is less able to care for aging parents than were the large immobile families of earlier times. The question that this new reality presents to us is who will fill in where the family cannot?

Incomplete Pension Coverage

A major failure of most Latin American pension systems is incomplete population coverage. Even experiments with government matching intended to encourage individuals to save for retirement have little effect and coverage rates remain low in all but a few countries, especially for women, workers in small firms, and those with the lowest incomes (Kritzer et al. 2011; Rofman and Oliveri 2011). Additional problems, including low rates of contribution among salaried workers, reduce the number of workers who qualify for benefits (Rofman and Oliveri 2011). Given the growth in the proportion of populations in the older age ranges, social security and pension systems, no matter how reformed and rationalized, are unlikely to provide adequate financial support to all older individuals. Although more affluent families may be able to purchase eldercare services (Papadópulos and Falkin 2011), most will not. The fact that both the state and families face financial and practical limitations in their ability to provide eldercare services requires a serious examination of possible alternatives that combine the services available and employ new options in ways that optimize the level of care in a sustainable manner.

Major Challenges and Systemic Structural Problems

Ultimately old age security, just as the security and quality of life of younger generations, depends on national economic performance in addition to enlightened social policy. For Latin American nations the aging of the population presents problems that will be very difficult to address given current economic performance. Although inequality in Latin America has been decreasing in recent years, those nations remain among the most unequal in the world (Gasparini 2005; Gasparini and Lustig 2011). These high levels of inequality are due to many factors, among which are ineffective taxation and transfer policies that fail to benefit the poor. Unlike European nations, in which income inequality is reduced considerably by government taxes and transfers, among Latin American nations fiscal policy has little impact on income inequality (Goñi et al. 2011). Tax and transfer policies in Latin American nations fail to substantially reduce inequality produced by the labor market. Pension systems can be seen as part of this policy and institutional dilemma. One of the major objectives of public pensions systems is to transfer income from those with higher lifetime earnings and more resources to those older individuals with lower incomes and fewer resources (Goudswaard and Caminada 2010; Mahler and Jesuit 2006). Private pension systems eliminate that transfer, except to the extent that they retain a non-contributory component.

Related to the lower efficiency of tax and transfer policies is the fact that a large fraction of the labor forces of most Latin American nations are employed in the informal sector. The concept of informality remains somewhat ambiguous and imprecisely measured. It can refer to wage labor in small unregistered enterprises or self-employment, both of which are associated with low levels of productivity (Maloney 2004), or it can refer to employment in enterprises that do not include labor contracts, that do not pay taxes or abide by labor regulations, and that offer no benefits (Tornarolli et al. 2012). However it is measured, though, despite some recent improvement, rates of informality remain high in many Latin American countries. One major indicator of informality is the fraction of the salaried labor force that does not have a pension to provide retirement income. Evidence based on a large number of household surveys places that figure at over 60% in Mexico, over 30% in Argentina, 22% in Chile, and 19% in Uruguay (Tornarolli et al. 2012).

Chile

Although we do not have the space to describe all Latin American retirement systems in detail, four comparisons help illustrate the problems involved in pension systems generally and the differential success of various Latin American nations in addressing them. Since 1982 all new Chilean workers have been required to set up individual retirement accounts with one of the government regulated private pension investment companies (*Administradoras de Fondos de Pensiones*, AFPs) of their choice. The old public system is expected to close by 2050. Workers are required to contribute 10% of their monthly earnings, up to a preset amount, into the account. Workers can contribute more to their regular account, or open a new voluntary savings account. They are free to change AFPs whenever they wish. The AFPs charge an administrative fee and premium, which initially were quite substantial. At the age of 65 men and at 60 women can make programmed withdrawals, purchase an immediate annuity, or purchase a deferred annuity, in which case the retiree makes programmed withdrawals until a specific date at which the annuity takes force. One can also choose to combine an annuity and programmed withdrawals. As is the case with private investment accounts generally, the retiree's well-being depends on how much he or she was able to save and the performance of the investments. The government guarantees a basic pension if the company that provides the annuity goes bankrupt. Other aspects of the private plans include disability and survivors benefits (Kritzer 2008).

The 2008 pension re-reforms addressed many of the shortcomings that remained even after interim reforms (Farías 2012; Mesa-Lago 2008, 2009). As in other countries, many of the reforms were highly technical and aimed at improving coverage, equity, and long-term sustainability. We will not go into the specific details but focus on the larger systematic reforms. Initially, labor force coverage was lower than before the introduction of the private system. This was partly due to the fact that the self-employed were not required to participate and their participation rate was extremely low (5%). Insured individuals who had not contributed for the required 20 years did not qualify for a pension. Non-contributory pensions for the poor were hampered by long waiting lists and financial problems (Mesa-Lago 2009). In addition, the armed forces were excluded and received higher pensions paid for by the treasury.

Three of the major objectives of the 2008 reforms were improvements in the administration and return to individual accounts, the extension of coverage to previously uninsured individuals, and greater gender equity. The reforms greatly increased competition in the pension fund industry, reduced risk to individual investors, encouraged voluntary savings, and more (Kritzer 2008). Since 1981 major changes had liberalized investment rules, increased the number of and type of funds offered, and addressed the excessively high administrative fees that had plagued the program (Farías 2012; Kritzer 2008). The 2008 reforms continued to address problems with the system. Although a lower fraction of the Chilean labor force is informal than is the case in other Latin American nations, informality remains significant and coverage and low and discontinuous contributions remain major problems. The new reform requires that the self-employed participate in the private system and over time increases the portion of their earning on which they must contribute. In addition, in 2008 previous means-tested income plans for retirees were replaced by a noncontributory basic pension, called the Basic Solidary Pension (Pensión Básica Solidaria, PBS) that provides a basic pension to a large fraction of the elderly (Kritzer 2008).

In Chile, as elsewhere, gender inequity remains a serious problem. Women are at particularly high risk of inadequate retirement income for several reasons (Arza 2012; Joubert and Todd 2011). Because of the fact that they have other family responsibilities and must care for children and the elderly, women on average contribute to their individual plans for less than half of their potential working lives (Kritzer 2008). When they do work, women receive lower wages than men and like others in the informal sector are at risk of not contributing enough to qualify for adequate pension benefits. The reforms increase a woman's pension by providing her a supplement for each child she has had. Other technical changes related to survivors and disability insurance, divorce, and other life events are also aimed at increasing gender equity.

Mexico

Mexico is an interesting case of the problems that the old pay-as-you-go systems faced. The Mexican Institute for Social Security (Instituto Mexicano de Seguridad Social), IMSS was founded in 1944 to administer several public programs including old age security, severance, disability, and life insurance. Until 1992 IMSS subsumed two major functions: (1) retirement, retirees' health benefits, and disability pensions funded by a payroll tax of 8.5 % charged to formal sector workers, and (2) a housing fund administered by the National Workers' Housing Fund Institute (Instituto del Fondo National de la Vivienda de los Trabajadores, INFONAVIT) funded by employers at 5 % of wages (Grandolini and Cerda 1998).

Since its beginning IMSS operated as a pay-as-you-go system, which functioned adequately until the 1970s because of Mexico's young workforce. There were enough workers per pensioner to support the retirement and the health systems. To receive an old age pension a worker was required to be 65 and to have contributed to the system for 500 weeks, or approximately 10 years. Pensions were indexed to changes in the minimum wage in 1989. The system included a guaranteed minimum payment equal to the minimum wage. The pension was based on a formula that included the average wage for the last 5 years plus an addition for each year over ten.

Other complex rules limited the amount of the payment. Problems in the design of the system, though, meant that many workers did not qualify for any pension.

Two major long-term problems plagued the system. First, the fact that given high levels of informality and women's need to leave the work force to deal with family issues, many individuals did not contribute for the full 10 year contribution period and would not qualify for a pension. Other inequities in payment favored higher-paid workers. Second, the benefit computation essentially penalized contributions for longer than the required 10 years since the additional years did not significantly increase what one received. This structural problem, along with many other aspects of Mexican labor law and business practices, encouraged workers to leave the formal sector after 10 years of participation. In short, the Mexican pension system was plagued by financial distortions, inadequate pensions, and evasion (Grandolini and Cerda 1998).

During the 1990s, the pension debate in Mexico, as in the other nations of the region, focused on the long-term sustainability of the pay-as-you-go aspects of the system. Even though over half of the work force was not in the formal system, given the demographic changes we have outlined, the long-term viability of the system was questionable. In 1997, Mexico made Individual Retirement Accounts mandatory for all private sector employees. The military, workers for PEMEX, the national oil company, and public sector workers continued in their traditional systems, although public sector workers not covered by special programs could choose the private system. The individual retirement accounts are funded by contributions from employers, employees, and the federal government.

The government guarantees a minimum pension. The retirement age is 65 and one must contribute for 1250 weeks, or nearly 25 years to receive full benefits. Those who fall short can take a lump-sum payment or continue contributing. The private funds are administered by pension fund managers (AFORES: Administra-doras de Fondos para el Retiro) that are regulated by a national retirement saving system (CONSAR: Comisión Nacional del Sistema de Ahorro para el Retiro). There are four groups of AFORES for individuals of different ages. The older and closer to retirement one is the more conservative the investment strategy of the fund. Since its inception there have been several reforms to the original plan to increase coverage, control administrative costs, and increase participation (Bertranou et al. 2009).

Uruguay

Uruguay began its individualized retirement system in 1996. It differs from Mexico's in that only high income workers were required to join the private system. Workers below a certain income threshold had the option of splitting their contributions between the new private system and the old public system. This new mixed system is mandatory for all workers born after 1956. Uruguay also differs from Mexico in that it has higher rates of pension coverage and greater participation. In

2009, 34.2 % of the labor force in Uruguay and 29.2 % of the labor force in Mexico had individual accounts (Kritzer et al. 2011). These numbers are lowered by the fact that certain groups such as the police and public employees have their own systems. The two countries differ in other regards as well. Among those with individual retirement accounts, 64.5 % have made recent contributions in Uruguay, while only 34.1 % have done so in Mexico. This pattern makes it likely that a far larger number of retirees in Uruguay will have adequate savings than in Mexico. In addition to easing pressure on the national budget, reforms were expected to encourage workers in the informal sector to participate in the formal sector and, therefore, the pension system, thereby increasing labor force coverage. The reform did not accomplish that goal, and currently only 27 % of the economically active population participate (Carranza et al. 2013).

Argentina

Unlike Chile which completely replaced its public pay-as-you-go system in 1981, Argentina kept the old public system and in 1994 introduced a new "mixed system" that consisted of two "pillars", one which provided a public basic pension, and another private supplementary option (Mesa-Lago 2009; Repetto and Masetto 2012). The privatized Argentine system suffered from many of the same problems that plagued the Chilean system and introduced new ones related to the continuation of the old pay-as-you-go system. As in Chile coverage declined and non-contributory social pensions for the poor did not cover all of those in poverty. The armed forces were excluded, but so were municipal and civil servants. Thirty-five, rather than 20 years of contributions were required to qualify for a pension, so it was clear that a large fraction of workers would never qualify. Administrative costs were high and the mixed system created substantial fiscal costs related to the support of those who remained in the old system which was not closed but to which younger workers did not contribute (Mesa-Lago 2008, 2009).

The problems in Argentina were exacerbated by the 2001 economic crisis which drastically reduced coverage and the value of individual pension funds. As the worldwide recession of 2008 illustrated, private investment accounts can be decimated by economic downturns. Other financial shortcomings related to the organization and capitalization of the private funds added to their limited coverage and long-term negative outlook. Reforms to the system in 2007 addressed some of the problems, but they did not solve them all. For these reasons, and perhaps for other motivations related to the national debt, in 2008 Argentina's senate gave final approval for the nationalization of the private pension system with the ostensible objective of protecting the value of pensions from the global economic downturn. Although there were some objections, the private system was not popular enough to survive. The confiscation of 25 billion dollars in private pension assets solves certain problems and strengthens the public system in the short run, but in the long-run Argentina faces the same problems of a growing pension burden that motivated privatization in the first place (Mesa-Lago 2009).

Civil Society and the Welfare of the Elderly

In Mexico, as in other Latin American nations, many programs provide nutritional, medical, and other social services to elderly adults. The Secretariat of Social Development (Secretaria de Desarollo Social, SEDESOL) coordinates general social services, including services for the elderly. The program Opportunities (Oportunidades), provides nutritional and financial assistance to the poorest families, including individuals over the age of 70. One organization focused on the elderly is the National Institute of Mature Adults (Instituto Natioal de las Personas Adultas Mayores, INSEN). Another agency that addresses needs of the elderly is the System for the Integral Development of the Family (Sistema Nacional para el Desarrollo Integral de la Familia, DIF). Other organizations, including IMSS and ISSTEE that deal with physical and mental health also provide other services to the elderly.

Despite these governmental programs, though, in Mexico as elsewhere the magnitude and speed of the increase in old-age dependency, in addition to the structural and administrative barriers we have described, means that governments alone will never be able to address all of the needs of older citizens. Even Chile, which has experienced significant economic growth in recent years, will be hard pressed to provide the entire range of financial, medical, and other support that a large retirement-age population will need. Nations with less dynamic economies will be more seriously challenged. In countries with low social security and retirement coverage older individuals who are able to work will have no choice but to continue working, perhaps at lower wages. When they can no longer work, they will turn to children or charity as has traditionally been the case.

Yet the demographic and social changes mentioned above, including smaller families, marital disruption, the need for women to enter the work force, and high rates of migration make it increasingly unlikely that the family will be able to fully assume its traditional role, especially as life spans increase and older individuals spend many years functionally impaired. Given this new demographic and social reality, both developed and developing nations will be forced to experiment with new models of community and institutional care (Angel 2011; Qingwen and Chow 2011). These experiments, which are informally if not formally underway in most countries, include combinations of formal support by the state and informal support by family and other institutions, including faith-based and other non-governmental organizations (Angel 2011; Pereira and Angel 2009; Pereira et al. 2007, 2009).

Financial, instrumental, and emotional support can be provided by some combination of the family, the State, and civil society, which includes non-governmental and faith-based organizations (NGOs and FBOs) like OnLok in San Francisco which provides community based support to Chinese elders, HelpAge, which provides assistance and support to the elderly and many other organizations that address problems of nutrition, transportation, isolation, and more. The complex and extensive nature of the needs of the elderly make the potential role of civil society organizations clear. While the state can provide financial and medical care, providing companionship, assistance with basic activities of daily living, house cleaning, yard work, care of pets, meal preparation, and the rest of what daily life requires are tasks that are best dealt with by families or individuals familiar with the older person. As the family becomes less able to deal with the needs of elderly parents who are living longer, pseudo-families, including volunteers may serve to fill the void.

The core theoretical and practical question that our analysis raises relates to the roles of civil society organizations, including NGOs and FBOs, as complementary providers of care to the elderly. The recent rise in the desire to devolve responsibility for social services from higher levels of government to lower levels of government and even non-governmental organizations reflects the belief that local entities have certain advantages in dealing with the more routine and manageable needs of specific populations (Pereira and Angel 2009; Pereira et al. 2007). Although complex and expensive high-tech medicine can only be paid for or provided by the State, routine and relatively inexpensive services, such as assistance with activities of daily living and providing companionship are often more effectively and economically provided by local groups including churches and other local agencies. Such local care providers are more likely to be aware of the legal, transportation, nutritional, and other needs of the elderly in their communities. These capacities potentially make such organizations important allies in service provision and important actors as advocates for the elderly.

A large body of research documents that social support and social networks are important in maintaining a high quality of life in old age (Krause 2006; Moren-Cross and Lin 2006). Such findings suggest the potential importance of local organizations. Social engagement and human contact are central to wellbeing at any age, and for the elderly non-family sources may in the future play an increasingly important role in providing social and instrumental support and preventing isolation. A recent study in Mexico showed that individuals with few resources and high needs receive significant amounts of support and assistance from neighbors, churches, and other voluntary organizations (Burcher 2008).

We end this section with a brief description of the eldercare functions of a few well-known non-governmental organizations as a general illustration of the possibilities. There are far too many such organizations to do more than mention a few. Many organizations include eldercare assistance and advocacy as part of larger social agendas. One of the major roles of such organizations is preventing isolation for which infirm elderly in the community are at high risk. For the elderly companionship and visitation are important needs that can be readily provided at little expense by volunteers. The *Meals on Wheels Association of America* (http://www.mowaa.org) is the oldest and best known non-governmental nutrition program for the elderly in the country and is dedicated to preventing hunger among the elderly. In addition to providing meals, though, volunteers provide important human contact to older persons. Although many governments provide funding for nutrition services, getting those meals to isolated elderly individuals in their homes and providing companionship are clearly tasks best suited to local organizations and volunteers.

The less well-know *Little Brothers—Friends of the Elderly (LBFE)* (http://www.littlebrothers.org) is an international network of non-profit, volunteer-based organizations that provide companionship to elderly people to reduce isolation and loneliness. The organization is a member of a larger international organization known as the Fédération Internationale des petits frères des Pauvres (International Federation of Little Brothers of the Poor, http://www.petitsfreres.org). A similar international organization, the *Fédération Internationale des Associations de Personnes Agées* (International Federation of Associations of Older Persons, FIAPA: http://www.fiapa.org), headquartered in Paris also focuses on the prevention of isolation and a better quality of life for older individuals. These international organizations make it clear that the risk of isolation affects all developed societies in which the old do not routinely live with their children.

A short perusal of the internet reveals many organizations in many countries that focus on support and advocacy for the elderly. India, like most of the rest of the world, is facing a serious problem related to the care of a growing elderly population. According to the Indian government in 2005 93 % of India's total labor force, and 82 % of its non-agricultural labor force was informally employed (Agarwala 2011). Consequently, very few Indians have a pension plan and older individuals are completely reliant on their own efforts or on family for support. In this environment NGOs are important advocates for and service providers to the elderly (Sawhney 2003). *Dignity Foundation* (http://www.dignityfoundation.com), an Indian organization that is a member of the American Association of Retired Persons (AARP) Global Network, provides housing, companionship, recreation and other services to elderly individuals in several Indian cities. *HelpAge India* (http://www.helpageindia.org) provides financial, medical, and emotional support to poor elderly Indians. This NGO has developed new programs to provide services to previously underserved areas. One example highlighted on the organization's website is a Mobile Medicare Unit (MMU) program that provides basic health care to underserved areas.

The cases of Dignity Foundation and HelpAge India are examples of eldercare NGOs that work with seriously impoverished populations. Another example in a nation that is far more developed is Christ's Home (*Hogar de Cristo*: http://www.hogardecristo.cl) in Chile (Pereira et al. 2007). Hogar de Cristo is a faith-based organization in a highly Catholic country, a fact that no doubt has contributed to its success. It began in 1944 when a Catholic priest named Alberto Hurtado decided to found an organization to provide for the needs of poor Chileans. Given high levels of poverty among the elderly and the seriously curtailed social services that were part of the neoliberal reforms introduced by the Pinochet dictatorship, Hogar's mission has expanded to provide such services as day care, nutritional programs, and even housing to impoverished elder persons with no other means of support.

These are only a few of thousands of organizations that operate in most countries of the world. While NGO assistance with food, medical care, and housing clearly benefit at least some of the most needy elders, NGOs are no substitute for adequate government sponsored social security programs (HelpAge International 2009; Willimore 2006). In order to do more than provide palliative assistance, NGOs must engage in advocacy for the elderly. In the United States the *American Association of Retired Persons* (AARP) is undoubtedly the most well-known and effective advocate for older persons (Binstock 2004). As is common knowledge, the AARP is a formidable political force. Other organizations such as the national *Committee to preserve Social Security and Medicare* (NCPSSM: http://www.ncpssm.org/), the *Alliance for Retired Americans* (ARA: http://www.retiredamericans.org), and the *National Hispanic Council on Aging* (NHCOA: http://www.nhcoa.org/) are examples of other organizations that advocate for changes in policy to preserve and enhance programs for the elderly.

Since it focuses on elderly Hispanics the NHCOA is particularly interesting in the context of our presentation. Its mission includes advocacy, the support of research, the funding of community-based projects, as well as the creation of support networks, capacity-building in Hispanic communities and the support and strengthening of Hispanic community-based organizations. The organization's core objective is to "empower Hispanic community organizations and agencies, as well as Hispanic older adults and their families." NHCOA offers educational programs focused on the major health risks that affect Hispanics, including diabetes, and it has developed an e-course on cultural competence that educates health care professionals concerning culture of their patients (http://edu.nhcoa.org).

Conclusion

This essay has addressed several highly complex and interrelated issues. Hopefully we have made the interconnection among various demographic, social, economic, and political factors a bit more obvious. The complexity and urgency of increasing old age dependency and its implications for the future of the welfare state define clear research and policy agendas. Old age economic security depends on vibrant economies, enlightened social policies, and a favorable demographic profile. Unfortunately, the nations of Latin America, like those of the developed world are facing serious challenges in providing financial, medical, social, and other support to rapidly aging populations in the context of fewer workers per retiree, what is likely to be permanent fiscal retrenchment, and growing political resistance to the welfare state. What is certain is that in the years to come the debate over the welfare state and the support of the elderly will only become more salient.

We began by mentioning the solidarity and its role in the justification of traditional public pension plans in which an endogenous tie between generation resulted from the fact that current retirees are supported by those currently employed. This solidarity remains a central value among supporters of public retirement systems. The neoliberal private approach deals with the fiscal problems of pension systems by explicitly rejecting the notion of solidarity. Indeed, as we noted, José Piñera believes that public systems ultimately undermine solidarity. There can be little doubt that given current demographic trends pension systems will come under increasing fiscal strain, even as they fail to cover large segments of the population. All nations face economic realities, but the optimal choice of how to divide the economic pie is not completely objective nor completely economic. These choices reflect basic cultural values and political processes.

As we have suggested, the reality is that new options in the support of the elderly that combine the efforts of the State, the family, and civil society will in all likelihood be necessary. No single social institution can bear the full burden. A greater reliance on non-governmental organizations certainly has its potential drawbacks and might be viewed as letting the state off the hook. Neoliberal and structural adjustment policies clearly take their greatest toll on the most vulnerable citizens of poor and middle income nations. Structural adjustment policies and the reduction of state budgets translate directly into the loss of the educational, health, and other services that the citizens of developing nations desperately need. While we call for exploration into the role of non-governmental and faith-based organizations in improving the quality of life for older and infirm people, we must reaffirm the basic social rights upon which the modern welfare state was built. States clearly cannot do everything, but they must provide for the basic needs of citizens in an equitable and just manner. Civil society organizations can only augment, but they can never replace the state in guaranteeing basic citizenship rights (Angel et al. 2012).

Solidarity, then, is a complex and multifaceted concept and reality. It is based on the indisputable fact that humans are not solitary creatures and are highly interdependent. Contemporary communitarian theories and politics recognize that interdependence and harken back to what were supposedly ideal communities in the past. Such calls are probably overly romantic, but ongoing debates concerning the nature of community and our mutual interdependence lie at the core of future discussion on the welfare state and the economic, physical, and social welfare of citizens of all ages.

References

Agarwala, R. (2011). India in transition: India's informal workers and social protection. Philadelphia: Center for the Advanced Study of India and the Trustees of the University of Pennsylvania. http://casi.sas.upenn.edu/iit/agarwala. Accessed 15 Nov 2014.

Angel, R. J. (2011). Civil society and eldercare in post-traditional society. In R. A. Settersten & J. L. Angel (Eds.), *Handbook of sociology of aging* (pp. 549–581). New York: Springer.

Angel, R. J., & Angel, J. L. (1997). *Who will care for us? Aging and long-term care in multicultural America*. New York: New York University Press.

Angel, R. J., Bell, H., Beausoleil, J., & Lein, L. (2012). *Community lost: The state, civil society and displaced survivors of hurricane Katrina*. New York: Cambridge University Press.

Arza, C. (2012). Pension reforms and gender equality in Latin America. In *Gender and Development*. Geneva: United Nations Research Institute for Social Development.

Barrientos, A. (2006). Poverty reduction: The missing piece of pension reform in Latin America. *Social Policy & Administration* (Vol. 40, pp. 369–384). Brighton: Institute of Development Studies.

Bertranou, F., Calvo, E., & Bertranou, E. (2009). *Is Latin America retreating from individual retirement accounts?* Boston: Center for Retirement Research at Boston College. http://www.ilo.org/wcmsp5/groups/public/---ed_protect/---secsoc/documents/publication/wcms_secsoc_12756.pdf. Accessed 5 Aug 2013.

Bharmal, N., Tseng, C.-H., Kaplan, R., & Wong, M. D. (2012). State-Level variations in racial disparities in life expectancy. *Health Services Research, 47*(1pt2), 544–555.

Binstock, R. H. (2004). Advocacy in an era of neoconservatism: Responses of National Aging Organizations. *Generations, 28*(1), 49–54.

Burcher, J. (Ed.). (2008). *México solidario: Participación cudadana y voluntaiado.* Mexico: Limusa.

Carranza, L., Melguizo, Á., & Tuesta, D. (2013). Matching contributions in Colombia, Mexico, and Peru: Experiences and prospects. In H. Richard, H. Robert, T. David, & T. Noriyuki. *Matching contributions for pensions: A review of international experience* (pp. 193–213). Washington, DC: The World Bank.

CEPAL. (2007). Demographic trends in Latin America in Observatorio demográfico N 3: Proyección de población. CEPAL: La Comisión Económica para América Latina. http://www.eclac.cl/publicaciones/xml/0/32650/OD-3-Demographic.pdf. Accessed 25 Jan 2014.

Cotlear, D. (Ed.). (2011). *Population aging: Is Latin America ready?* Washington, DC: The World Bank.

Easterlin, R. A. (2000). The worldwide standard of living since 1800. *The Journal of Economic Perspectives, 14*(1), 7–26.

Elo, I. T. (2009). Social class differentials in health and mortality: Patterns and explanations in comparative perspective. *Annual review of sociology* (pp. 553–572). Palo Alto: Annual Reviews.

Farías, C. R. (2012). *Social protection systems in Latin America and the Caribbean: Chile.* Santiago: ECLAC. http://www.eclac.org/publicaciones/xml/8/48988/SPS_Chile_ing.pdf. Accessed 29 July 2013.

Ferro, G., & Castagnolo, F. (2010). On the closure of the Argentine fully funded system. *Pensions, 15*(1), 25–37.

Gasparini, L. (2005). Income inequality in Latin America and the Caribbean: Evidence from household surveys. *Económica, La Plata, LI*(1–2), 29–57. http://economica.econo.unlp.edu.ar/documentos/20081127035039PM_Economica_542.pdf. Accessed 5 Aug 2013.

Gasparini, L., & Lustig, N. (2011). The rise and fall of income inequality in Latin America. ECINEQ: Society for the Study of Economic Inequality. http://www.ecinEq.org/milano/WP/ECINEQ2011-213.pdf. Accessed 5 Aug 2013.

Goñi, E., López, J. H., & Servén, L. (2011). Fiscal redistribution and income inequality in Latin America. *World Development, 39*(9), 1558–1569.

Goose, N., & Looijesteijn, H. (2012). Almshouses in England and the Dutch Republic circa 1350–1800: A comparative perspective. *Journal of Social History, 45*(4), 1049–1073.

Goudswaard, K., & Caminada, K. (2010). The redistributive effect of public and private social programmes: A cross-country empirical analysis. *International Social Security Review, 63*(1), 1–19.

Grandolini, G., & Cerda, L. (1998). *The 1997 pension reform in Mexico: Genesis and design features.* Washington, DC: The World Bank. http://elibrary.worldbank.org/docserver/download/1933.pdf?expires=1375106885&id=id&accname=guest&checksum=CD466D08B56FC462040075D2D2F1260F. Accessed 29 July 2014.

HelpAge International. (2009). *Working for life: Making decent work and pensions a reality for older people.* London: HelpAge International.

Joubert, C., & Todd, P. E. (2011). *The impact of Chile's 2008 pension reform on labor force participation, pension savings, and gender equity.* Santiago: Labor Ministry, Chilean Government. http://www.consejoprevisional.cl/documentos/articulos/impacto-reforma-2008-fza-laboral-ahorro-genero-joubert-todd.pdf. Accessed 8 Aug 2013.

Krause, N. (2006). Social relationships in later life. In H. R. Binstock & K. G. Linda (Eds.), *Handbook of aging and the social sciences* (pp. 181–200). New York: Academic.

Kritzer, B. E. (2008). Chile's next generation pension reform. *Social Security Bulletin, 68*(2), 69–84. http://www.ssa.gov/policy/docs/ssb/v68n2/v68n2p69.html. Accessed 6 Aug 2013.

Kritzer, B. E., Kay, S. J., & Sinha, T. (2011). Next generation of individual account pension reforms in Latin America. *Social Security Bulletin, 71*(1), 35–76.

Link, B. G., & Phelan, J. (1995). Social conditions as fundamental causes of disease. *Journal of Health and Social Behavior, 35*(Special Issue), 80–94.

Lloyd-Sherlock, P. (2000). Population ageing in developed and developing regions: Implications for health policy. *Social Science & Medicine, 51*(6), 887–895.

Mahler, V. A., & Jesuit, D. K. (2006). Fiscal redistribution in the developed countries: New insights from the Luxembourg income study. *Socio-Economic Review, 4*(3), 483–511.

Maloney, W. F. (2004). Informality Revisited. *World Development, 32*(7), 1159–1178.

Mesa-Lago, C. (2008). *Reassembling social security: A survey of pensions and health care reforms in Latin America*. Oxford: Oxford University Press.

Mesa-Lago, C. (2009). Re-reform of Latin American private pensions systems: Argentinian and Chilean models and lessons. *The Geneva Papers on Risk and Insurance—Issues and Practice, 34*, 602–617.

MIDES. (2012). Plan Nacional de Envejecimiento y Vejez, 2013–2015. Montevideo: Unidad de Informacíon y Comunicación. MIDES—Consejo Consultivo del Instituto Nacional del Adulto Mayor—INMAYORES.

Morbidity and Mortality Weekly Report. (1999). Ten great public health achievements—United States, 1900–1999. *The Journal of the American Medical Association, 281*(16), 1481.

Moren-Cross, J. L., & Lin, N. (2006). Social networks and health. In R. H. Binstock & L. K. George (Ed.), *Handbook of aging and the social sciences* (pp. 111–126). New York: Academic.

Müller, K. (2000). Pension privatization in Latin America. *Journal of International Development, 12*(4), 507–518.

Myles, J. F. (1983). Conflict, crisis, and the future of old age security. *The Milbank Memorial Fund Quarterly Health and Society, 61*(3), 462–472.

Myles, J. (2002). A new social contract for the elderly. In G. Esping-Andersen & D. Gallie (Ed.), *Why we need a new welfare state* (pp. 130–172). New York: Oxford University Press.

Orenstein, M. A. (2008). *Privatizing pensions: The transnational campaign for social security reform*. Princeton: Princeton University Press.

Orenstein, M. A. (2011). Pension privatization in crisis: Death or rebirth of a global policy trend? *International Social Security Review, 64*(3), 65–80.

Papadópulos, J., & Falkin, L. (2011). *Documento conceptual: personas adultas mayores y dependencia. Dimensionamiento de necesidades en materia de cuidados y alternativas de incorporación de servicios y población*. Montevideo: Sistema Nacional de Cuidados—Presidencia de la República.

Pereira, J., & Angel, R. (2009). From adversary to ally: The evolution of non-governmental organizations in the context of health reform in Santiago and Montevideo. In S. Babones (Ed.), *Social inequality and public health*. Bristol: Polity Press.

Pereira, J., Angel, R. J., & Angel, J. L. (2007). A case study of the elder care functions of a Chilean non-governmental organization. *Social Science & Medicine, 64*, 2096–2106.

Pereira, J., Angel, R. J., & Angel, J. (2009). A Chilean faith-based NGO's social service mission in the context of neoliberal reform. In T. L. Hefferan, J. Adkins, & L Occhipinti (Eds.), *Bridging the gaps: Faith-based organizations, neoliberalism, and development in Latin America and the Caribbean* (pp. 151–164). Lanham: Lexington Books.

Phelan, J C., Link, B. G., & Tehranifar, P. (2010). Social conditions as fundamental causes of health inequalities. *Journal of Health and Social Behavior, 51*(Suppl 1), S28–S40.

Pierson, P. (2001). Coping with permanent austerity: Welfare state restructuring in affluent democracies. In P. Pierson (Ed.), *The new politics of the welfare state*. Oxford: Oxford University Press.

Piñera, J. Fall/Winter. (1995/1996). Empowering workers: The privatization of social security in Chile. *Cato Journal, 15*(2–3), 155–156.

Preston, S. H. (1996). *American longevity: Past, present, and future*. Maxwell School of Citizenship and Public Affairs, policy research paper 36. Syracuse University, Syracuse, NY.

Qingwen, X., & Chow, J. C. (2011). Exploring the community-based service delivery model: Elderly care in China. *International Social Work, 54*(3), 374–387.

Repetto, F., & Masetto, F. P. D. (2012). *Social protection systems in Latin America and the Caribbean: Argentina*. Santiago: Economic Commission for Latin America and the Caribbean (ECLAC). http://www.eclac.org/publicaciones/xml/4/48984/SPS_Argentina_ing.pdf. Accessed 30 July 2012.

Rofman, R., & Oliveri, M. L. (2011). La covertura de los sistemas previsionalesen América Latina: conceptos e indicadores. In *Serie de Documentos de Trabajo sober Políticas Sociales No 7*. Buenos Aires, Argentina: Banco Mundial Argentina, Chile, Paraguay y Uruguay.

Sawhney, M. (2003). The role of non-governmental organizations for the welfare of the elderly: The case of HelpAge India. In P. S. Liebig & S. I. Rajan (Eds.), *An aging India: Perspectives, prospects, and policies*. Binghamton: The Hayworth Press.

Thane, P. (Ed.). (2005). *The long history of old age*. London: Thames & Hudson.

Tornarolli, L., Battistón, D., Gasparini, L., & Gluzmann, P. (2012). Exploring trends in labor informality in Latin America, 1990–2010. Documento de proyecto LaborAL, CEDLAS y IDRC.

United Nations. (2002). *Report of the Second World Assembly on ageing*, Madrid, 8–12 April 2002. New York, United Nations. http://www.c-fam.org/docLib/20080625_Madrid_Ageing_Conference.pdf. Accessed 1 May 2013.

United National Economic Conference for Europe. (2008). A society for all ages: Challenges and opportunities. Report of the UNECE. *Proceedings of the UNECE Ministerial Conference on Ageing*, 6–8 Nov 2007, León, Spain. New York and Geneva: United Nations. http://www.unece.org/index.php?id=10834. Accessed 1 May 2013.

WHO and Milbank Memorial Fund. (2000). Towards an international consensus on policy for long-term care of the ageing. Geneva, Switzerland. http://www.milbank.org/uploads/documents/000712oms.pdf. Accessed 1 Feb 2013.

Willimore, L. (2006). Universal age pensions in developing countries: The example of Mauritius. *International Social Security Review, 59*(4), 67–89.

Wilmoth, J. R. (1998). The future of human longevity: A demographer's perspective. *Science, 280*(5362), 395.

World Bank. (1994). *Averting the old-age crisis: Policies to protect the old and promote growth*. New York: Oxford University Press.

World Health Organization. (2000). *Home-based long-term care*. Geneva: World Health Organization. http://whqlibdoc.who.int/trs/WHO_TRS_898.pdf. Accessed 2 Jan 2013.

World Health Organization. (2002). *Lessons for long-term care policy*. Geneva: World Health Organization. http://whqlibdoc.who.int/hq/2002/WHO_NMH7CCL_02.1.pdf. Accessed 1 Feb 2013.

World Health Organization. (2003). *Key policy issues in long-term care*. Geneva: World Health Organization. http://www.who.int/chp/knowledge/publications/policy_issues_ltc.pdf. Accessed 1 Feb 2013.

Zaidi, A., Gasior, K., & Manchin, R. (2012). Population aging and intergenerational solidarity: International policy frameworks and European public opinion. *Journal of Intergenerational Relationships, 10*, 214–227.

Chapter 24
The Politics of Aging in a Majority-Minority Nation

Courtney M. Demko and Fernando M. Torres-Gil

Introduction

As Chap. 1 shows, the United States is aging and becoming more diverse, and these demographic trends are increasingly ensconced in mainstream discussion about the needs of older persons and how to respond to elders, their families and their caregivers. In particular, the emerging Latino population and the increased life expectancy of Hispanics (i.e., Mexicans, Puerto Ricans, Cubans, Central Americans) in the United States is of paramount importance, because these groups will collectively become the nation's largest minority group. The Aging in the Americas conference and the research agenda have made important contributions to understanding the physical health, mental health, long-term care and social conditions facing older Latinos in the U.S., Mexico and Latin America and provided directions for expanded research agendas in these important scholarly areas. Yet, as we examine the issues further, posing questions and offering programmatic and policy responses to these areas, we find that the mainstream debates increasingly have an overtly political overlay that may influence how the United States, and potentially Mexico, respond to the aging and growing diversity of their respective populations. Specifically, the "politics of aging of a majority-minority nation" may determine the political agenda and the public policy response to this demographic transformation and perhaps shape the narrative of how the public views these issues prior to its understanding of the facts, data and analyses that are currently underway by the impressive group of scholars in this volume.

This paper will explore the nuances and complexities of politics, aging, ethnicity and public policy within the prism of one subset of aging and diversity: Latino baby boomers. We ask: What are the questions and concerns that intertwine with aging, Latinization and politics? What does it mean that older persons are an influential segment of the electorate, and how do current controversies over entitlement

C. M. Demko (✉) · F. M. Torres-Gil
University of California, Los Angeles, CA, USA
e-mail: cbleecher@ucla.edu

© Springer International Publishing Switzerland 2015
W. A. Vega et al. (eds.), *Challenges of Latino Aging in the Americas*,
DOI 10.1007/978-3-319-12598-5_24

reforms and budgetary politics influence the future aging of the Latino population? And what might a deeper examination of Latino baby boomers reveal vis-á-vis the politics of aging in a majority-minority nation? This chapter explores these issues and suggests directions for future policy research and legislative ideas for a more proactive response to the reality of a nation that is becoming majority-minority as it grows older. First, however, we briefly review the demographic trends discussed in the first section of this volume.

Demographic Overview

As Murdock describes in ample detail, two major demographic changes in the United States are anticipated to have a profound impact on the social, economic and political landscape in the next 40 years, as the nation becomes older and more diverse. The U.S. Census Bureau projects that the population of people aged 65 and older will more than double between 2010 and 2050, from 40.2 to 88.5 million (U.S. Census Bureau 2010). This significant numerical increase is due in large part to the aging of the baby boomers, a large cohort born between 1946 and 1964. The data presented in the Murdock chapter shows that the boomers will have their maximum impact—culturally, economically, politically—between 2012 and 2035. The end point in 2060 creates another milestone question: who replaces elderly boomers? What trends lead us to a new politics of aging?

Several trends are operating simultaneously: the aging of the baby boomer co-hort, the increasing share of the elderly population, and increased life expectancy. While these trends are well known, together they create a complex demographic mosaic of a nation experiencing the full effects of longevity and a population where older persons will dominate.

Yet, there is another demographic change that will add a new variable, further complicating this profile: diversity and the increase in minority, ethnic and immigrant populations. Between 2012 and 2060, we can expect minority populations such as blacks, Asians and Latinos to account for a larger proportion of the U.S. population in general and to be represented as an increased share of the U.S. older population. Again, this is not a surprise; demographers and the mainstream media are increasingly aware that the United States is both older and more diverse. A new phenomenon is the juxtaposition of these two key trends—aging and diversity—and what it signifies in terms of the next 20–40 years of a nation undergoing a demographic transformation.

If this was not enough of a challenge, one more key variable will come into play, a trend that has not fully inserted itself into the national debates about how best to respond to a nation of more older persons and more ethnic/immigrant groups: replacement rates and changing fertility patterns. How might the basic question, "Who is having the babies?" illuminate the "politics of aging in a majority-minority nation," and what might this say about demographics of aging in the Americas?

The total fertility rate (TFR) refers to the average number of children born to a woman who is in her childbearing years (OECD Family Database 2011). Assuming there are no migration flows and that mortality rates remain unchanged, a TFR of 2.1 children per woman insures that a population remains stable: the number of deaths equals the number of births. This can also be called a replacement fertility rate (RFR), and this chapter will refer to both terms. Throughout modern history, and in nations that are more agrarian and developing toward first-world nationhood, replacement rates have remained above 2.1. Thus, many nations, and the world in general, have witnessed increased populations. That all began to change in the latter part of the twentieth century. With increased global aging and increased economic prosperity for many nations, women have chosen to have fewer children. In key parts of the world, such as Europe (1.6) and East Asia (1.5), the TFR is less than 2.1. North and Central America (2.2) and South America (2.2.) are closer to that magic mark (New York Times Magazine 2008; National Geographic 2011). Low TFR appears to be a global trend, although certain regions, such as Africa (4.7) and Western Asia (3.1) continue to have high fertility rates. The United States and Mexico, however, are not in that first category.

In 2010, for the first time, the Hispanic replacement rate (2.4) exceeded that of whites, blacks and Asians and remained above the requisite 2.1. Moreover, the answer to "who is having the babies" can be visually answered in many K-12 public school playgrounds throughout the Southwest, Great Lakes, East Coast and Florida, where Latino kids comprise the majority of public school children. Even in formerly black neighborhoods, like Watts and South Central Los Angeles, the majority of children are Hispanic. In a recent conversation (at a Milken Global Institute event in 2014) with this co-author, Ron Brownstein made a prescient commentary: that the high school graduating class of 2014 will be the last such group to be majority white! Of course, acculturation of Hispanic immigrants may change and represent a lower TFR, and evidence abounds that Latina women, with education and professional opportunities, will choose to have fewer than 2.1 children (New York Times 2013). But for the present and into the next two decades, Latinos/Hispanics and immigrants will account for the net population growth in the United States. Even Mexico, a nation long considered youthful, is undergoing similar trends albeit a generation or two behind the U.S. For example, life expectancy at birth for Mexicans increased from 36 to 74 years between 1950 and 2000. In the same period, its fertility level decreased from 7.0 in 1960 to 2.4 in 2000. And it continues to decline. Mexico will soon face a dramatic increase in its older adult population while its youthful population decreases. In fact, by 2013 it was reported that Mexico's fertility rate had reached 2.1. What might this ultimately mean?

It demonstrates that the nation is becoming not just older but also more diverse. The U.S. Census Bureau projects that the Hispanic population is set to more than double from 53.5 million in 2012 to 128.8 million in 2060 (U.S. Census Bureau 2012). This means that, in 2012, one in six U.S. residents was Hispanic. But by 2060, one in three residents will be Hispanic (U.S. Census Bureau 2012). Populations of other ethnic groups, such as African Americans and Asians, are also projected to increase in the United States between 2012 and 2060. The Asian population is projected to more than double, from 15.9 to 34.4 million; and the African American population is expected to increase from 41.2 to 61.8 million (U.S Census Bureau 2012).

The U.S., then, is projected to become a majority-minority nation for the first time in 2043, meaning that, while the non-Hispanic white population will remain the largest group, not one ethnic group will be the majority (U.S. Census Bureau 2012). Overall, the minority population is set to more than double from 116.2 million in 2012 to 241.3 million in 2060 (U.S. Census Bureau 2012).

This new demographic reality will engender a mix of uncertainty, surprises and discomfort. For the first time in this country's history, we must respond to three concurrent demographic trends: growing diversity, continued immigration—legal and illegal—and increasing longevity. In some ways, this is not a completely new experience. The United States has a 300-year history of successfully responding to constant waves of immigration and majority-minority changes. Although often fraught with conflicts, divisions and even tragedies (e.g., race riots, incarceration of Japanese Americans, violence against Native Americans and blacks, deportations of Mexicans), the United States has survived these crises and become a stronger and more united political entity, in large part because of its political institutions (democracy), its founding documents (Bill of Rights, U.S. Constitution, Declaration of Independence) and its historic willingness to socialize all newcomers with the common identity of "American." What is unprecedented is that this constant wave of immigration and shifting majority-minority populations is occurring while people in this country (like those in much of the world) are living longer, with an elderly population that is increasing in numbers. This new reality, which is unique to the twenty first century, must now account for a politics of aging: the inordinate influence of older persons in the electorate, while emerging Hispanic and other ethnic populations seek political empowerment. What new realities does this create for an aging nation while it becomes majority-minority?

Politics of Aging

With the twin phenomena of longevity and diversity, we can surmise that the political institutions of the United States and the electorate that is composed of older persons and ethnic groups/immigrants will have an impact on registration, voting, electoral politics, public policy (at the local, state and federal levels), and the ongoing debates about national priorities (e.g., public budgeting) and social policy (e.g., immigration reform). For example, the literature is replete with research and analyses of the political participation and voting behavior of older persons. There is also a growing body of articles and studies about political participation and voting behavior of ethnic groups, particularly the emerging Hispanic population (Logan et al. 2012).

The literature and surveys demonstrate that, on average, as one gets older, he or she is more likely to be registered to vote. In most voting cycles, then, especially in primary elections and state and local elections, persons over 65 years of age, particularly retirees, are more likely to vote than younger groups and thus have inordinate influence in electoral politics (Campbell and Binstock 2011; Binstock and Day 1996; Day 1990). But do they vote as a block? And how influential might they be in public policy? Our understanding of the politics of aging is heavily influenced by a leading scholar in this area: Robert Binstock. He was a pioneer in asking, "Why do older persons vote more often than younger groups?" Do they constitute a voting block? Is their electoral and political influence disproportionate to their numbers? In his seminal article, "Older People and Voting Participation" (2000), Binstock explores their voting behavior and concludes that, in large part, they have not proven to be a cohesive block and that the inherent differences within the older voter cohort mitigates against age as the predominant variable. On the other hand, Binstock et al. (2012) do suggest the possibility that an age cohort particularly, aging baby boomers, may coalesce around issues of self-interest, especially if this cohort perceives a direct and personal stake in policy and political debates. And herein lies the potential for a politics of aging that coalesces the baby boomer voting cohorts around issues of self-interest and perceived threats to their priorities.

Brownstein suggests a potential generational and ethnic/racial divide in his prescient article, "Brown Versus Gray" (National Journal 2011). In this and other essays (2012), he describes the growing intergenerational and interethnic divide in American politics with an electorate that is increasingly dividing by age, race and ethnicity. For the next 20 years, for example, the over-50 electorate will be primarily white and English-speaking, while the under-50 population segments will become increasingly minority, immigrant, ethnic, and younger. This age-race-ethnicity stratification has led to voting behavior based on different public priorities, with older whites focused on fewer taxes, less government intervention and greater public safety and younger, more ethnically diverse voters focused on enhanced public benefits and investments in jobs, employment and health care. One can see this divide in places like Arizona, with white retirees enjoying the "snowbird" life while relying on Hispanic workers for cooking, house cleaning, car washing and other low-wage services while supporting laws to restrict immigration. This may be an overly dramatic view of the "Brown Versus Gray," but it points out a change in the politics of aging in a majority-minority nation: the growing role of age and ethnicity in voting behavior and in the allocation of public benefits.

There is, however, a public policy issue that may highlight a potential age-based block of voters and also serve as a potential intergenerational and interethnic common agenda: the reliance on entitlement programs for retirement and health security. Social Security and Medicare are the bedrock of subsistence for older persons, particularly Tea Party conservative retirees. They depend heavily on these two programs for a modicum of security in their old age. Minorities, particularly Hispanics, depend on Social Security and Medicaid and expect that Medicare will provide their health security as they experience greater longevity (Brownstein 2012, 2013).

What might this mean for a better understanding of aging in the Americas and the future policy implications of a nation that is becoming older and more diverse? What might be the analytical and research directions for scholars in the politics of aging and ethnicity? And how might the subgroup of Latino boomers give us insights into these issues?

Several themes emerge that may serve as roadmaps for further exploration on these matters. First, the contemporary policy debates about national budgets, entitlement reform and immigration reform are occurring now and will be pressing political concerns for the foreseeable future. Thus, attention to mitigating any potential divide between the emerging Hispanic population, a younger workforce and the current cohort of older persons and retirees is needed. Promoting intergenerational coalitions benefits a diverse population in an aging society. Reframing the debates and narratives about intergenerational and interethnic tensions becomes imperative. And in this vein, focusing on the "canary in the mine shaft" gives us important clues for future directions in policy analysis and research.

Intergenerational Coalitions

This politics of aging in a majority-minority nation reveals an opportunity instead of potential conflict for intergenerational coalitions (Binstock 2010). Older adults and younger, more diverse populations share the same need for security and thus may come to depend on one another. As the older population increases, the solvency of entitlement programs such as Social Security and Medicare are being questioned. Younger, more diverse populations can keep entitlement programs such as Social Security and Medicare solvent by working and paying into the system. A recent study in *Health Affairs* by Harvard Medical School researchers states that immigrants may be "disproportionately subsidizing the Medicare Trust Fund" and shows how immigrants have contributed a surplus of $ 115.2 billion to the fund between 2002 and 2009 (Zallman et al. 2013). This research illustrates the need for an intergenerational coalition by demonstrating that the younger, more diverse generations that account for the majority of the workforce can give older adults the financial security they may need.

There is evidence of a potential bipartisan agreement on these divisive matters. Three notable conservative Republicans have spoken about the stake that older Americans have in immigration reform. Senators John McCain and Charles Schumer argue that "immigration reform would stimulate economic growth and create jobs" and improve the solvency of Social Security and Medicare (AARP Bulletin 2013). Congressman Paul Ryan, a notable Tea Party leader, also argues that immigration reform can lift the economy (Cook 2013). These conservative supporters of immigration reform lend credence to a potential ideological agreement on the most complex issue of aging in a majority-minority society.

Alexis de Tocqueville (1945) provides another reframing context: "self-interest rightly understood." In his seminal visit to the United States in 1831, he noted how Americans of that era would support government actions and their new nation's federal directives even if they disagreed or felt it was against their personal interests. They had, as he noted, a sense of the common good, and that what was done for all would eventually benefit an entire society, including those who might not support a particular action. This "self-interest rightly understood" would add resonance to Binstock and Schulz's reframing and allow, perhaps, for such controversial legislation as the Dream Act, as well as comprehensive immigrant reform, given the growth of a diverse workforce and taxpayer base that is replacing the declining number of non-Hispanic whites in the United States.

At the same time, the younger, more diverse population is in need of employment for its own security and well-being. Research by the Latino Economic Security (LES) shows that Latinos will comprise 20% of the workforce by 2030 but are the least likely demographic to have a high school or college degree, which may imply lower wealth and incomes (Gassoumis et al. 2011). Gassoumis et al. (2011) describe how investing in the education of young Latinos will have a positive impact on the older adult population due to the generation of much higher tax revenues. Latinos will need the education and employment to maintain their own security and overall well-being. In addition, research shows that Latinos rely on entitlement programs more than the overall U.S. population (Haliwell et al. 2007). This further illustrates the need for an interdependent relationship among older adults and the growing Latino workforce.

The Need to Reframe

With these findings, there is a need to reframe and show that an intergenerational coalition would be beneficial to the country as a whole, regardless of age, race or ethnicity. Scholars have suggested ways to reframe and adopt a more opportunistic view. Binstock and Schulz (2006) propose a "we're all in this together" approach, in which the frame does not pit old against young but instead highlights the benefits that can be reaped by all generations, regardless of age. In order to do this, Binstock and Schulz suggest constructing a scenario that would show what life would be like without Social Security and Medicare not just for baby boomers but also for their families (Binstock and Schulz 2006). By reframing these old-age issues into "family issues" that will affect the "daily lives of persons of all ages," there would be the potential for intergenerational coalitions that would, in effect, garner more widespread political support (Binstock and Schulz 2006).

Latino Baby Boomers: The Canary in the Mine Shaft

Into the vortex of aging, diversity and a nation that is undergoing demographic transformation lies a group that may serve as both a prism of future changes and as a "canary in the mine shaft" for what to do and not to do, given these trends. The Latino baby boomer population can help assess the future of a nation that is growing older and more diverse. With the growth of the Latino and older populations in America, Latino baby boomers mark the convergence of both groups. This subset shows the nexus of aging and diversity, and for this reason, Latino baby boom-

Fig. 24.1 According to the
U.S. Census Bureau, the
aging baby boomer popula-
tion is estimated at 76.4 million http://www.census.gov/
newsroom/releases/archives/
population/cb12-243.html.
The data provided by Gas-
soumis et al. used in this
figure incorporated additional
immigrants (minus deaths in
that cohort) making the aging
baby boom population rise
to 80 million. (Source: Gassoumis et al. (2008). "Latino
Baby Boomers: A Hidden
Population.")

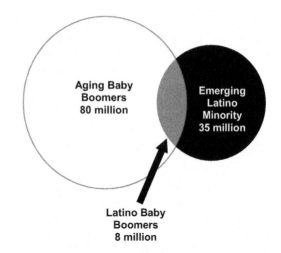

ers can be seen as the "canary in the mine shaft." Latino baby boomers show that
the baby boomer generation is composed not just of older white adults but also of
an increasingly diverse group of people. Because of this, lessons can be learned
from this relatively unknown population. Latino boomers comprise the cusp of the
next generation of Latino elders but also came of age during the civic and political
empowerment of the 1960s and 1970s (civil rights advances, Chicano movement,
farmworker organizing) and now represent the political, economic and social leadership of an increasingly influential Hispanic population.

As seen in Fig. 24.1, Latinos account for 8 million people, or 10 % of the baby
boomer population (Gassoumis et al. 2008). This statistic emphasizes the need for
policymakers to consider the social, economic and political implications of a nation
that is not just growing older but also more diverse.

One implication policymakers should recognize is that Latino baby boomers are
not as prepared for retirement as non-Latino elders. One reason is that, as a group,
Latino baby boomers have lower incomes. In addition, research shows that, compared with the U.S. population as a whole, Latinos are less likely to have savings
and retirement accounts (Ibarra and Rodriguez 2006; Orszag and Rodriguez 2005).
It is important for policymakers to be aware of this research when forming new
policies now and in the future, as the Latino and older adult populations continue
to escalate.

Latino baby boomers as "the canary in the mine shaft" also give clues about immigration reform. The discussion on immigration reform and whether citizenship
should be granted for 11 million undocumented immigrants in the United States is
high on the political agenda. To the extent that the politics of immigration continue
to motivate Hispanic voters, it may influence the outcome of national and state
elections.

Implications and Future Directions

Where then might we go with all that we understand about longevity, diversity, an emerging Latino population and the reality of a majority-minority nation. And how do these issues inform the future directions of "Aging in the Americas?" Firstly, we choose to view this demographic transformation as a set of opportunities for research, for policy analysis, for advocacy and public policy. As importantly, we believe that these issues and the conference series on Aging in the Americas provides a forum for enhancing bilateral relationships between the United States, Mexico and Latin America. The chapters in this series related to Mexico, Chile and other Latin American nations illustrate the varied and divergent paths individual nations are taking as they respond to the universal reality of populations growing older and declining replacement rates. Future directions for bilateral partnerships can include forums throughout Latin America bringing together scholars and policy researchers as well as government officials to learn from and to share divergent experiences and best practices. Perhaps the next such forum could occur in Mexico City, a location representing these demographic phenomena, but also a geographic case study of a local governmental entity providing a basic social safety net for elderly residents (e.g., a minimum pension). This conference has served to bring together the experts in gerontology and geriatrics from throughout Latin America and thus provides an opening for future collaborations on a topic that engender goodwill and a common agenda.

The Aging in the Americas series of conferences and the related volumes and papers also give insights into the methodological, analytical and substantive areas for future research. Data gathering and the creation and availability of comparative data sets is a crucial first step in understanding the complexity and scope of demographic challenges facing the U.S., Mexico and Latin America. Hereto, the availability of data, information and benchmarks for measuring progress and outcomes as each nation grapples with its aging population varies and a common Latin American forum on aging can provide a "sharing tool" for creating universal benchmarks of data points for understanding and responding to demographic challenges. Supporting the training and education of professionals in gerontology and geriatrics and expanding a trained workforce supporting the needs of elders throughout Latin America becomes the next important step in this effort. As such, the development of governmental institutions focused on gerontology and geriatrics (as we see in the United States and Mexico), provides the public policy ability for agenda-setting and public sector responses to the needs of an aging population. These suggested responses have short term, interim and long –term implications for analyzing the impacts of various legislative, regulatory and political actions at all levels of government and public and private sectors. In this regard, the role of elders, families and caregivers raises the spectrum of potential politics of aging throughout the Americas. Granted, this level of civic engagement and empowerment rests on the nature of the polity of each nation but we believe that in time, older persons themselves, merit a level of engagement, influence and control over their destinies as elders in their respective societies.

The measure of our response to aging and its affect on the emerging Latino populations throughout the hemisphere will be the level of dignity and quality of life provided to the elderly, their families and caregivers. The real test and outcomes of the analyses and recommendations throughout this volume may not be known for several decades; the demographic transformations are just beginning. Yet, we hope that the efforts of the scholars and researchers making important contributions to the "Aging in the Americas" scientific conferences and the products from these efforts, will be the seminal starting point for the dramatic and exciting realities of aging and longevity.

References

AARP Bulletin. (2013). A fix for immigration. September, 2013.

Binstock, R. H. (2010). From compassionate ageism to intergenerational conflict? *The Gerontologist, 50*(5), 574–585. doi:10.1093/geront/gnq056.

Binstock, R. H., & Schulz, J. H. (2006). *Aging nation*. Baltimore: The Johns Hopkins University Press.

Binstock, R. H., & Day, C. L. (1996). Aging and politics. In R. H. Binstock & L. K. George (Eds.), *Handbook of aging and the social sciences* (4th edn., pp. 362–387). San Diego: CA Academic.

Brownstein, R. (2012). White like me. The National Journal. Retrived from http://www.national-journal.com/columns/political-connections/white-like-me-20120315.

Brownstein, R. (2013). What the GOP fears. *Los Angeles Times*. October 11, 2013, p. A17.

Campbell, A., & Binstock, R. (2011). Politics and aging in the United States. In R. Binstock & L. George (Eds.), *Handbook of aging and the social sciences* (7th edn., pp. 265–280). London: Academic.

Cook, N. (2013). Paul Ryan: Immigrants "Bring Labor to Our Economy So Jobs Can Get Done." National Journal.com. Retrieved November 19, 2014, from http://www.nationaljour-nal.com/magazine/paul-ryan-immigrants-bring-labor-to-our-economy-so-jobs-can-get-do-ne-20130725.

Day, C. L. (1990). *What older Americans think: Interest groups and aging policy.* Princeton: Princeton University Press.

De Tocqueville, A. (1945). *Democracy in America*. New York: Vintage Books.

Gassoumis, Z., Wilbur, K., Benavidez, M., & Noriega, C. (2008). Latinos and Economic Security Policy Brief, Number 3 July 2008. Retrieved from http://www.chicano.ucla.edu/publications/report-brief/latino-baby-boomers-0.

Gassoumis, Z. D., Wilber, K. H., & Torres-Gil, F. (2011). The economic security of Latino boomers & beyond: The role of citizenship among generational cohorts. Gerontological Society of America Annual Meeting, Boston, MA.

Halliwell, P. A., Gassoumis, Z. D., & Wilber, K. H. (2007). Social security reform: How various reform options will affect Latino retirees (Research Report No. 2). Los Angeles: UCLA Center for Policy Research on Aging. from http://www.spa.ucla.edu/lss/documents/reports/Research%20 Report%202.pdf. Accessed 7 Jan 2009.

Hudson, R. B., & Gonyea, J. G. (2012). Baby boomers and the shifting political construction of old age. *The Gerontologist,* 1–11. doi:10.1093/geront/gnr129.

Ibarra, B., & Rodriguez, E. (2006). Closing the wealth gap: Eliminating structural barriers to building assets in the Latino community. *Harvard Journal of Hispanic Policy, 18,* 25–38.

Logan, J. R., Darrah, J., & Oh, S. (2012). The impact of race and ethnicity, immigration, and political context on participation in American electoral politics. *Social Forces; A Scientific Medium of Social Study and Interpretation.* doi:10.1093/sf/sor024.

National Geographic. (2011). The infant formula. pp. 120–121, September, 2011.

New York Times Magazine. (2008). Childless Europe: No babies? pp. 34–68. June 29, 2008.

New York Times. (2013). U.S. birthrates dips as hispanic pregnancies fall. p. A1 and A3. Jan 1, 2013.

OECD Family database. (2011). www.oecd.org/social/family/database-SF2.1: Fertility rates. Accessed 26 Oct 2011.

Orszag, P., & Rodriguez, E. (2005). Retirement security for latinos: bolstering coverage, savings and adequacy. The retirement security project. retrieved from: http://www.cfsinnovation.com/sites/default/files/imported/managed_documents/retirement_security_for_latinos.pdf.

U. S. Census Bureau. (2012). U.S. "Census bureau projections show a slower growing, older, more diverse nation a half century from now"-Population-Newsroom-U.S. Census Bureau. http://www.census.gov/newsroom/releases/archives/population/cb12-243.html. Accessed 4 April 2014.

Vincent, G., & Velkoff, V. (2010). "The Next Four Decades: The Older Population in the United States: 2010 to 2050 Population Estimates and Projections," U.S. Department of Commerce Economics and Statistics Administration U.S. CENSUS BUREAU.

Zallman, L., Woolhandler, S., Himmelstein, D., Bor, D., & McCormick, D. (2013). Immigrants contributed an estimated $ 115.2 billion more to the medicare trust fund than they took out in 2002-2009. *Health Affairs, 32*(6), 1153–1160. doi:10.1377/hlthaff.2012.1223.

Epilogue

Conference Series on Aging in the Americas (CAA):
Past, Present, and Future Directions
Prepared for
Challenges of Latino Aging in the Americas
Jacqueline L. Angel, Ph.D.
E-mail: jangel@austin.utexas.edu
August 7, 2014

The Conference Series on Aging in the Americas (CAA) was launched in 2001 to promote interdisciplinary collaboration on the complex issues associated with the health of the growing population of older Hispanics in the United States and Mexico. The conference organizers understood that the time was ripe for such a collaboration: the burgeoning population of elderly Mexicans and Mexican Americans was a healthcare crisis in the making, and a growing number of social scientists and policy analysts were turning their attention in that direction. Thirteen years and seven conferences later (the 2014 installment will be held this fall), the serial approach to identifying trends and proposing solutions has proved to be an effective one. Each meeting has built on the new information that arose from the previous ones, and much progress has been made in understanding the unique health status and health care needs of this important subgroup of the elderly population in both countries. It is vital that we continue to build on this momentum.

The following summary describes some of the issues that have driven the conference agendas and that continue to be of critical importance in formulating viable solutions. It also provides an overview of the conference series and its satellite mentoring program as well as a list of peer-reviewed publications and conference proceedings. Perhaps most importantly, the summary offers a look at future agenda themes that will carry the work of the conferences forward.

© Springer International Publishing Switzerland 2015   425
W. A. Vega et al. (eds.), *Challenges of Latino Aging in the Americas,*
DOI 10.1007/978-3-319-12598-5

Background

The population of the United States, like that of most other nations of the world, will age rapidly well into the twenty-first century (Bloom et al. 2011, 2013). The older population is also anticipated to become more ethnically and racially diverse than ever before. As a result of high immigration and fertility rates, as well as improvements in life expectancy, the U.S. Census Bureau projects the steepest growth in older Hispanics. The number of older Hispanics (65 years and over) is expected to increase by more than six times by 2050 to 17.5 million (Vincent and Velkof 2010). The oldest and by far the largest segment of the Latino population in the United States is of Mexican heritage, accounting for almost two-thirds of all Hispanics. The coming nexus of aging and diversity faced by the United States will require a greater level of scrutiny and analysis if we are to provide policy solutions to aging and health care in the United States and, by example, in Mexico and throughout Latin America.

Clearly, we are entering a new era in human history in which aging populations present unique problems to both developed and developing nations and to local economies. The challenges in the Americas are particularly daunting. Improved nutrition and living conditions have increased life expectancies at all ages. By 2025, at least one-fifth of the population in 15 countries in the Americas is likely to be age 60 and over (Bravo 2013). The number of elderly 65 and over in Latin America will triple as a share of the population by 2050 (Jackson et al. 2009). As the result of a precipitous decline in the fertility rate, Mexico's population has aged significantly in the past two decades. There will be more Mexicans in their sixties than their twenties by 2050 (Jackson et al. 2009). Like Mexico, the United States will age swiftly and more than double its population aged 65 and older to 88.5 million in 2050 (Vincent and Velkof 2010). The Mexican-origin population will become the largest minority group in the United States. Both nations face similar problems with the growing aging population.

Despite improvements in general health levels in both the U.S. and Mexico, access to preventive and acute care remains problematic for many older people of Mexican-origin in both countries. Although longer life is desirable, people of Mexican-origin tend to have low levels of education, low rates of health insurance coverage, low incomes, and little wealth in both countries (Angel et al. 2012). In Mexico, high rates of poverty and a fragmented health care system place poor elderly individuals at risk of inadequate care (Pagán et al. 2007). Only about 50% of adults aged 70 and over had health insurance coverage: 54% of return migrants and 52% of those who had no history of U.S. migration (Wong and Espinoza 2007). In the United States, despite nearly universal Medicare coverage, poor Mexican-origin elders without supplemental Medigap policies often lack full access to high-quality health care (Angel and Angel 2009).

At all ages, the Mexican-origin population is the most inadequately insured subgroup in the United States (Angel et al. 2007). A large body of research shows that adequate coverage is clearly associated with better health and, on the other hand, a lack of coverage results in negative health outcomes (Institute of Medicine 2001). For example, Mexican Americans are substantially more likely than other groups to develop diabetes within their lifetime (Langa et al. 2002), but they also have drastically lower rates of insurance coverage and use of nursing facilities (Fennell et al. 2010).

Because of relatively longer life spans and poorer health, older Hispanics, especially Mexican-origin immigrants (Hayward et al. 2007), spend a larger number of years with chronic health problems than non-Hispanics (Cantu et al. 2013). This longer period of frailty and infirmity means a greater need for assistance. Low levels of institutional care use suggest that in the future Latino communities will need and demand more and better community-based long-term care services (Herrera et al. 2012). For individuals of Mexican descent in the United States, then, the possibility of a longer life characterized by compromised health and material hardship raises serious questions about the potential burden on government and families.

Aging populations in the U.S and Mexico have implications for all aspects of social policy including financial security. The education and income trends of this population will have great impact on future tax revenues, with low levels of education and income resulting in lower levels of income available for taxation (Torres-Gil 2005). Education has a direct impact on income for Mexican immigrants in particular. For example, 63 % of male Mexican immigrants have not completed high school by age 18 compared to 8.7 % of native-born workers in the U.S. (Borjas and Katz 2007). Consequently, male Mexican immigrants earn 53.3 % less than their native-born counterparts even after accounting for differences in age, education, and state of residence. The coming age wave, then, poses serious economic problems. Both countries will need to craft retirement policies that are capable of providing adequate levels of support for the elderly without imposing an untenable dependency burden on the young. To the extent that the care of elderly populations undermines development and educational and employment opportunities for younger generations, political stability is threatened and the lure of extremist political positions heightened. In addition, given high levels of immigration to the United States, many Latin Americans will age in the United States and draw upon social services here.

As older Hispanics and Mexican Americans become increasingly impaired, the demands encountered by the family and society at large will become more serious. Today's Hispanics will be tomorrow's majority in many states, and they will make up a large share of the U.S. labor force that supports tomorrow's retirees. If these groups are confined to the low-paying service sector, which in addition to low wages offers poor access to health insurance, yet another disparity based on race and ethnicity will be introduced into our structures of social inequality. The fact that the productive potential of a large segment of the future labor force might be undermined by poor health and low educational levels has profound implications for older, as well as younger, Americans. Everyone's welfare, therefore, depends upon the productivity of minority groups. Local, state, and federal governments are confronted with increasing challenges related to the allocation of limited resources for the care of vulnerable citizens — both young and old.

Notwithstanding the gravity of these issues, we currently lack a sophisticated understanding of the impact of rapid aging on minority populations and the consequences of age differentials among minority and majority groups for the nation as a whole. It is imperative that we identify and develop the most effective and equitable approaches to meeting the unique financial and health care needs of Latino families in later life, within the context of what is likely to be long-term fiscal retrenchment. A series of scholarly conferences focused specifically on aging in the Mexican-origin population and the unique social, political, and economic ties between Mexico and the United States is an essential vehicle for understanding how aging affects the well-being of both societies in an increasingly globalized economy. Rubén Rumbaut underscored this point at the 2013 ICAA closing keynote lecture at The University of Texas at Austin. He added that "planning and policymaking for the aging and well-being of the rapidly growing Mexican-American population need to be contextualized within the tangled 'tango' of U.S.-Mexico interconnectedness, specifically with respect to international migration flows, the putative demographic 'fit' of U.S.-Mexico economic and population needs, the incorporation of immigrants and their descendants, and the political contexts which mold it."

History of Conference Installments

In recognition of these challenges, the Conference Series on Aging in the Americas (CAA) has brought together leading senior scholars and emerging scholars to develop a critical mass of theoretical and practical work aimed at greatly expanding our knowledge base concerning the consequences of population aging in the Americas. The first meeting was held at The University of Texas at Austin in 2001, and was followed by six additional conferences in 2005, 2009, 2010, 2012, 2013, and 2014 http://lbjschool.austin.utexas.edu/caa/. Each meeting is convened during National Hispanic Heritage Month and Mexico's Independence Day. The overarching goal of the conference series is to address vitally important issues that affect not only the Hispanic population of the United States, but those of all Latin American nations in which the demographic transition that took a century or more in Europe is occurring in a few decades. The rapid demographic transition in Mexico has immediate practical implications for both that country and the United States. New policies dealing with education, health care, housing, transportation, and more will be necessary to deal with changing population age compositions.

The ICAA series has three major aims: (1) to facilitate the exchange of ideas aimed at addressing key issues confronting the aging populations in the Americas; (2) to promote an interdisciplinary collaboration by researchers in the fields of Hispanic health, health care policy, and behavioral and social aspects of aging (the CAA installments are unique in their focus on the aging populations of the United States and Mexico); and (3) to set a research agenda for understanding and improving the health and well-being of older Latino adults and their families.

One of the many attractive aspects of this conference series is that it supports interdisciplinary collaboration on state-of-the art research on Hispanic health and aging developed from the sharing, analysis, and dissemination of ideas. The series has a logical sequential order aimed at building upon previous work of each presenter through integration of knowledge, interest, and research findings on substantive and methodological issues.

Each installment in the series has a unique and individual signature theme, but follows a similar format. In the last three installments, topics covered the methodological problems of conducting cross-national comparative studies in context: Mexico and the U.S.; implications of changing demographics in Mexico for Mexican-origin health and economic well-being; and research designed to improve understanding of the influences of the immigrant family experience on Hispanic (Mexican-origin) healthful aging. Each meeting lasts for two days and consists of two keynotes, ten presentations, and a poster session. In addition, there is a consensus building session at the end of each conference where participants review key issues raised during the conference and identify new directions for research.

Future Directions

Future installments of the Conference Series on Aging in the Americas will address many critical issues that identify and inform an agenda for cutting-edge research that addresses disparities in the health and human services systems of Mexico and the United States. The impetus for additional installments come at a time when Latino health care for older adults is emerging as an important state budget issue and public health initiative in Mexico and the U.S. As a result of the seven meetings, the intellectual rapport among scholars in the field of Mexican-origin health and behavioral and social aspects of aging research now exists at the national level. This cross-disciplinary collaboration has fostered scholarship that significantly contributes to research, training, and informs policy analysis of key health issues affecting the aging Hispanic population of Mexican descent. Although the scholarship on Hispanic aging in the United States and Mexico is growing, there is a significant dearth of knowledge and information about the unique strengths and characteristics that underrepresented minority groups possess and experience as they age. Further examination of these understudied groups from a cross-national perspective offers the possibility to promote healthy aging for the entire nation. Local, state, and federal governments are faced with increasing challenges related to the care of vulnerable citizens, both young and old. As older individuals become impaired the caregiving demands faced by the family become more difficult to meet. There is also profound uncertainty over how the potential dependency burden affects government's ability to project and plan for future long-term care services.

Currently we lack a sophisticated understanding of the impact of rapid aging on minority populations and the consequences of age differentials among minority and majority groups for the nation as a whole. It is imperative that we identify and develop the most effective and equitable approaches to meeting the unique financial and health care needs of Hispanic families in later life, within the context of what is likely to be long-term fiscal retrenchment. This meeting provides a unique opportunity to increase scholarly research on aging Hispanics in the U.S. and Mexico. The knowledge is critically needed, and we believe that it promises to make a substantive contribution to inform social policy with respect to health care in the Americas.

The longer term plans include meetings on various topics of relevance to Hispanic aging with more of a comparative focus. The following are some general directions and new themes along these lines.

A. Work on the role of randomized control studies and experimental interventions in policy studies to improve the quality of life among older Hispanics on both sides along the U.S./Mexico border.
B. Studies assessing the impact of immigration processes on the health and socio-behavioral outcomes of migrant populations.
C. In-depth investigations of personal health services and their effect on public health and geriatric social services in both the U.S. and Mexico. This could include mixed-method studies of temporal patterns of caregiving and effect on formal and informal support systems.
D. A session devoted to research resources for questions on ethnicity and elder health. These might include access to data sets, biobanks, informatics platforms and data sharing capacities to inform trainees and others. Particularly the resources at NIA and the equivalents in Mexico and other Latin American nations would be of value.
E. Examination of the history of Hispanic populations and culture in the US since the American Revolution. This would enable the humanities to become more involved in the research process and provide additional opportunities to support bi-national studies of aging in the Americas.
F. A theme suggested by one of the policy sessions of this meeting is addressing the tensions and difficulties of merging policy and politics with social and behavioral science, including economics of aging in the Americas. This is currently an important issue for both the National Science Foundation and the National Institutes of Health, and there could be an important benefit for trainees in hearing cogent analyses of the problem. Resources at the National Research Council may be able to provide assistance.
G. The growing emphasis on biomarkers of health and psychosocial behavior has been supported in part by the National Institute on Aging. It may be worthwhile to introduce new investigators to the methods and directions of such research to entice some to pursue multi-disciplinary career pathways that are likely to be productive in the future.
H. Methodological research focusing on variation in acculturation across culture and ethnicity, and how this variation influences and is associated with health. An important, and related issue, is the extent of variation in the acculturation experience and what it means for the use of an overarching "Latino" designation. This last is particularly relevant in light of the U.S. Census Bureau's proposal

to eliminate the question of specific Hispanic national origin, such as Mexican, Cuban, and Nicaraguan and instead designate Hispanic/Latino as a race category of its own.

I. There is a dearth of contemporary studies on the consequences of aging populations and improved medical technology on the health care systems of the Americas and the world. Aging populations and medical innovation contribute to high rates of medical inflation and soaring hospital costs that raise serious challenges to governments and raise important public policy concerns. In particular, understanding the process and impact of implementation of recent health reform in Mexico and comparisons to the U.S. is crucial in developing comparative frameworks for investigating global aging and sustainability of support systems. Future fiscal crises will require all nations to ration increasingly expensive health care services. This rationing raises serious questions of equity related to the treatment of Hispanic populations, the poor, minorities, immigrants, and native populations.

The University of Texas at Austin has hosted the CAA since its inception in 2001. The Population Research Center in collaboration with the LBJ School of Public Affairs provides the institutional support to administer meeting activities. Other future venues include Mexico City in the not too distant future and San Francisco, California to be held concurrently with the 2017 International Association for Gerontology and Geriatrics.

Conclusion

In summary, by now a large body of research points to a dramatic demographic transformation in Mexico. While Mexico's population is relatively young today, with a median age of 27, it will age rapidly in coming years and will grow to 25 million by the middle of this century. The number of older Latinos (65 years and over) in the United States is also expected to increase by more than six times in 2050 and reach almost 18 million. Since Mexico and the United States are swiftly becoming old, both countries face problems of caring for the elderly and often frail and disabled populations. The CAA provides a unique opportunity to engage with invited speakers, keynote lectures and emerging scholars to address these issues in panel discussions, peer-reviewed posters, and a consensus building session. In addition to the presentations, the CAA sponsors a mentoring program for emerging scholars in minority aging with a special emphasis on Hispanic health in the Americas. Our hope is that this collaboration will continue in the future to produce new insights and to offer solutions to emerging social problems.

References

Bloom, D. E., Boersch-Supan, A., McGee, P., & Seike, A. (2011). *Population aging: Facts, challenges, and responses*. PGDA Working Paper. Cambridge: Harvard University.

Bravo, J. (2013). *World population ageing report 2013*. New York: United Nations, Department of Economic and Social Affairs.

Jackson, R., Strauss, R., & Howe, N. (2009). Latin America's aging challenge: Demographics and retirement policy in Brazil, Chile, and Mexico. *Global Aging Initiative*. csis.org/files/media/csis/pubs/090324_gai_english.pdf.

Vincent, G. K., & Velkof, V. A. (2010). The next four decades: The older population in the United States: 2010–2050. *Current Population Reports, P25*(1138), 1–14.

CPSIA information can be obtained at www.ICGtesting.com
Printed in the USA
BVOW08*0820210415

397030BV00004B/6/P

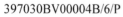